P9-EEG-967

RADIATION BIOPHYSICS

RADIATION BIOPHYSICS

Second Edition

HOWARD L. ANDREWS
Professor of Radiation Biology
and Biophysics (retired)
University of Rochester

PRENTICE-HALL, INC.

Englewood Cliffs, New Jersey

Library of Congress Cataloging in Publication Data

ANDREWS, HOWARD LUCIUS. (date)
Radiation biophysics.

Includes bibliographies.
1. Radiation—Physiological effect.
2. Radiobiology. I. Title. [DNLM: 1. Biophysics.
2. Radiobiology. WN610 A566r 1974]
RA569.A5 1974 612′.01448 73-15870
ISBN 0-13-750075-0

10 9 8 7 6 5 4 3 2 1

Printed in the United States of America

PRENTICE-HALL INTERNATIONAL, INC., *London*
PRENTICE-HALL OF AUSTRALIA, PTY. LTD., *Sydney*
PRENTICE-HALL OF CANADA, LTD., *Toronto*
PRENTICE-HALL OF INDIA PRIVATE LIMITED, *New Delhi*
PRENTICE-HALL OF JAPAN, INC., *Tokyo*

Contents

Preface

The uses of radiation in all of its known forms have increased considerably since the first edition of this book. New and sophisticated instruments and a number of previously unavailable radioactive isotopes are now in daily use in biological research and in diagnostic and therapeutic medicine. The field of nuclear medicine has come of age. Our understanding of the genetic code and the role played by DNA in the scheme of human biology is due in no small measure to the intensive use of radioactive tracers. It appears that our ever-increasing demands for energy will require the extensive utilization of nuclear sources, under conditions which are acceptable to our threatened environment.

Conceptual and technological advances have not been confined to the ionizing radiations of modest energy. On one hand, physicists have been pushing toward higher and higher quantum energies as they investigate the ultimate properties of matter. At the other extreme, low-energy electromagnetic waves are being increasingly exploited in communications, in industrial and domestic food preparation, and as an indispensable aid to safe transport by sea or air.

The student interested in obtaining a working knowledge of radiation, its effects and its potential uses, is faced today with a field of vastly greater scope than was the case a few years ago. Along with the expanded uses of radiation has come the recognition that each use alone, and all combined, must be compatible with the maintenance of the earth as a habitable planet. The opposing forces of economics and environmental protection require detailed analyses of all proposed applications of radiation in order to maximize the benefit : risk ratio at an acceptable cost. Fortunately, today's student comes to his task with a good foundation in the basic sciences upon which to build.

The present edition of this book reflects some of the changes in the subject matter and in the student. The main aim is to present to the student entering the field the physical principles that lie behind the production of radiation, its transmission, its absorption, and some of the physical aspects of its biological action. Authoritative texts are now available in radiation biology and measurements, and the present treatment drastically reduces the space devoted to them. This edition, like the first, is designed for the student with an incomplete mathematical background. The results of the calculus are used here without rigorous, detailed derivations. Emphasis is placed on physical concepts and on the computational methods that are required for the safe and effective use of radiation.

Space limitations prevent individual acknowledgements of my gratitude to all who have contributed to this volume. I am particularly grateful for the constructive criticisms received from those who used the first edition in the classroom. I hope that the response to those suggestions has been favorable. My special thanks go to Dr. J. Newell Stannard whose continued interest and guidance have made this edition possible.

H.L.A.

RADIATION BIOPHYSICS

1

Mass, Energy, and Radiation

1.01 Radiation and Living Systems

A considerable mass of evidence indicates that our earth, including all living systems thereon, has been exposed to various forms of radiation for an extremely long period of time. Radiation from our sun, consisting of infrared, visible, and ultraviolet light has made possible life as we know it on earth. High-energy radiations from extraterrestrial sources and from radioactive materials on earth are capable of penetrating through substantial thicknesses of any absorber and can react throughout any living system. Radiation may have played an important or even an essential role in the production of simple molecules from primordial matter and in the later incorporation of these structures into living organisms.

Man has lived in essential equilibrium with his radiation environment up until the very last years of the nineteenth century. He had already made use of the chemical effects of a wide variety of substances but he had produced almost no sources of radiation beyond the fires used in cooking, heating, and industry.

Man-made sources of radiation became widely used in medicine following the discovery of X rays by Roentgen in 1895. It soon became evident that these radiations could have a profound effect on living tissues. Deleterious as well as beneficial effects were observed. In spite of these observed effects relatively little research was carried out on the biological consequences of radiation absorption until the United States undertook the development of nuclear explosives in the 1940's. Since that time large programs in radiation biology have been supported, with the emphasis on the ionizing radiations.

Much has been learned and nuclear energy has been exploited on a large and increasing scale.

Today man is generating ever-increasing quantities of all forms of radiation, from the low-energy radio waves used in communication and in heating to the high-energy radiations which have many applications in medicine and industry. We can understand the actions of these radiations only from a thorough appreciation of the physical properties of the radiations and the nature of their interactions with chemical and biological systems. We conceive of two main modes of energy transmission: moving particles and wave motions. Each of these modes will be dealt with in detail in the following sections.

1.02 Elementary Particles

The production of X rays in 1895 and the discovery of radioactivity in 1896 inaugurated an era of intensive research into the physical nature of matter and energy and the interactions between them. Atoms were shown to consist of electrically neutral assemblies of nuclei and electrons, and subnuclear particles were discovered and identified. As research became more sophisticated and more refined equipment became available, more particles were discovered. Today the list of identified particles is a formidable one, and still other particles may remain to be found.

The roles played by some of these particles are quite clear but the functions of others are obscure. Some interact only weakly with any matter through which they pass; some appear as free particles only fleetingly, and only under unusual circumstances. These latter types play no direct, major role in the interactions with chemical systems or with living tissues. We shall be intimately concerned here with only eight particles, listed with some of their pertinent properties in Table 1-1.

Size has no unique definition in the domain of the elementary particles. Fortunately, exact sizes are of little consequence for present purposes. It is sufficient to think of the particles as spheres with radii of the order of 10^{-15} m (meter).

Mass, on the other hand, can be precisely defined and measured and is a most important parameter. Masses measured in grams form one of the basic units of the centimeter–gram–second, or CGS, system of units which has now been largely replaced by the meter–kilogram–second–ampere, or MKSA, system. The latter is sometimes called the Systeme International, or SI.

Although there are a number of multiplicative prefixes in common use (Appendix 1), the gram and the kilogram are inconveniently large for use in calculations involving the elementary particles. The *unified mass unit* (umu or u) is defined as exactly one-twelfth of the mass of a single carbon atom of the type that contains 12 particles or nucleons in its nucleus. This unit supersedes the older *atomic mass unit* (amu) which was based on oxygen instead of carbon. The unified mass unit can be readily expressed in terms of the basic CGS units. Exactly 12 g (grams) of ^{12}C will consist of 6.02×10^{23} atoms* and hence

$$1 \text{ umu} = \frac{12}{12 \times (6.02 \times 10^{23})} = 1.66 \times 10^{-24} \text{ g} \tag{1-1}$$

Classical, or Newtonian, mechanics found several parameters useful for describing bodies in motion. One of these, *momentum*, p, is the product of the mass and velocity: $p = mv$. Momentum is a quantity that is constant, or *conserved*, in any interparticle collision. *Energy*, consisting of two components, potential and kinetic, has also proved to be an important parameter of any mechanical system. According to Newtonian mechanics, kinetic energy† $T = \frac{1}{2}mv^2 = p^2/2m$. Whenever a body is located in a field of force, it will have a potential energy which is capable of conversion to kinetic energy. Thus a body in the earth's gravitational field can convert some of its gravitational potential energy to kinetic energy if it is allowed to fall toward the center of the earth.

1.03 Special Relativity

By the beginning of the twentieth century physicists had discovered that the measured velocity of light had a constant value regardless of the relative motions of the source, the detector, and the postulated medium through which the light was propagated. The explanation of this observation involves the behavior of bodies moving with high relative velocities. Einstein's studies led to a new or *relativistic* mechanics that was applicable to bodies moving with either high or low relative velocities. Newtonian mechanics was shown to be a special case, adequate for bodies moving at slow speeds but failing at the high velocities that are usual with subatomic particles and the ionizing radiations. Experiments have amply demonstrated the validity of the relativistic mechanics.

Einstein postulated that the velocity of light in a vacuum, as measured in any laboratory coordinate system, would indeed have a constant value, c. Ponderable bodies could have any velocity from zero up to c but could not equal nor exceed it. This requirement is in sharp contrast to classical mechanics, which puts no upper limit on velocity or any other parameter.

*Avogadro's constant: more precise values of this and other constants are found in Appendix 2.

†To avoid confusion, T will be used to denote kinetic energy when E designates an energy state or level. When the meaning is unambiguous, E may denote kinetic energy.

Relativity theory recognizes that the mass of a moving body, m, depends on its rest mass m_0 and also on its velocity v through the relation

$$m = \frac{m_0}{\sqrt{1 - (v^2/c^2)}} \tag{1-2}$$

For small values of v, m and m_0 are almost identical, but as v approaches c, the mass increases without limit.

A most important prediction of the theory is that mass and energy are really only two different manifestations of a single entity and are interchangeable one into the other. The quantitative relation between the two forms is

$$E = mc^2 \tag{1-3}$$

where the units will be either

$$\text{joules} = \text{kilograms (meters per second)}^2 = \text{kg (m s}^{-1})^2$$

or

$$\text{ergs} = \text{grams (centimeters per second)}^2 = \text{g (cm s}^{-1})^2$$

Combining Eqs. (1-2) and (1-3) we have for the *total* energy of a moving mass

$$E = \frac{m_0 c^2}{\sqrt{1 - (v^2/c^2)}} \tag{1-4}$$

In the special case when $v = 0$, Eq. (1-4) becomes

$$E_0 = m_0 c^2 \tag{1-5}$$

where E_0 represents energy that is inherent in mass itself. This term has no counterpart in Newtonian mechanics but its existence has been repeatedly demonstrated.

From Eq. (1-4) the relativistic expression for kinetic energy becomes

$$T = E - E_0 = m_0 c^2 \left[\frac{1}{\sqrt{1 - (v^2/c^2)}} - 1 \right] \tag{1-6}$$

The first term inside the brackets can be expanded in a power series of terms which leads to

$$T = m_0 c^2 \left[\frac{1}{2} \left(\frac{v}{c} \right)^2 + \frac{3}{8} \left(\frac{v}{c} \right)^4 + \cdots \right] \tag{1-7}$$

When $(v/c) \ll 1$, all terms beyond $(v/c)^2$ can be neglected and we have just the classical expression for kinetic energy. As (v/c) becomes larger, more terms in the series must be used.

When Eq. (1-4) is squared to remove the radical and is rearranged, we have

$$E^2(c^2 - v^2) = m_0^2 c^6 \tag{1-8}$$

When $v = c$, this relation can only be satisfied by $m_0 = 0$. Thus any transfer of energy that takes place at the velocity of light can only be made through an entity which has a zero rest mass. Conversely, if energy is transferred by an agent with zero rest mass, the transfer must be carried out at the velocity of light.

In the relativistic mechanics, momentum is still defined as $p = mv$, which is then related to kinetic energy by

$$p = \frac{1}{c}\sqrt{T(T + 2m_0c^2)} \tag{1-9}$$

The inverse relation is frequently useful.

$$T = m_0c^2\left[\sqrt{1 - \left(\frac{p}{m_0c}\right)^2} - 1\right] \tag{1-10}$$

At low velocities Eqs. (1-9) and (1-10) become equivalent to the simpler forms of the older mechanics.

1.04 Electrical Units

All matter is composed of an enormous number of positive and negative electric charges, with these charges so accurately balanced in the normal undisturbed state that there is no macroscopic evidence of any charge. When sufficient energy is added to an electrically neutral system, as by a collision, high temperature, electrical discharge, or some other means, some of the positive and negative charges may be separated and then both kinds of charge become manifest.

A stationary electric charge q_1 is surrounded by an *electric field* because a second charge q_2 brought into the vicinity will experience a force. The magnitude of this force follows an inverse-square law of the distance between the charges.

$$F = \frac{q_1q_2}{\epsilon r^2} \tag{1-11}$$

In Eq. (1-11) r is the distance between the charges and ϵ is the *permittivity*, a constant whose value depends on the properties of the medium separating the charges and on the units in which the other quantities are measured. When F is expressed in dynes and r in centimeters and ϵ is set equal to 1.00 for a vacuum, Eq. (1-11) can be used to define the unit of charge in the CGS electrostatic or esu system. In the MKSA system used here, F will be in newtons, r in meters, and each q in coulombs, whence ϵ for a vacuum will have a value of 8.85×10^{-12} farad.

Electric and magnetic phenomena are closely related. A moving electric charge produces a magnetic field and a changing magnetic field sets up an electric field in the surrounding space. Although an isolated magnetic pole of either sign has never been observed, separation can be achieved in principle by using a long slender magnet whose two poles are separated by a relatively great distance. Then an equation similar to Eq. (1-11) can be written for the magnetic poles and this gives rise to a third system, the CGS electromagnetic or emu system.

Many careful measurements have demonstrated that electricity is not infinitely divisible but exists only in integral multiples of a fundamental unit of charge. The ubiquitous electron, which is involved in all electric current flows, has just one basic charge unit, 1.6×10^{-19} coulomb, of negative sign.

When a unit charge moves through a difference of potential V, the amount of work done will be numerically equal to V. In general, for a charge q,

$$W = V \times q \tag{1-12}$$

$$\text{joules} = \text{volts} \times \text{coulombs}$$

For our purposes it is convenient to use a hybrid system of units, taking V in volts from the MKSA system and the fundamental quantity of charge as the unit for q. When this is done, the work will be expressed in *electron volts*, or eV. Obviously 1 eV $= 1 \times (1.6 \times 10^{-19})$ joule or 1.6×10^{-12} erg. In addition to the new unit of work itself we shall find the multiples keV and MeV most useful.

1.05 Excitation and Ionization

When sufficient energy is absorbed by an electrically neutral atom or molecule, an electron may be ejected from it, leaving behind a positively charged residue. This process, known as *ionization*, leads to the formation of an *ion pair*. Ionizations may be produced by direct mechanical collisions with neutral particles. Electrically charged particles can produce ionizations without a direct-contact collision. A moving charge will be accompanied by an electric field which extends out from the particle itself. As this field sweeps past a neutral atom, the electric force may be sufficient to eject an electron and form an ion pair.

In some cases the energy received either by collision or by radiation absorption may not be sufficient to form an ion pair. The atom may then be left in a neutral, un-ionized but *excited* electronic state. We shall have much more to say about ionizations and excitations. It is sufficient now to note that an excited atom may radiate the excitation energy and return to its original, undisturbed state of lowest energy, the *ground state*. A molecule in an excited state may radiate and return to its ground state or it may dissociate into two fragments.

At any temperature above absolute zero a molecule will rotate about its center of mass and the component atoms will vibrate with respect to each other. Each mode of mechanical motion will have an average energy content that is determined by the temperature. Energy transfers of less than about 1 eV may not even excite the molecular electrons but are quite capable of increasing the vibrational and rotational energies. Only a fraction of an eV is required to increase the vibrational energy and an even smaller amount is required in the case of rotations. The absorption and release of these forms of mechanical energy are discussed in detail in Chapters 16 and 17.

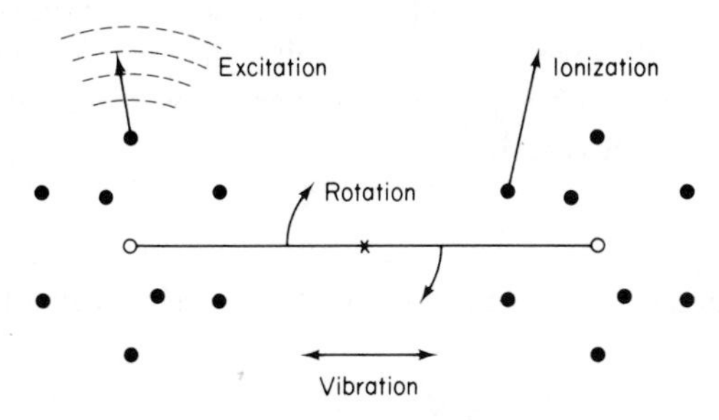

Figure 1-1. A molecule may respond to a sudden energy input by releasing an electron to form a pair of charged ions, raising an electron into an excited state, increasing the amplitude of vibration, or increasing the speed of rotation. Absorbed energy may divide between two or more modes.

Figure 1-1 is a schematic representation of the various results of an energy transfer to a molecule. In ionization an electron is ejected from the neutral structure to form a pair of charged ions; in excitation the electron involved remains attached to the parent structure but has an increased energy. Vibrational amplitudes and rotational velocities are increased when energy is absorbed.

1.06 General Properties of Waves

Wave motions have long been recognized as one of the means by which energy can be transferred from one system to another. The general concept of wave generation and transmission is simple. Some type of source vibrating at a frequency ν Hz disturbs and sets up stresses in a transmitting medium. The medium, in returning to its original unstressed state, propagates the stress, which moves outward from the source at a velocity v which is a characteristic of the propagating medium. Eventually the disturbance may reach and activate a suitable receiver.

Many types of sources and waves are known. A vibrating string, as in a violin, creates a series of pressure waves in the adjacent air. Pressures above and below normal move outward as the air tends to return to its normal constant pressure. When the pressure waves strike an eardrum, they set it in vibration and the brain interprets the disturbance as sound.

Whatever the type of wave, at the end of 1 second there will be ν cycles spread uniformly over a distance v from the source. Each cycle will occupy a distance λ where

$$\lambda = \frac{v}{\nu} \tag{1-13}$$

The wavelength λ is the distance between consecutive points of equal phase in the wave train. Spectroscopists frequently find it more convenient to use the *wave number* $\tilde{\nu}$, which is just the number of cycles in a unit length of the propagating wave. Obviously

$$\tilde{\nu} = \frac{1}{\lambda} \tag{1-14}$$

Another parameter of a wave motion is the *period* or time of one oscillation, τ.

$$\tau = \frac{1}{\nu} \tag{1-15}$$

When two waves of equal frequency and opposite phase are superposed, crest will fall on trough and trough on crest; the summed effect is a cancellation and the waves are said to *interfere destructively*, *D*, Fig. 1-2. If the two waves are in phase, crest will fall on crest and an increased amplitude will be produced by *constructive interference*, *C*, Fig. 1-2. Interference appears to be a phenomenon exclusively associated with wave motions. It is extremely difficult to conceive of discrete particles interacting to produce interferences.

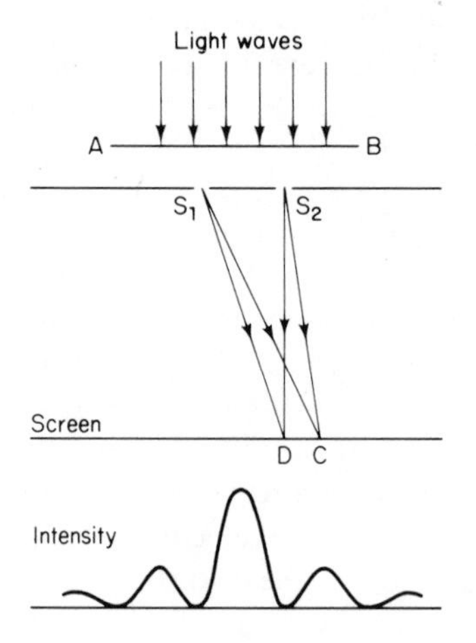

Figure 1-2. When the path lengths from the two slits S_1 and S_2 differ by $\lambda/2$, as at *D*, there will be destructive interference and no light will be seen. At *C* the path lengths differ by λ and the interference is constructive.

1.07 Electromagnetic Waves

Consider a simple dipole antenna, Fig. 1-3, so connected to a radio transmitter that electric charges are oppositely and alternately forced onto the

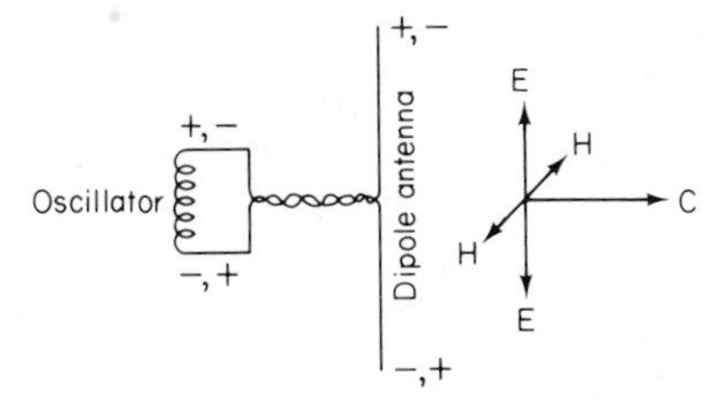

Figure 1-3. Electric charges driven out along the dipole radiator by the vacuum-tube oscillator set up an electromagnetic field which is propagated through space with a velocity c.

two sections of the dipole. The charge movements in the wires produce an oscillating electric field in the immediate vicinity. This electric field will produce a magnetic field oscillating with the same frequency and this magnetic field will, in turn, create an electric field. Thus an *electromagnetic field* originates at the antenna and is propagated outward from it. An electromagnetic wave can be initiated equally well by an oscillating magnetic field. In this case the changing magnetic field sets up electric stresses in the surrounding medium and a two-component disturbance is propagated as before.

An electromagnetic wave consists of an oscillating electric field and an oscillating magnetic field, inextricably connected, with each depending on the other for its existence. Each component vibrates with a common frequency and in space the two components are in phase. The two amplitudes are at right angles to each other, Fig. 1-4, and since the amplitudes are also perpendicular to the direction of propagation, the waves are *transverse.*

In 1865, Maxwell predicted that visible light consisted of an electromagnetic wave; experiments have amply verified his ideas. Later developments showed that visible light is just one special example of electromagnetic radiation, generated by a wide variety of methods and exhibiting a wide variety of properties. Whatever the mode of generation, all electromagnetic waves travel in empty space with a common velocity, denoted by c. Careful measurements give a value of $c = 2.997925 \times 10^8$ km s^{-1}. For all but the most precise calculations a value of 3×10^8 km s^{-1} may be used. When electro-

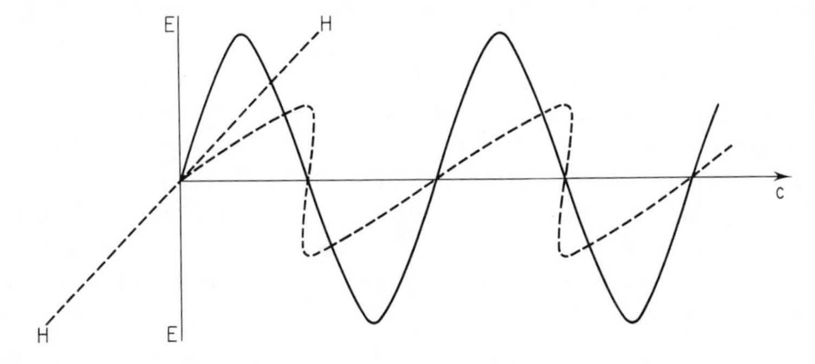

Figure 1-4. In space the two component fields E and H are in phase and are perpendicular to each other and to the direction of propagation.

magnetic waves travel in a ponderable medium, the velocity may be substantially less than c and it may vary with the wave frequency.

All wave motions involving mechanical displacements travel in a readily recognizable medium such as air, water, or a stretched string. Elaborate investigations have attempted without success to identify the medium which propagates electromagnetic radiation. The transmitting medium is of secondary present interest since we shall be primarily concerned with energy exchanges rather than with details of the energy transmission.

1.08 Quantum Nature of Energy Exchanges

If an electromagnetic wave has sufficient energy, it is quite capable of producing an ionization when it is absorbed. Consider specifically a photoelectric cell consisting of a photosensitive cathode surface and a collecting electrode or anode in an evacuated bulb. The circuit from the cathode, which usually consists of an alkali metal such as cesium, is externally completed through a battery and some sort of current-measuring device. Under proper conditions visible light falling on the cathode will eject electrons from it. These electrons will be attracted to the anode by the polarizing potential and a current will flow in the external circuit.

According to the classical wave model of light the oscillating fields in the incident wave set the electrons in the cathode into vibration. The amplitude of these oscillations will increase as energy is absorbed until a critical amplitude is exceeded, when the electron will fly away from the cathode surface. If this model is correct, a strong light should produce some free electrons instantly, whereas an appreciable time should be required for a dim light to bring the electrons up to the critical energy. A quite different response is actually observed. With an alkali metal cathode even the dimmest blue light will instantly produce a few photoelectrons, while a strong red light will produce none even after a prolonged exposure. The explanation of this behavior lies in the quantum ideas of Planck and Einstein rather than in the wave nature of the radiations.

By the beginning of the twentieth century many experiments had shown light to be a wave motion and wavelengths had been measured with high precision. In spite of the demonstrated wave nature of light, all attempts to use it to explain the spectral distribution of the energy radiated from hot bodies had failed. Planck rejected the wave model and proposed that light behaves in energy transfers like a concentrated particle or *photon* with a definite energy content available for the transfer. The *quantum* of energy in a photon was postulated to be

$$E = h\nu \tag{1-16}$$

where h is Planck's constant and ν is the frequency of vibration in hertz.

The universal constant h has a value of 6.6×10^{-34} joule s or 4.14×10^{-15} eV s.

When a photon strikes a photocathode, the energy exchange will be given by

$$h\nu = T + \phi \tag{1-17}$$

where ϕ is the energy or *work function* required to eject an electron from the cathode surface. Energies from separate photons are not additive in the process described by Eq. (1-17) so even a large number of photons will not eject an electron if for each photon $h\nu < \phi$. When the photon energies exceed ϕ, photoelectrons will be produced with kinetic energies given by Eq. (1-17). The predictions of the equation have been precisely confirmed and, in fact, this relation provides one of the best methods for evaluating the constant h.

Further evidence for the need to think of electromagnetic radiation as a series of quanta as well as a wave motion comes from the quantitative relations involved in the production of X rays. Neither the wave nor the particle model provides a complete explanation of the observed properties of the radiation. Both models are needed and it seems evident that each of them merely represents our attempt to describe an entity whose exact nature is unknown. Because of the quantum or particulate nature of light, we are justified in including the photon in the list of elementary particles in Table 1-1.

TABLE 1-1
PROPERTIES OF SOME ELEMENTARY PARTICLES

Name	*Symbol*	*Rest Mass* *umu*	*Rest Mass* *MeV*	*Charge*	*Comments*
Photon	$h\nu$			0	
Neutrino	ν			0	
Electron	e^-	0.000549	0.511	-1	Stable
Positron	e^+	0.000549	0.511	$+1$	Stable*
Muon	μ	0.11320	105.659	$\pm 1, 0$	Unstable
Pion	π	0.14990	139.578	$\pm 1, 0$	Unstable
Proton	p	1.007277	938.256	$+1$	Stable
Neutron	n	1.008665	939.550	0	Life, 12 min

*Stable in free space; in matter, lifetime $\simeq 10^{-6}$ s.

1.09 The Electromagnetic Spectrum

The known frequency range of electromagnetic radiation is enormous, Fig. 1-5. Commercial power lines radiate at the generator frequency, usually 60 Hz, but it is quite easy to produce much lower frequencies so the lower

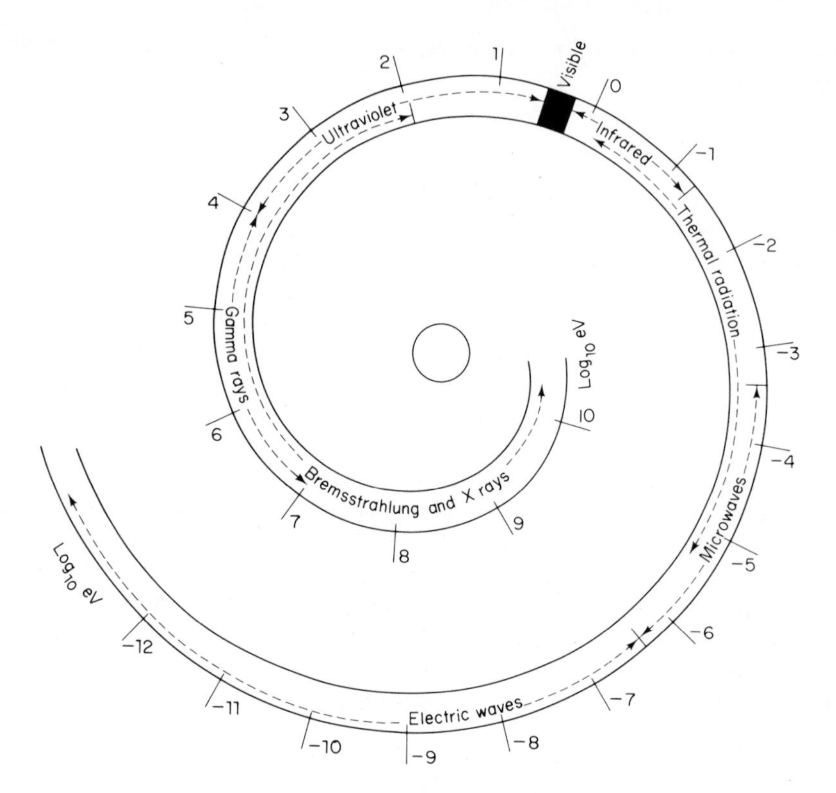

Figure 1-5. A logarithmic display of the electromagnetic spectrum. Most of the divisions between regions are quite arbitrary.

limit is practically zero. Vacuum tubes and other electronic devices provide sources up to about 10^{11} Hz and this limit may be extended with future technical developments. This frequency region contains all the applications to radio communication, radar, diathermy, and commercial dielectric heating. Radiations in the range of 10^8 to 10^{11} Hz are sometimes known as *microwaves* because they represent about the shortest wavelengths that can be generated with electronic equipment presently available.

All matter not at absolute zero temperature will emit electromagnetic radiation produced by the rotations and vibrations of the constituent electric charges. There will be a characteristic broad emission spectrum at each temperature, the most probable photon energy being given by

$$E = kT \quad \text{eV} \tag{1-18}$$

if k, which is the Boltzmann constant, is taken as 8.62×10^{-5} eV/°K. Known absolute temperatures range from a fraction of a degree above zero to 10^7

degrees or so and the radiated frequencies cover a correspondingly wide range. Thermal radiations overlap almost all known regions of the electromagnetic spectrum, from the lowest power frequencies up into the X-ray region.

The infrared region, extending down in energy from the red end of the visible spectrum, is defined by methods of detection rather than by source types. Here, and in the higher-energy regions, the frequency designations used at lower energies are usually abandoned in favor of energy or wavelength designations. At these higher energies, frequency cannot be determined directly but must be calculated from other parameters.

A variety of units are in common use for specifying wavelengths, chiefly the meter with appropriate prefixes and the angstrom. The micron and the millimicron are preferably replaced by the micrometer and the nanometer, respectively. A useful, readily derivable conversion of Eq. (1-16) is

$$E = \frac{12{,}400}{\lambda\ (\text{angstroms})}\ \text{eV} \tag{1-19}$$

The low-energy limit of the infrared may be set at a wavelength of 25 μm (0.05 eV), which is the limit of useful transmission by a potassium bromide prism. The infrared extends to perhaps 7000Å (1.8 eV), the red limit of human vision. This continues to 4000Å (3.1 eV) where the visible violet merges into the invisible ultraviolet.

Although photon energies increase through the ultraviolet, they are strongly absorbed. The 1900–10 Å subregion is known as the vacuum ultraviolet since here even a very short air path will absorb the UV photons.

Bremsstrahlung and X-ray production start at about 100 eV and extend upward to the highest photon energies produced by man, about 2 GeV. Over this range the photons gradually change from highly absorbable to the strongly penetrating radiations used for flaw inspection of large castings or for deep radiation therapy. Gamma rays, the photons emitted by radioactive nuclei, occupy a small portion of this energy region, from about 10 keV to 10 MeV.

The demands and accomplishments of modern technology are pushing practical applications of photon radiation outward at each end of the energy scale. Much work has been done in some regions toward an understanding of the biological effects. In other areas little or nothing is known about biological effects at the power levels now in use or contemplated for future use.

1.10 Wave Properties of Particles

In 1924, the dichotomy already established for photons was extended to entities previously considered to be strictly particulate. De Broglie suggested that a wave structure of wavelength

$$\lambda = \frac{h}{p} = \frac{h}{mv} \tag{1-20}$$

$$\text{cm} = \frac{\text{erg s}}{\text{g cm s}^{-1}}$$

was to be associated with any moving particle. By substituting appropriate values into Eq. (1-20) we see that for the usual range of macroscopic masses and velocities the associated wavelengths are far too small to be detected. In the domain of atomic particles, however, the wavelengths are comparable to atomic dimensions and diffraction effects should be observed.

Experimental results quickly confirmed the theoretical predictions of de Broglie. Electron beams directed at crystals showed diffraction patterns similar to those seen with X rays and the calculated wavelengths were in accord with Eq. (1-20). Electron diffraction was soon put to practical use in the electron microscope; this has permitted the resolution of objects much closer together than can be separated by optical microscopy. When beams of neutrons became available, they too were shown to exhibit diffraction. Today neutron diffraction spectrometers are important tools in the study of atomic and molecular structure.

We are again faced with the fact that neither a wave model nor a particle model alone can adequately describe the behavior of particles in the atomic domain. As with electromagnetic radiation, the true unified model is not available and the two apparently incompatible models must be used, each where it is appropriate.

It must be understood that the de Broglie waves are not to be associated with electromagnetic waves. The de Broglie waves have no vectors comparable to the electric and magnetic fields that can directly actuate detectors and thus demonstrate their presence.

The pioneer work of de Broglie led to the development of entirely new concepts, replacing both the old classical mechanics and the quantum theory of Planck. According to the *wave* or *quantum, mechanics*, each particle has an associated set of *wave functions* which can have only a set of discrete energy values. These wave functions have both real and imaginary parts and so they cannot directly represent a real particle. The square of the wave function will, however, have only real and positive values and these are interpreted as expressing the probability of the particle's location. According to the wave mechanics it is not possible to say with complete assurance that a particle *is* at a precise location. Certainty in the older mechanics must be replaced by probabilities. The probability function may have a maximum value at a precise point but there will be a finite probability that the particle is not exactly at that point. If the initial position of a particle is not precisely known, its subsequent positions also can not be established with certainty, even though it moves in a precisely defined field.

A theory which is not subject to absolute proof and which departs so drastically from such basic ideas as precise location and causality must prove its validity by verified predictions. In this the quantum mechanics has been remarkably successful. The mathematical difficulties of calculating even the simplest systems are profound but progress has been spectacular and predictions are in good accord with experiment.

1.11 The Indeterminacy Principle

Even though the position of a particle cannot be determined exactly, there is a most probable position and the probability of existence in another position falls off very rapidly with distance from the most probable point. Consequently the wave functions, whose squares give the probabilities, cannot be the extended, constant amplitude waves that we have come to recognize in more familiar wave motions. A localized wave structure can be obtained by combining a group of constant amplitude waves if each component wave has a slightly different wavelength. Such a group can be in phase and additive at one point but it will rapidly get out of phase and diminish in amplitude away from the in-phase point. Such a wave group might, then, serve to describe a particle.

The wavelength spread of the constituents of the group will be due to slight differences in the velocities of propagation. The in-phase or most probable position will move through space with what we call the velocity of the particle. It has been necessary to introduce a velocity spread, however, and so we are unable to define precisely the particle's velocity. Because of the inherent nature of things, it is not possible to establish precisely either the location or the velocity of a moving particle.

Heisenberg put these considerations into quantitative form in his *indeterminacy principle*, which is usually written as

$$\Delta x\, \Delta p \simeq \frac{h}{2\pi} = \hbar \tag{1-21}$$

$$\text{m kg m s}^{-1} = \text{J s}$$

The Δ symbols in Eq. (1-21) are to be read *the uncertainty in*. x is a position coordinate and p is the corresponding momentum. Equation (1-21) states that either the coordinate or the momentum may be determined with any desired degree of precision but any increase in the certainty of one parameter will inevitably result in a decreased certainty of the other. Note that these uncertainties are conceptual; they are not the result of any imperfection or inadequacy in the measuring equipment. Note also that there is a slight uncertainty in the product of the Δ's. This product will be very nearly but not necessarily exactly equal to $h/2\pi = 1.05 \times 10^{-34}$ J s. The factor $h/2\pi$ occurs so frequently that it is given a special symbol $\hbar$.

Another useful relation is obtained by expressing the Heisenberg principle in terms of energy and time. A simple conversion of Eq. (1-21) leads to

$$\Delta E \, \Delta t \simeq \hbar \tag{1-22}$$

where E is the energy of a given state and t is the time of existence of that state.

Illustrative Example

A particular excited state is found to exist for 10^{-5} s before it goes to the ground state with the emission of a gamma ray. What will be the range of energies of the emitted gamma rays?

The uncertainty in t cannot be greater than t itself. If we set them equal, we shall be maximizing Δt and minimizing ΔE. For calculations of this sort it is useful to take $\hbar$ in electron volt seconds (eV s) rather than in joule seconds (J s). Then

$$\Delta E \times 10^{-5} = \frac{4.14 \times 10^{-15}}{2\pi} = 6.56 \times 10^{-16} \text{ eV s}$$

$$\Delta E = 6.56 \times 10^{-11} \text{ eV}$$

Although this is only an approximation, the value calculated is a good representation of the width of the energy state. The emissions are obviously nearly monoenergetic.

REFERENCES

FEYNMANN, R. P., R. B. LEIGHTON, and M. SANDS, *The Feynmann Lectures on Physics*, Vol. I, II, and III. Addison-Wesley Pub. Co., Reading, Mass., 1965.

LAPP, R. E. and H. L. ANDREWS, *Nuclear Radiation Physics*, 4th ed. Prentice-Hall, Inc., Englewood Cliffs, N.J., 1971.

LEIGHTON, R. B, *Principles of Modern Physics*, International Series in Pure and Applied Physics. McGraw-Hill Book Co., New York, N.Y., 1959.

SHORTLY, G. and D. W. WILLIAMS, *Elements of Physics*, 5th ed. Prentice-Hall, Inc., Englewood Cliffs, N.J., 1971.

2

Atomic Structure

2.01 The Rutherford Model

For many years after the electrical constituents of matter were recognized, there was uncertainty as to the way in which the charges were distributed. Some visualized each electrically neutral atom as a sort of gel with the charges distributed rather uniformly throughout it. Others believed the charges to be more concentrated. The definitive experiments and the accompanying theory were reported by Rutherford in 1911.

Rutherford used the energetic, positively charged alpha particles emitted by radium and other radioactive substances as probes, directing these beams at very thin metal films. As an alpha particle approaches close to an atom, it no longer sees electrical neutrality, as at a distance, but rather interacts individually with the component charges. The resulting trajectory of the alpha particle will depend strongly on the spatial distribution of the charges.

Figure 2-1 shows the situation on the assumption as made by Rutherford that the negative charge is spread over a thin shell at a relatively great distance from a highly concentrated positive charge. The magnitude of the latter is taken to be Z times the fundamental charge unit e. Once inside the negative shell the incoming alpha particle will be repelled by the positive charge on the *atomic nucleus*. The alpha particle will be deflected or *scattered* from its original trajectory by an amount which depends on the relation of this trajectory to the nucleus, Fig. 2-1. Although most of the atomic mass is assumed to be in the central nucleus, it will not remain stationary and its recoil must be taken into account in exact calculations.

Rutherford's theoretical treatment of the scattering situation showed that

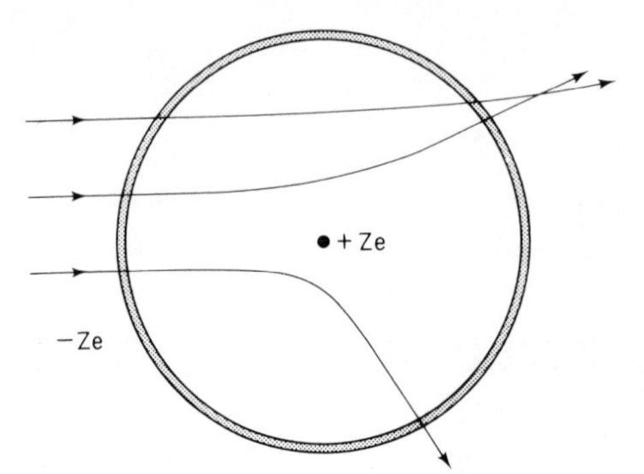

Figure 2-1. The Rutherford scattering of alpha particles, charge $+2e$, by an atomic nucleus of charge $+Ze$.

the relative number of particles scattered in each direction would be given by

$$f_\phi = N\left(\frac{Zze^2}{4T}\right)^2\left(\frac{1}{\sin^4(\phi/2)}\right) \tag{2-1}$$

where f_ϕ = fraction of the incident particles scattered through angle ϕ
N = number of target atoms per unit area
Ze = charge on the nucleus
ze = charge on the incoming particle, $2e$ for the doubly charged alpha particles
T = kinetic energy of the incident particles

The form of Eq. (2-1) predicts that small angle scattering will predominate with relatively few particles being deflected through large angles. An experimental confirmation of the predictions of Eq. (2-1) would not only furnish strong support to the Rutherford concepts but would also permit a determination of the Z values for each scattering element.

The experimental test of the model consisted in the tedious counting, as a function of ϕ, of the number of tiny visible scintillations produced when the scattered alpha particles struck a zinc sulfide (ZnS) screen. Experimental results confirmed the model in every way and permitted the determination of Z for a series of metallic elements. In every case Z was found to be an integer, with a value nearly one-half of the atomic weight of the element.

The Rutherford scattering experiments demonstrated unequivocally that an atom consists of a nucleus of about 10^{-15}-m radius, containing almost all the atomic mass and an integral number of positive charges. Much later the distance of 10^{-15} m became known as the fermi, or f, named after the famous Italian nuclear physicist. The neutralizing electric charges are electrons, distributed at a radius of the order of 10^{-10} m. Most of the atomic volume is empty space.

The Rutherford scattering experiment was of the utmost importance in establishing the basic structure of the atom. It is fitting that our first direct knowledge of the chemical constitution of the moon came from similar scattering measurements, with sophisticated modern detectors replacing the ZnS screen, the microscope, and the human eye.

2.02 Spectral Emissions of Excited Elements

Spectroscopy in the visible and the ultraviolet regions was a highly developed science by the time the Rutherford model of the atom was demonstrated. Each element was known to emit spectral lines at characteristic wavelengths when the element was excited in an electric arc or by other suitable means. Some spectra were simple, with only a relatively few lines. Others, notably iron, contained thousands of lines in a seemingly random distribution. No trace of the harmonic overtones seen in some other types of wave motions was observed in the spectral emissions.

In some spectra, lines could be grouped in series according to their appearance in the spectroscope or by their persistence as the level of atomic excitation was reduced. Thus came the designations of *sharp*, *principal*, *diffuse*, and *fundamental* series. Empirical relations were developed to fit the observed wavelengths but no satisfactory theoretical explanation or model was available to account for them. It remained for Bohr, building on the results of the Rutherford results, to provide a satisfactory explanation of the emission spectra.

2.03 The Bohr Theory of Hydrogen

A giant step forward was taken in 1913 when Bohr announced his model for the emission of spectral lines by atomic hydrogen. Bohr assumed that the hydrogen atom consisted of a relatively massive, singly charged nucleus, called the *proton*, with a single electron revolving around it in a closed orbit,

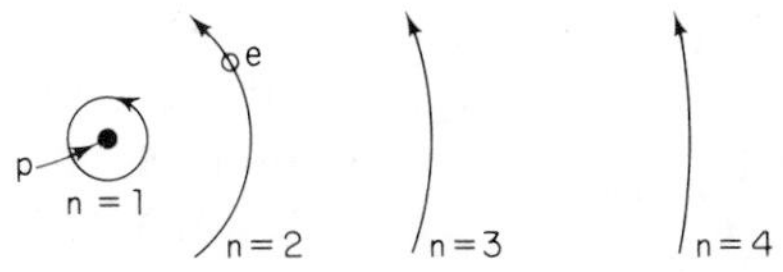

Figure 2-2. Schematic representation (not to scale) of the allowed orbits with principal quantum number designations. The electron shown is in the first excited state.

Fig. 2-2. Any object moving in a circular orbit is undergoing a continuous acceleration toward the center of rotation. One of the basic requirements of electromagnetic theory is that an accelerated electric charge radiate energy in the form of electromagnetic waves. According to classical theory the electron in the hydrogen atom should rapidly radiate its energy and spiral into the nucleus. This obviously does not take place.

Bohr proposed that the electron could exist without radiating in a particular series of orbits known as *stationary states*. Each stationary state was assumed to have a definite energy value and the atom could absorb or radiate energy only when the electron moved from one stationary state to another. Under normal, unexcited conditions the electron will remain in the state of lowest energy, the *ground state*. When energy is absorbed, the electron is forced into a higher-energy or *excited* state. Residence times in excited states will be, in general, very short. After 10^{-8} s or so the electron will return to a lower-energy state with the emission of a photon. The photon energy will be related to the state energies by

$$h\nu = E^* - E \tag{2-2}$$

where E^* is the energy of the excited state involved in the transition. The amount of energy given by Eq. (2-2) is precisely the amount needed to create E^* from E in the first place but this energy was not necessarily obtained by the absorption of a photon. If E^* represents a high level of excitation, the return to the ground state may be by any one of a series of alternative routes of energy degradation through intermediate stationary states rather than by dropping in one step to the ground state.

2.04 Energy of a Stationary State

Bohr applied the laws of nonrelativistic mechanics to the circular motion of the electron around the nucleus and obtained some simple relations among the electron mass m, velocity v, orbital radius a, and energy E of the system. In the classical mechanics there was no rule for selecting the particular values of the parameters that would yield the proper stationary states. Bohr added the requirement that for a stationary state

$$mv(2\pi a) = nh \tag{2-3}$$

where n can take on only integral values 1, 2, 3, ... The Bohr requirement *quantized* or restricted the system to specific values of a and v and hence of E. The running integer n was called the *principal quantum number*.

Although the quantization relation is dimensionally correct, there were no basic, irrefutable considerations that led to its choice. Justification for the quantization rule came quickly from the remarkable ability of the Bohr theory to predict the wavelengths of the lines emitted by the hydrogen atom.

The Bohr quantization leads to energy values:

$$E_n = -\frac{me^4}{8\epsilon^2 h^2}\left(\frac{1}{n^2}\right) \text{ joules} \tag{2-4}$$

when all quantities are taken in the MKSA system. The minus sign in Eq. (2-4) appears because the zero point of the energy scale was chosen for an electron radius $a = \infty$. When the electron moves into an orbit with a finite radius, work will have been done *by* the system because of the force of attraction between the two charges. This energy will have been radiated out of the system; the final system energy must be less than the original energy and hence negative.

The energy radiated in a transition is

$$E_1 - E_2 = h\nu = -\frac{me^4}{8\epsilon^2 h^2}\left(\frac{1}{n_1^2} - \frac{1}{n_2^2}\right) = -R\left(\frac{1}{n_1^2} - \frac{1}{n_2^2}\right) \qquad (2\text{-}5)$$

where the n values are related to the orbital radii as in Fig. 2-2. Equation (2-5) has precisely the form developed empirically by the experimental spectroscopists. They had determined the value of the constant multiplier outside the parentheses, the Rydberg constant R, from experimental data. The Bohr theory permitted a calculation of this constant from basic constants that were not evaluated from spectroscopic measurements. The triumph of the Bohr theory was immediate. The calculated wavelengths agreed very closely with those observed in the laboratory.

The Rydberg constant has the dimensions of an energy since each n is a dimensionless number. Spectroscopists find it convenient to express R in wave numbers but for present purposes the electron volt is more useful. When evaluated, $R_\infty = -13.6$ eV where the subscript ∞ denotes that the value applies to the case of an infinitely massive nucleus which does not move as the electron revolves. The proton does indeed take part in the motion but this needs to be accounted for only in precise calculations or in the case of hydrogen-like structures where the nuclear mass is comparable to that of the electron.

Figure 2-3 shows some of the possible energy transitions in hydrogen. The lines empirically classified into series by the early spectroscopists can now be identified with specific energy states and transitions. For example, all transitions ending at the state $n = 1$ produce lines in the Lyman series; transitions to $n = 2$, the Balmer series; and so on. Each radiated energy will be less than 13.6 eV. If this energy, or more, is added to a hydrogen atom in the ground state, the electron will be freed from the proton to produce an ion pair. Thus 13.6 eV is the ionization potential of hydrogen.

2.05 Additions to the Theory

Although the Bohr theory correctly predicted most of the observed emissions of atomic hydrogen, there were some disturbing discrepancies well outside any experimental error. Several extensions of the theory were made in an attempt to eliminate the lack of agreement.

When an inverse-square law of force is operating, the laws of mechanics permit elliptical as well as circular orbits around a central body. Long before

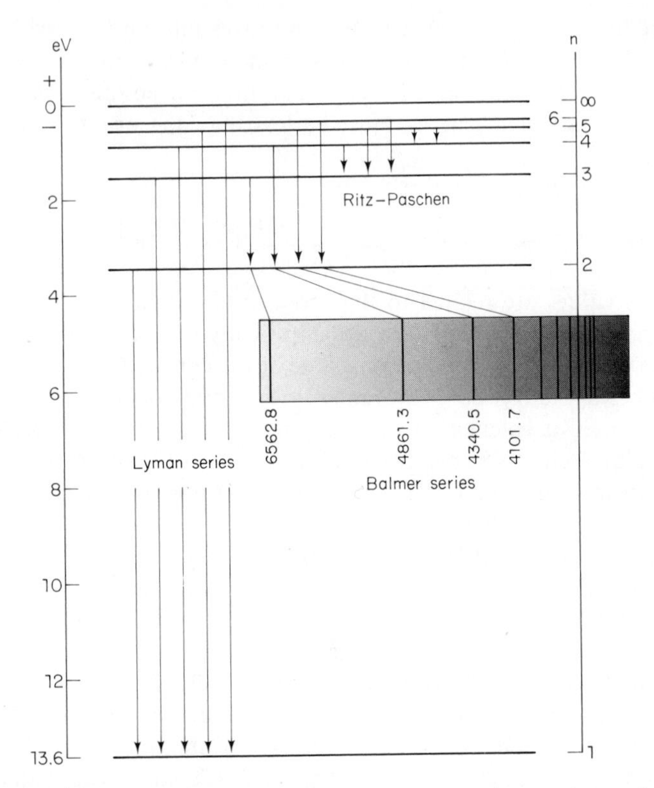

Figure 2-3. Partial energy-level diagram for atomic hydrogen. Some of the spectral lines of the Balmer series are shown as they are actually observed in a spectrograph, with wavelength designations in angstrom units.

Bohr, Kepler had shown that the planets in our solar system move in elliptical orbits around the sun with the latter at one focus of the ellipse. When elliptical orbits are introduced into the atomic case, a second quantum number is required to restrict the orbits to certain degrees of ellipticity. It is more precise to say that the *orbital* quantum number l quantizes the angular momentum of the electron. An extreme ellipse, a straight-line orbit through the proton, will have a zero angular momentum, while a circular orbit will have the maximum value allowed by energy restrictions, Fig. 2-4. The angular momentum of each state must be compatible with the momentum of the system allowed by the principal quantum number n. This requirement restricts the orbital quantum number to the integral values 1, 2, 3, . . . , $(n - 1)$.

An electron moving in its orbit is equivalent to an electric current in a closed circuit and such a current produces a magnetic field or is said to have a *magnetic dipole moment* μ, Fig. 2-5. Such a dipole moment will orient itself in an external magnetic field H. The energy of the system will then depend

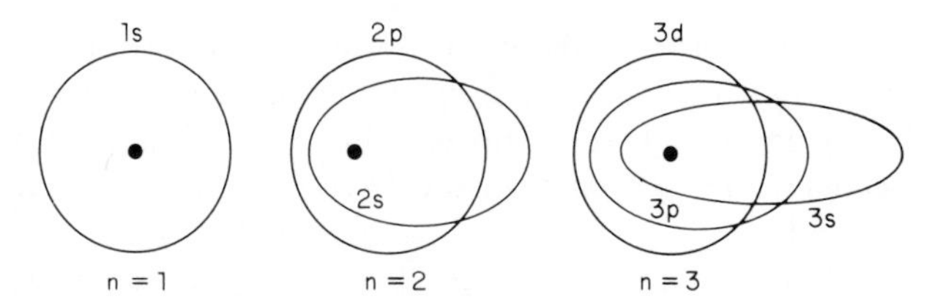

Figure 2-4. Some allowable elliptical orbits with the letter designations developed from spectroscopic observations.

on its degree of alignment with the field; hence this orientation must also be quantized. The *magnetic* quantum number m can take on integral values $0, \pm 1, \pm 2, \pm 3, \ldots, \pm l$.

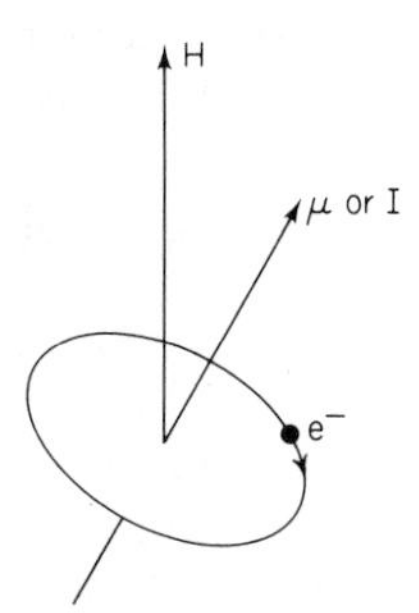

Figure 2-5. An electron in orbital motion produces a magnetic moment which interacts with an external magnetic field H.

The electron appears to spin about its own axis and this spin gives rise to an angular momentum in addition to that in the orbit that is quantized by l. Electron spin, like the magnetic orientation, can have two different directions referred to the orbital motion. Only one numerical value of electron spin is observed, however, so the *spin* quantum number, s, can have only the two possible values $\pm\frac{1}{2}$.

Four quantum numbers n, l, m, and s are required to designate a state of hydrogen. The energy level is determined primarily by n with the other numbers specifying small variations around the level. Agreement with experiment was improved by the use of the more complex theory but some discrepancies remained that were to be removed only by the introduction of the wave mechanics.

2.06 Wave Mechanical Concepts

A generalized wave equation, applicable to any type of wave motion, furnished the starting point for the wave mechanical treatment of the hydrogen atom. The de Broglie relation $\lambda = h/mv$ was then introduced to make the equation apply specifically to the particle waves. Solutions of the resulting *Schrödinger equation* exist for only specific integral values of four variables. Thus the four quantum numbers, introduced rather arbitrarily into the older theory, appeared naturally as mathematical necessities for the solution of the differential equation.

Wave mechanics retains the concept of stationary states but replaces the picture of precisely localized orbits by probability functions. In the old theory the radius of the ground-state orbit in hydrogen, 5.29×10^{-11} m, is known as the *Bohr radius*. This same value appears in the wave mechanical calculation but only as the radius of greatest probability when $n = 1$. There is now a small but finite probability that the electron will be found at a slightly different radius than that predicted precisely by the original theory.

The predictions of the wave mechanics agree almost exactly with experimental results and the theory must be regarded as a very accurate representation of atomic affairs. The Bohr treatment is a good approximation and has the advantage of a simpler physical picture; so it will be used freely, recognizing that precise calculations will be made from the wave-mechanical model.

2.07 Applications to Heavier Atoms

Atoms beyond hydrogen in the periodic table will have Z values greater than 1 and each will require Z orbital electrons in the neutral state. Helium, an atom with two orbital electrons, has an emission spectrum distinctly different from that of hydrogen. When one electron is completely removed by an ionization, the remaining structure He^+ is hydrogen-like in that a single electron has a set of energy levels in a centrally acting, inverse-square-law force field. The energy levels are

$$E_n = -\frac{Z^2 m e^4}{8\epsilon^2 h^2}\left(\frac{1}{n^2}\right) \tag{2-6}$$

where $Z = 2$ for He^+, $Z = 3$ for Li^{2+}, and so on. Note that the energies increase rapidly, as Z^2.

In any atomic configuration each orbital electron has an energy designated by the four quantum numbers. The values that can be assumed by these numbers are not completely independent but are governed by the *Pauli exclusion principle*. According to this principle only one electron in each atom can be in any energy state, which is equivalent to saying that no two electrons can be described by exactly the same set of quantum numbers. The Pauli principle has many far-reaching consequences.

Consider the situation at the lowest energy level where $n = 1$. Then l can only have the value $0 = (n - 1)$ and m is necessarily zero since it cannot exceed l. There are two possible spin values and so for $n = 1$ there can be a maximum of two electrons. These will have oppositely directed spins, $+\frac{1}{2}$ and $-\frac{1}{2}$. The two electrons have nearly equal energies and form a *closed shell*, known as the K shell.

When $n = 2$, l can have values of 0 or 1. As before, only two electrons can exist in the subshell $n = 2$, $l = 0$. In the $l = 1$ subshell, m can have the values 0, +1, or −1. Each of these m states can be occupied by two electrons, with oppositely directed spins, and so the $l = 1$ subshell has space available for six electrons. Eight electrons will completely fill the L shell.

Electrons with $l = 1$ are known as s electrons (distinguish this s from that of the spin quantum number); p electrons are those with $l = 2$; and thence $d, f, g, h, \ldots$, in order of increasing l values. The peculiar sequence of the first four designations comes from the original spectroscopic descriptions as *s*harp, *p*rincipal, *d*iffuse, and *f*undamental. These series types can now be identified with the corresponding values of the orbital quantum number.

Atomic electron configurations are given by first specifying the principal quantum number, then the letter description of the orbital number, and finally the number of electrons in that configuration as a superscript. In the case of argon with 10 orbitals, the designation is $1s^2 2s^2 2p^6$. The next element in the periodic table is sodium, $Z = 11$, and now one electron must go into the third energy level, or M shell. The complete designation will be $1s^2 2s^2 2p^6 3s^1$. This is frequently written as Ar $3s^1$ since the structure consists of an argon core with one more electron added.

The chemical properties of each element are determined primarily by the electron configuration of the outer, or *valence*, shell. The filled-shell sequence He, Ne, Ar, . . . , are all noble gases, essentially inert chemically. Atoms with exactly filled orbital shells have a greater stability (less reactivity) than the neighboring structures which either lack shell completion or are just starting to fill a new shell. The noble metals Cu, Pd, and Pt are found at Z values where the $3d$, $4d$, and $5d$ subshells, respectively, are just closed. Each element in the sequence immediately following the noble gas sequence has one electron in the valence shell. Each of these elements is a reactive alkali metal Li, Na, K, ...

As we move upward through the periodic table with increasing values of Z, there is a tendency for one orbital shell to fill completely before electrons start to enter the next higher shell, each configuration assuming the arrangement of lowest total energy. At the higher quantum numbers the energy differences between successive n values becomes small and one shell may be incomplete when entry into the next higher shell begins. Thus the elements from Ce, $Z = 58$, to Lu, $Z = 71$, have nearly constant chemical properties because each has a constant outer electron configuration of $5d6s^2$, while the inner $4f$ shell is filling from $4f^1$ to $4f^{14}$, Fig. 2-6.

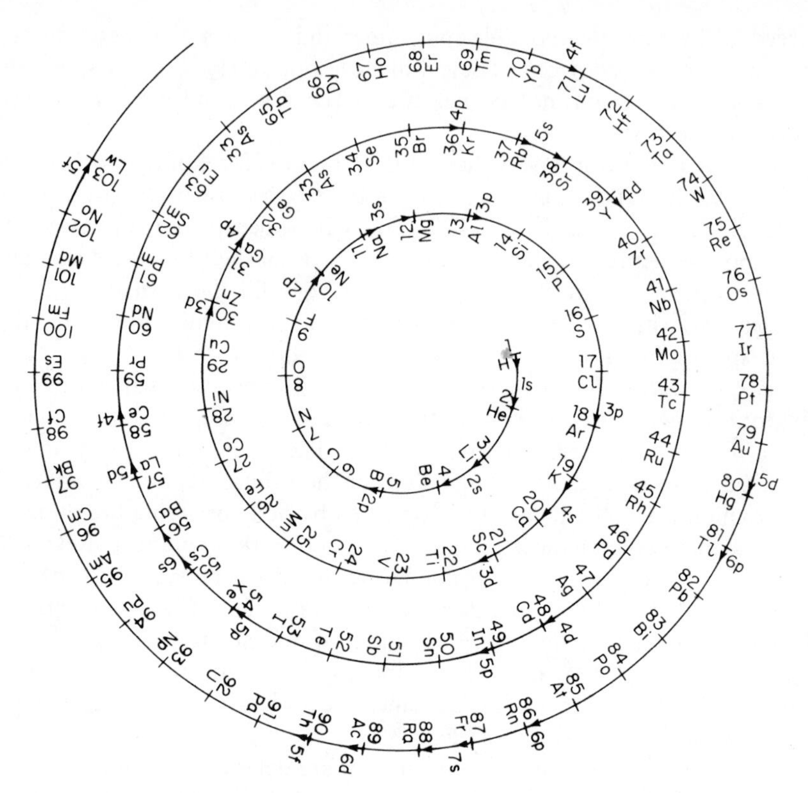

Figure 2-6. The periodic table of the elements showing electron shell filling.

2.08 Energy-Level Multiplicity

An atom of hydrogen will have an angular momentum arising in part from the movement of the 1*s* electron in its orbit and in part from the spin of the electron. The two angular momenta will combine, or *couple*, to form the total angular momentum of the atom. Angular momentum is a vector and consequently direction as well as magnitude must be taken into account in combining the two components into one. The two spin possibilities $+\frac{1}{2}$ and $-\frac{1}{2}$ will have different relations to the orbital angular momentum and so two slightly different energy levels are to be expected for each value of n and l. Each spectral line would then be expected to be a closely spaced *doublet* rather than a single monoenergetic emission.

Any closed shell or subshell will have an even number of electrons with all possible combinations of spin and orbital motions represented. Oppositely directed vectors will cancel and the net angular momentum of any closed shell will be zero. In any atomic configuration the total angular momentum and hence the energy will be determined by the angular momenta of the electrons in the nonclosed shells, usually the valence electrons in the outermost shell. As an example, the argon core in the sodium atom has closed K and L shells, and these make no contribution to the total angular momentum. This quantity is determined only by the single $3s$ electron in the M shell. As in the case of hydrogen this electron can have either spin direction and again the spectral lines should be doublets. The well-known D emission lines of sodium at 5890 and 5896 Å form the doublet produced by transitions from the first excited state.

Two possibilities arise when there are two electrons in the valence shell outside a closed-shell core. In either configuration the total orbital angular momentum will be $L = l_1 + l_2$. In the simplest case the spins of the two electrons will be oppositely directed, or *antiparallel.* Then the total spin $S = s_1 + s_2 = +\frac{1}{2} - \frac{1}{2} = 0$, Fig. 2-7A. The total angular momentum $J = L + S$ can now have only one value for each value of L and so each energy level is a *singlet.*

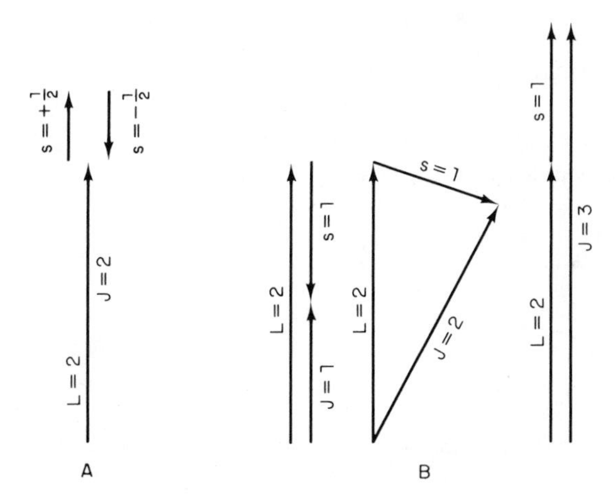

Figure 2-7. Spin–orbit coupling of angular momenta to form (A) a singlet energy state and (B) a triplet.

When the two electrons have equally directed, or *parallel*, spins, $S = +\frac{1}{2} + \frac{1}{2} = 1$ and $J = L + 1$. Vector addition permits three integral values of J for each integral value of L, Fig. 2-7B, and so each energy level is a *triplet*. In general the multiplicity of a level will be $(2S + 1)$.

The summing method described here is known as Russell–Saunders or LS coupling. When this type of coupling applies, all the l values and the s values are summed separately and the two sums are then added to obtain J. In jj coupling, which applies in some cases, each individual electron is summed, as $j_1 = l_1 + s_1$, and the j values are then summed to obtain J. Both types of coupling lead to the same multiplicity values and the details of the summations need not concern us here.

2.09 Magnetic Spin Resonance

The magnetic dipole moment of an electron can be represented by a vector having the same direction as that which represents the orbital angular momentum, Fig. 2-5. It can be shown that the relation between the dipole moment μ and the angular momentum I is

$$\mu = \frac{eI}{2m} = \frac{n\hbar e}{2m} \text{ joules tesla}^{-1} \tag{2-7}$$

where m is the mass of the electron and not the magnetic quantum number. If we set $n = 1$ in Eq. (2-7), we obtain an expression for the dipole moment of the hydrogen electron in the ground state.

$$\mu_B = \frac{he}{2m} = 9.27 \times 10^{-24} \text{ J T}^{-1} \tag{2-8}$$

μ_B is the *Bohr magneton*, a convenient unit of pole strength in the atomic domain.

A constant external magnetic field interacts with the dipole vector in a manner exactly analogous to the action of the earth's gravitational field on a

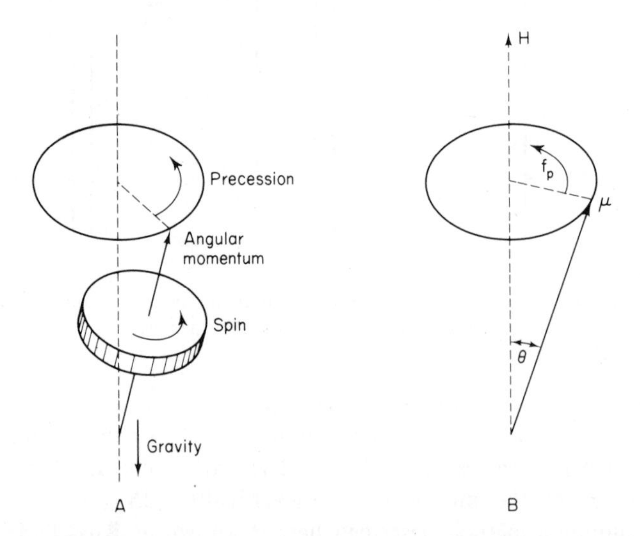

Figure 2-8. (A) A spinning gyro precesses rather than falls under the pull of gravity. (B) The magnetic moment due to orbital and spin angular momenta precesses around the external field vector.

spinning gyroscope. A gyroscope with its axis off-vertical, Fig. 2-8A, will experience a torque due to gravity which will tend to cause it to fall. Instead of falling the gyro responds by revolving or *precessing* around a vertical axis. In the magnetic analog, Fig. 2-8B, the dipole vector makes a quantized angle θ with the external field. This angle is determined by the value assumed by the magnetic quantum number. Like the gyro, the dipole vector, and hence the plane of the orbiting electron, precesses around the external field vector. The frequency of this *Larmor precession* can be shown to be

$$f_p = H\frac{e}{4\pi m} \text{ Hz} \tag{2-9}$$

Note that the frequency of precession is independent of all quantum numbers, and hence of the angle θ. The precessing electron will absorb energy from an external field of frequency f_p superposed on the steady field H. As it absorbs energy, the electron will go into higher quantum states but the frequency of precession will not change.

Precession frequencies are readily attainable with ordinary electronic techniques.

Illustrative Example

Calculate the frequency of precession of the hydrogen electron in a steady field of 1 gauss $= 10^4$ webers m^{-2}.

From Eq. (2-9) by direct substitution

$$f_p = \frac{10^4 \times (1.6 \times 10^{-19})}{4\pi \times (9.1 \times 10^{-31})} = 1.4 \times 10^6 \text{ Hz}$$

Sophisticated equipment has been developed to measure the frequencies of precession. In making these measurements it is usually most convenient to maintain the external oscillating field constant at some value close to the expected value of f_p and to vary H slowly through the resonance absorption.

When atoms are bound into molecular structures, the precession frequencies will not be given precisely by Eq. (2-9) because each electron will be moving in the fields of its neighbors as well as in the field H. Values of f_p can, however, be identified with known chemical groupings and the technique of *electron spin resonance* has become a very powerful analytical tool.

REFERENCES

EVANS, R. D., *The Atomic Nucleus*, International Series in Pure and Applied Physics. McGraw-Hill Book Co., New York N.Y., 1955.

FEYNMAN, R. P., R. B. LEIGHTON, AND M. SANDS, *The Feynman Lectures on Physics*, Vol. I and II. Addison-Wesley Pub. Co., Reading, Mass., 1965.

LAPP, R. E. AND H. L. ANDREWS, *Nuclear Radiation Physics*, 4th ed Prentice-Hall, Inc., Englewood Cliffs, N.J., 1971.

SNIPES, W., ed., *Electron Spin Resonance and the Effects of Radiation on Biological Systems*. National Academy of Science, Washington, D.C., 1966.

WHITE, H. E., *Introduction to Atomic Spectra*, International Series in Physics. McGraw-Hill Book Co., New York, N.Y., 1934.

3

Bremsstrahlung and X-ray Production

3.01 Roentgen and the Early Discoveries

During the early 1890's many physicists had been studying electrical conduction in gases at low pressures. The high voltages necessary to produce the gaseous discharge were obtained from crude spark coils or from electrostatic generators known as "influence machines." By applying these voltages to electrodes in partially evacuated glass tubes, Crookes, Lenard, and Hittorf, among others, had discovered many of the basic phenomena associated with gaseous conduction.

Wilhelm Conrad Roentgen was also working with gas discharge tubes; in December, 1895, he announced the discovery of a new type of radiation,* which he called X rays because their nature was then almost completely unknown. This report aroused great scientific interest and many laboratories were able to repeat and extend Roentgen's experiments promptly. Studies of these new radiations proceeded so intensively that the first case of human injury from X rays appears to have been recorded early in 1896.†

It is impossible to evaluate accurately the importance of Roentgen's discovery to science and indeed to all mankind. Prior to this work many scientists believed that there was little left for physics to accomplish, save the more accurate determination of numerical constants. Today, with radiation

*Communication to the Physikalisch Medicinschen Gesellschaft of Wurzburg, Dec. 28, 1895. Translated by A. Stanton, *Science* **III**, 227, 726, 1896.

†Brown, P., *American Martyrs to Science Through the Roentgen Rays*. Charles C. Thomas, Pub., Springfield, Ill., 1936.

both a powerful servant and a potential destroyer of man, it is foolhardy to predict where future developments will lead.

Early research soon led to the type of discharge tube shown in Fig. 3-1 as most suitable for the production of X rays. The electrical discharge between the electrodes depended on the presence of residual gas left in the tube by incomplete pumping. As the tube was used, some of the residual gas became adsorbed on the glass walls and the pressure decreased. At the lower pressure more voltage was required to start the discharge and because of

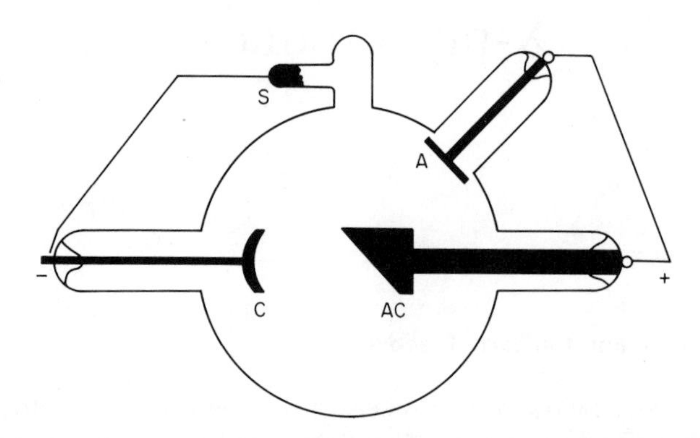

Figure 3-1. Early type of gas discharge X-ray tube. Positively charged gas ions bombard cathode *C* and liberate electrons. These bombard the "anticathode" *AC* and produce bremsstrahlung and characteristic X rays.

this difficulty the tube was said to be *hard.* To maintain the tube at proper working pressure, a gas-evolving material was heated by either an electric current or a small flame, and the evolved gas *softened* the tube by increasing the pressure. Today, X rays are not produced in gas-filled tubes but we still refer to *hard* X rays, meaning those produced by high voltage and having good penetrating power, as contrasted to the *soft* radiations produced at lower voltages.

Quality or the penetrating ability of X-radiation is determined by the voltage applied to the tube. The *quantity* of radiation produced at any given voltage is directly proportional to the number of electrons striking the target, which is to say, the current through the tube. In the old gas tubes, current and voltage were related through the gas pressure and only by changing this pressure could the current–voltage ratio be altered. Frequent checks were required to assure that the tube was in proper operating condition.

Tube operation was customarily checked by examining the bones in the left hand, held close to the tube, with a fluorescent screen held in the right hand. This practice was responsible for the large number of malignancies, which originated almost universally in the left hands of early radiologists.

3.02 The Coolidge X-ray Tube

In 1913, W. D. Coolidge announced the development of a hot cathode X-ray tube in which current and voltage were independently variable and which did not depend on an erratic gas discharge for its operation. By this time effective vacuum pumps had been developed and in the new tubes great care was taken to obtain the highest possible vacuum. Modern tubes are pumped for many hours and during this time the entire tube is enclosed in a furnace for "baking out" at temperatures just below the softening point of the glass. Before the tube is sealed off from the pumps, it is operated at abnormally high power so that all interior metal parts reach a red heat. These procedures drive off most of the occluded and adsorbed gas so that residual gas plays no role in the sealed-off tube.

Current–voltage independence is achieved by obtaining the electrons from a hot filament instead of from cathode bombardment by gas ions. When a filament is heated in a high vacuum, the most energetic electrons leave the wire to form an electron cloud, or *space charge*, around the filament. If a second electrode has a sufficiently high positive potential, all the electrons emitted by the filament will flow to the *target*, or *anode*. This *saturation current* varies extremely rapidly with the filament temperature, which is determined by the amount of current allowed to flow through it from a low-voltage source. Great care must be taken to maintain the heating current and the filament temperature constant in order to obtain a constant radiation output from the tube.

Although there are many types of X-ray tube designs for specific applications, nearly all modern tubes use a hot filament as the source of electrons. Radiation *quality* is varied by making appropriate adjustments to the high voltage, which accelerates the electrons toward the target. Radiation *quantity* is independently varied by changing filament temperature. By setting the filament in an appropriately shaped focusing cup the electron stream can be brought to a *focal spot* of the desired dimensions at the target surface.

Tubes for voltages up to about 250 kV are made with *reflection targets*, Fig. 3-2A. Above 250 kV, *transmission targets* are customarily used, Fig. 3-2B. At the higher voltages it has been found desirable to sectionalize the tube so that the total accelerating voltage is applied in a series of steps. Sectionalizing maintains a more uniform voltage gradient along the tube and reduces tube failures caused by high-voltage flashovers.

3.03 Wave Nature of X Rays

In the years immediately following Roentgen's discovery much effort was devoted to determining the nature of the new radiation. It was suspected that

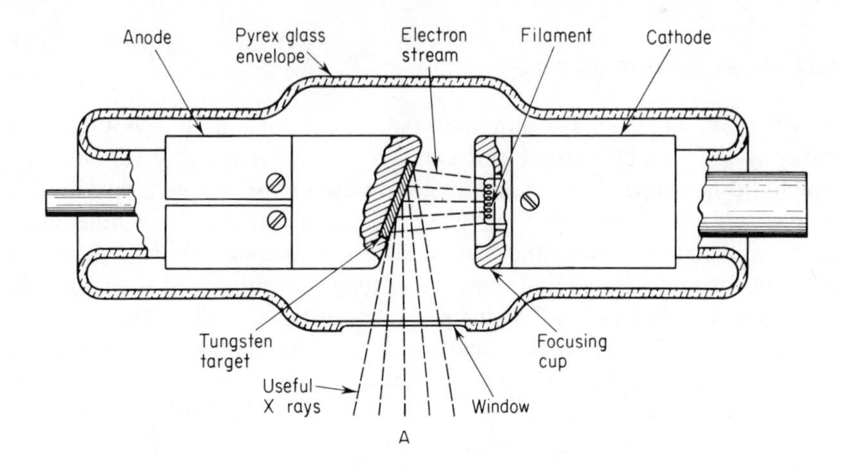

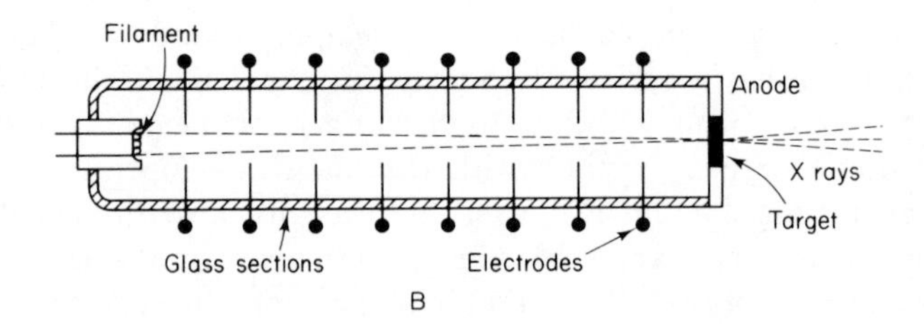

Figure 3-2. Modern types of X-ray tubes. (A) Reflection target. (B) Multisection tube with a transmission target. Each filament is set in an electrode shaped to provide electron beam focusing.

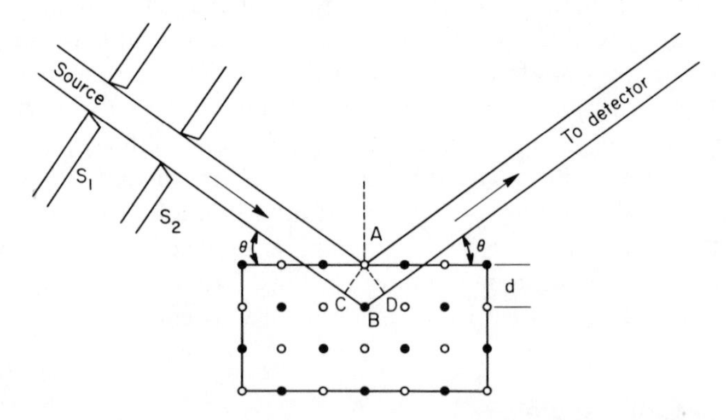

Figure 3-3. Diffraction, or coherent scattering, of X rays by the orderly arrayed atoms in a crystal.

X rays were electromagnetic waves but wave motions universally exhibit *interference* and *diffraction*; early attempts to demonstrate these phenomena with X rays failed. We now know that the failure resulted from the very short wavelengths involved.

In 1912, Von Laue succeeded in demonstrating X-ray diffraction, using a crystal as the diffracting material. In a simple cubic crystal such as NaCl or KCl the atoms are arranged in a regular alternating manner with an interatomic spacing or *grating constant d*. These recurring atoms serve as scattering or diffracting centers for an X-ray beam with a grating constant calculable from crystallographic data.

Illustrative Example

Calculate the grating constant for a sodium chloride crystal, density 2.163 g cm^{-3}; M. W., 58.45.

Molecules in 1 g NaCl	$6.02 \times 10^{23}/58.45$	$= 1.03 \times 10^{22}$
Atoms per gram	$2 \times 1.03 \times 10^{22}$	$= 2.06 \times 10^{22}$
In a cube of 1 g, atoms per edge	$= \sqrt[3]{2.06 \times 10^{22}}$	$= 2.74 \times 10^{7}$
Volume of 1 g	$= 1/2.163$	$= 0.4623$ cm^3
Length of cube edge	$= \sqrt[3]{0.4623}$	$= 0.7732$ cm
Interatomic spacing	$= 0.7732/2.74 \times 10^{7}$	$= 2.820 \times 10^{-8}$ cm

The experimental arrangement is shown in Fig. 3-3. Lead slits S_1 and S_2 produce a *collimated* or essentially parallel beam of X rays, which strike the face of the diffracting crystal. The crystal is mounted on an X-ray spectrometer in such a way that the angle of reflection into some sort of radiation detector is always kept equal to the angle of incidence θ. The ray diffracted at B and entering the detector must travel a distance $CBD = 2d \sin \theta$ farther than a similar ray diffracted at atom A. When the difference in path is equal to an integral number of wavelengths $n\lambda$, the diffracted waves will add constructively, radiation will reach the detector in phase, and the detector will respond. The condition for constructive interference is

$$n\lambda = 2d \sin \theta \tag{3-1}$$

With d known, λ can be determined from measurements of θ.

Equation (3-1) shows that d must be of the same order as λ if reasonable values of θ are to be obtained, and this explains the failures of the early workers to demonstrate X-ray diffraction. The coarse diffraction gratings originally used resulted in undetectably small values of θ. With diffracting crystals the angles become large enough for accurate measurement. In practice, calcite crystals, $CaCO_3$, are used for precision measurements since these crystals can be grown with fewer imperfections than most other crystals.

Studies with the X-ray spectrograph show a *continuous spectrum* having a sharp lower wavelength limit and a maximum energy emission at a considerably longer wavelength. When the voltage on the tube is sufficiently high, the continuous spectrum has superposed on it narrow bands of relatively intense emission, the *characteristic spectrum*.

3.04 Bremsstrahlung Production

Classical theory predicts that electromagnetic energy will be radiated whenever an electric charge is accelerated. Radiation produced in this way is known as *bremsstrahlung* from the German for "braking radiation," and this forms the continuous emission spectrum.

X-ray targets are customarily made of metals of high atomic number. A fast-moving electron in an X-ray tube, entering the target, will experience violent, random accelerations as it interacts with the electric fields of the target atoms. We would expect all values of acceleration from zero up to some maximum value and this should result in radiated photons having energies from zero to some maximum value $h\nu_0$.

The maximum photon energy is readily calculated. With a potential difference V across the X-ray tube the external power source will do an amount of work Ve on each electron striking the target. If V and e are both expressed in emu or esu, the work done will be in ergs. If the electron now loses its entire kinetic energy in a single deceleration process, conservation of energy requires that

$$Ve = h\nu_0 \tag{3-2}$$

Equation (3-2) provides one of the best means for obtaining h since all the other quantities can be measured with considerable precision.

Relatively few photons will have an energy $h\nu_0$ because the probability of a complete transfer of energy, electron to photon, in a single encounter is very low. Kramers* has deduced an expression based on interaction probabilities which gives the intensity of bremsstrahlung production as a function of wavelength and the operating parameters.

$$I_\lambda = \frac{KiZ}{\lambda^2}\left(\frac{1}{\lambda_0} - \frac{1}{\lambda}\right)d\lambda \tag{3-3}$$

where I_λ = emitted intensity in wavelength interval λ to $\lambda + d\lambda$
K = constant
i = tube current
Z = atomic number of target
λ_0 = limiting wavelength as calculated from Eq. (3-2)

*Kramers, H. A., "On the Theory of X-Ray Absorption and the Continuous X-Ray Spectrum." *Phil. Mag.* **46**, 836–871, 1923.

According to Eq. (3-3) the radiated intensities at λ_0 and at very long wavelengths will be very low. With a given target material, tube current, and voltage the radiated intensity will pass through a maximum at $3\lambda_0/2$. Plots of the Kramers equation are shown in Fig. 3-4 compared with bremsstrahlung intensities actually observed.

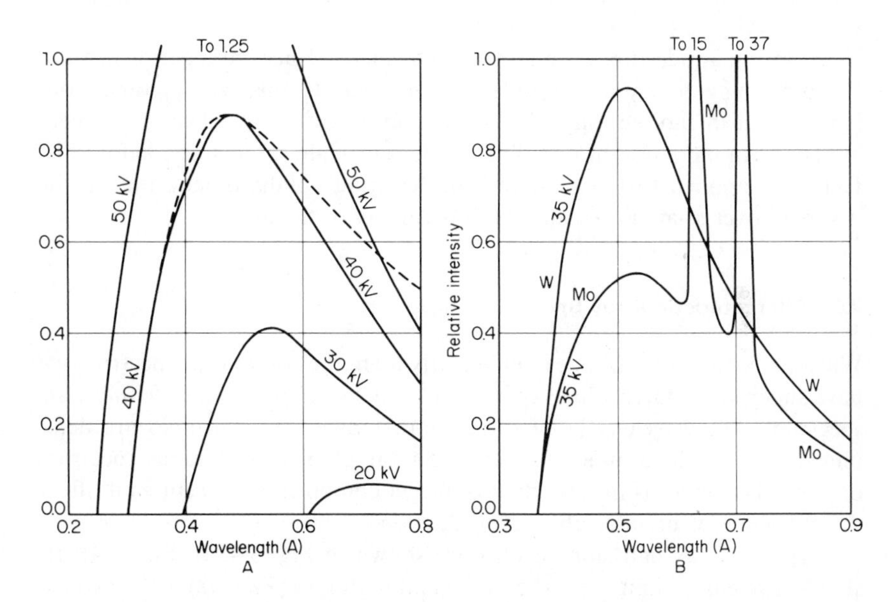

Figure 3-4. Total photon emission plotted as a function of wavelength. (A) Tungsten target, with variable kV. The dotted line is calculated from the Kramers equation. (B) Constant kV, tungsten ($Z = 74$) and molybdenum ($Z = 42$) targets.

An integration of Eq. (3-3) over all possible wavelengths will give the rate of total bremsstrahlung production I.

$$I = \frac{K_1 i Z}{\lambda_0^2} = K_2 V^2 i Z \tag{3-4}$$

The efficiency of bremsstrahlung production will be given by the ratio of total radiated power/total power input to the anode. Since the power input is Vi,

$$\text{efficiency} = K_2 V Z = V Z \times 10^{-9} \tag{3-5}$$

when V is expressed in volts. Equation (3-5) cannot be exactly correct since it leads to efficiencies greater than 100% at very high voltages. Up to a few MeV, however, the equation gives about the correct dependence of efficiency on V and Z.

X-ray tubes can be ruined quickly if the heat generated at the anode is not removed. Tungsten is widely used as a target material because of its high atomic number and high melting point. Tungsten is a poor conductor of heat, however, and tungsten targets are usually imbedded in copper blocks to improve the transfer of heat to cooling water or oil. Air cooling is satisfactory for moderate power and some tubes have anode cooling fins to increase the rate of heat dissipation.

Rotating anode tubes are also used to increase heat dissipation without attaining dangerously high anode temperatures. In this type of tube, commonly used in radiography, the anode consists of a disc of large diameter driven by an induction motor. The anode is brought up to speed before high voltage is applied to the tube and the resulting bombardment and heating is spread over a large area along the periphery of the disc.

3.05 Characteristic X-ray Spectra

When an X-ray tube voltage is sufficiently high, the continuous spectrum will have superposed narrow lines of relatively intense emission, the *characteristic* spectrum. The wavelengths of these characteristic emission lines are determined only by the atomic number of the target. The continuous spectrum, on the other hand, is independent of target composition, except as it affects the efficiency of bremsstrahlung production.

Typical X-ray emission spectra are shown in Fig. 3-4. In Fig. 3-4B the short-wavelength limit from the molybdenum target ($Z = 42$) is the same as that obtained from a tungsten target ($Z = 74$) at the same voltage. The molybdenum spectrum has a reduced intensity in accord with Eq. (3-4). In the molybdenum spectrum a group or *series* of characteristic lines appear which have no counterpart in the tungsten spectra.

The lines shown in Fig. 3-4B constitute the *K series* of molybdenum. In addition, molybdenum has a low-intensity L series at about 5.4 Å; in some elements, low-intensity, long-wavelength M and N series are known. In each series the high-intensity line of longest wavelength is given the subscript α as K_α with other subscripts following in order as L_α, L_β, L_γ. In 1914, Moseley established the relationship between the photon energy of a given line and the atomic number of the target. This relation, known as Moseley's law, is

$$\nu^{1/2} = K(Z - \sigma) \tag{3-6}$$

where K and σ are constants whose values depend on the line being considered. Moseley's law was an empirical fact in 1914 but can now be understood in terms of the Bohr theory and the Pauli principle.

3.06 Production of Characteristic Spectra

Figure 3-5 shows schematically the energy levels and electronic structure for copper ($Z = 29$) which is frequently used as a target in X-ray diffraction tubes. In accordance with previous discussion there will be 2 K, 8 L, 18 M, and 1 N electrons in each copper atom. No attempt is made in Fig. 3-5 to show the small energy differences for the different values of l, m, and s.

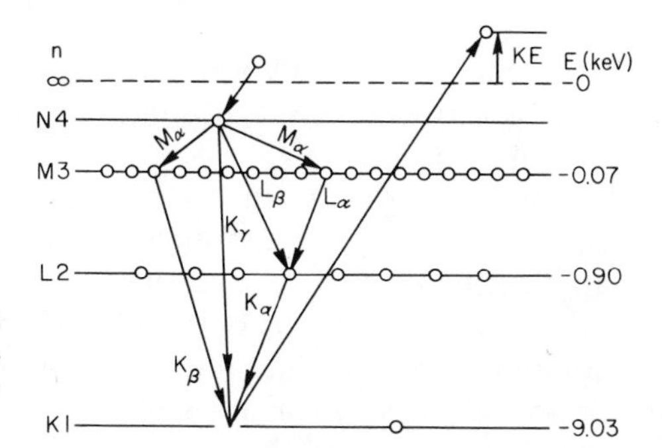

Figure 3-5. The electron transitions in copper that are responsible for producing some of the characteristic X-ray lines.

In Fig. 3-5 a bombarding electron has interacted with one of the K electrons, which has received enough energy to be ejected from the atom. Any excess energy remaining after the K electron is ejected goes into kinetic energy of the now free electron. The removal of the K electron leaves the target atom with a vacancy in the K shell which will be filled promptly.

An L, M, or N electron has an energy greater than that required for K-shell occupancy so any of these outer electrons can spontaneously drop into the K vacancy. An L electron will probably fill the K vacancy and when it does, the energy excess will be radiated as a K_α photon. The less probable transfer from the M shell results in the emission of a K_β photon, and so on. Following a K_α emission there will be an L-shell vacancy, which can be filled from above with an M or N electron. Finally the N vacancy will be filled with a free electron, the atom will again be electrically neutral, and the radiation process will be ended.

A K-shell vacancy can only be created if the bombarding electron has sufficient energy to remove an electron from the ground state and free it from the atom. An electron incapable of creating a K-shell vacancy may still have energy enough to remove an L electron. In this case all K lines would be missing from the spectrum but L and M lines would be present. Thus in Fig. 3-4B the K-series lines are missing from the tungsten spectrum since an energy of 70 keV is required to remove a K electron from tungsten.

The mathematical treatment developed by Bohr for the hydrogen atom can be extended to more complex nuclides. In the hydrogen atom the single electron moves in the electric field of the single proton composing the nucleus. In a nuclide of atomic number Z the two K electrons will move in the more intense electric field due to the Z protons in the nucleus. Electrons in outer shells will not experience the full nuclear field due to the positive charge Ze since part of this field will be *screened* by the negative charges of the inner electrons. This effect is taken into account by introducing a *screening constant* σ into the equation for the radiated energies. This is

$$E = 13.6(Z - \sigma)^2\left(\frac{1}{n_f^2} - \frac{1}{n_i^2}\right) \text{ eV} \tag{3-7}$$

where the factor 13.6 eV will be recognized as the ionization potential of the hydrogen atom and n_i and n_f are the numbers of the initial shell and the final shell, respectively.

The screening constant σ is a complicated function of the initial and final state involved in each transition. A σ value for a given transition is reasonably constant for all elements since σ is only slightly affected by electrons in outer shells. For example, since the K_α line arises from the drop of an L electron to the K shell, σ for this transition is little affected by the presence or absence of M- or higher-shell electrons. A table of characteristic X-ray emission energies will be found in Appendix 3.

3.07 Characteristics of the Useful Beam

The radiation emerging from an X-ray tube, the *useful beam*, differs considerably from that actually generated in the target. Absorption in the target itself and in the tube structures required to allow the photons to pass from the high vacuum inside the tube to atmospheric pressure introduces an *inherent filtration*. Since in general absorbers are most effective at the lower photon energies, Sec. 8.08, these will be attenuated preferentially and the useful beam will be harder or more penetrating than the originally generated beam. Inherent filtration and target absorption will also alter the spatial distribution of the radiation.

At electron energies below about 100 keV the emitted radiation will be nearly isotropic in space. As the accelerating voltage increases, emission in the forward direction (the direction of the original electron beam) tends to be favored. Figure 3-6 shows the forward peaking in the useful beam produced by voltages in the 1- to 2-MV range. At betatron or linac voltages (20–50 MV), Sec. 3.11 and 3.13, the useful beam may have an angular width at the half-maximum intensity points of less than 5°.

In some cases the size of the *focal spot*, which is the area of the target struck by the electron beam, introduces an undesirable lack of uniformity into the emitted beam. Consider a reflection target with a focal spot *FS*,

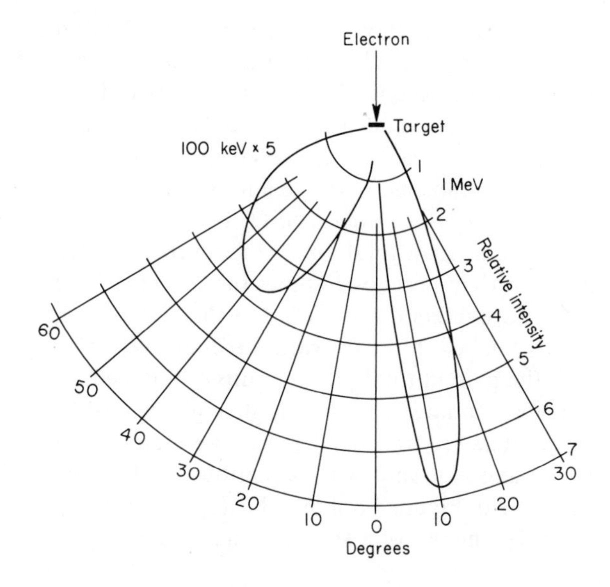

Figure 3-6. Angular distribution of radiation intensities generated at various constant potentials. From Buechner, Van de Graaff, Burrill, and Sperduto, *Phys. Rev.*, **74**, 1348, 1948.

Fig. 3-7A. The emergent beam will usually be defined by a diaphragm *D* to restrict the useful beam to the desired radiation field. Then radiation generated over the entire area of the focal spot will reach area 2-3 on the surface to be irradiated. Regions 1-2 and 3-4 "see" only a portion of the focal spot on the target and so receive a reduced amount of radiation. These regions of reduced intensity are known as the *penumbra*. Some decrease in the size of the penumbra can be achieved without reducing the target area struck by

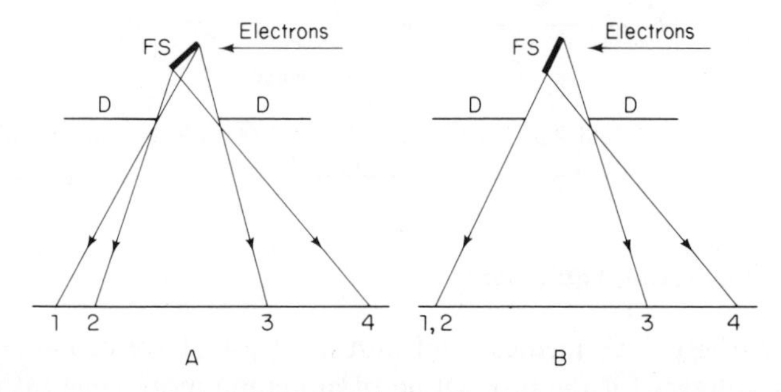

Figure 3-7. (A) The central radiation field 2–3 is surrounded by a penumbra 1–2 and 3–4 of lower intensity. (B) The size of the penumbra can be reduced somewhat by changing the angle of the target face.

the electrons by changing the target angle, Fig. 3-7B. Small focal spots can be obtained by introducing special focusing electrodes near the tube filament but very small spots may exceed the local heat capacity of the target and may result in local melting unless anode power is reduced to an undesirable level.

For many applications it is desirable to cover as large a field as possible with a constant radiation intensity. Large uniform fields can be obtained by increasing the *source–surface distance*, SSD, but only at the expense of a reduced intensity. The latter will fall off as the inverse square of SSD except at short distances less than perhaps 20 times the size of the focal spot. Large uniform fields are more difficult to produce at the higher energies where the emission is strongly peaked in the forward direction. When large irradiation fields are desired, the gain in total photon output obtained at high voltages, Eq. (3-4), is somewhat counteracted by the strongly peaked fields.

Field-flattening filters are frequently put in the useful beam to reduce the intensity of the on-axis hot spot without seriously affecting the intensities near the edge of the field. Figure 3-8 shows the improvement in the uniformity of a 2.5-MV beam by the use of a specially shaped absorbing filter.

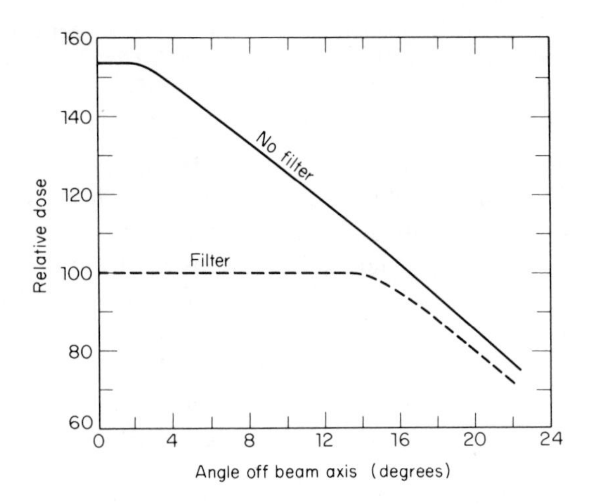

Figure 3-8. The field flattening achieved at 2.5-MV CP with a specially designed filter.

3.08 X-ray Generating Equipment

From the earliest days medical and industrial uses have demanded higher and higher voltages for the production of more and more penetrating radiation. Some truly mammoth induction coils and electrostatic machines were built but action was erratic and it became evident that other power sources were needed.

High-voltage transformers provide the simplest replacement. A step-up transformer with many turns of wire on the secondary coil furnishes the high voltage needed for the production of hard X rays. A step-down transformer with a few secondary turns gives the 2–10 V needed to supply the filament of the Coolidge tube. A special voltage-regulating transformer is usually connected in the filament circuit to reduce the effects of line voltage fluctuations.

In the simplest arrangement the high-voltage winding is connected directly to the X-ray tube in a *self-rectifying* circuit. Self-rectifying circuits put severe electrical strains on the X-ray tube and full output capabilities cannot be realized. For these reasons self-rectifying circuits have been largely replaced by some sort of rectifier circuit. Figure 3-9 shows a simple *half-wave* rectifier

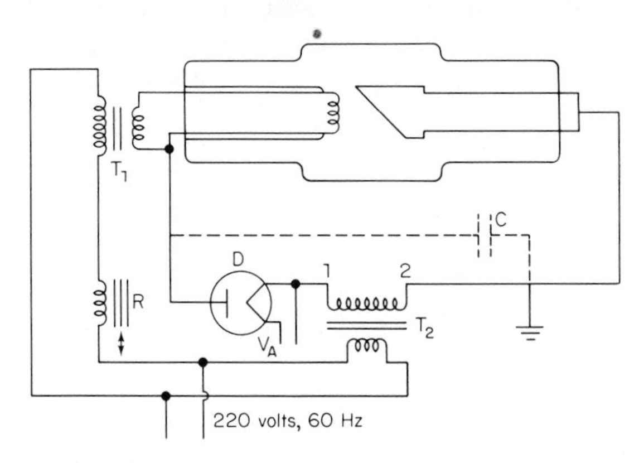

Figure 3-9. A half-wave rectifier circuit with an optional condenser *C*. The anode is at ground potential.

used either with or without the condenser *C*. Transformer T_1 supplies the low voltage needed to heat the filament of the X-ray tube. Smooth control of the filament current is achieved by the movement of the iron core in *reactor R*. In an AC circuit a reactor of this sort provides a stepless variation of the voltage supplied to T_1. T_2 supplies the high voltage to the X-ray tube through the high-vacuum diode or rectifier tube *D*. Control of the high voltage is through one or more sets of taps on the low-voltage winding of T_2, omitted from Fig. 3-9 for clarity. The filament of the diode is heated by a separate transformer, not shown. On the half-cycle when terminal 1 of T_2 is positive, the diode will permit current to flow to the X-ray tube and photons will be generated. When terminal 1 is negative, the diode will prevent the reverse potential from appearing across the X-ray tube.

It is very important to keep the filaments of all diode rectifier tubes at proper operating temperature. If the filaments are too hot, tungsten will evaporate rapidly from the filament and the life of the tube will be drastically shortened. If the filament is not hot enough to ensure a copious supply of electrons, the diode will act like a high resistance and there will be a high voltage across it instead of across the X-ray tube. As a consequence the diode anode may now become the chief source of bremsstrahlung and X rays. Dangerous radiation levels may develop unsuspected in the vicinity of diodes which under proper operating conditions produce only a negligible quantity of soft radiation. It must be remembered that bremsstrahlung will be generated whenever high-energy electrons strike a target, particularly one of high atomic number.

The half-wave circuit in Fig. 3-9 is shown with the anode of the X-ray tube grounded. This is usually desirable since it permits locating a patient close to the anode, which is the source of the radiation, without the danger of an electric shock. With this arrangement the X-ray tube filament will be at a high negative potential and the filament transformer must be insulated to withstand the maximum voltage that may be applied to the anode.

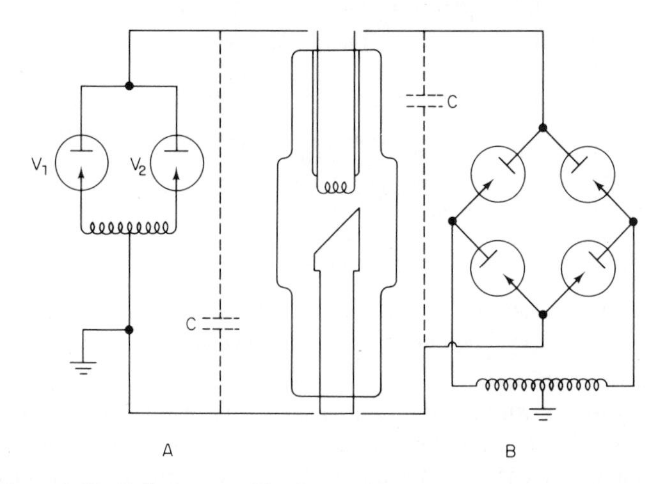

Figure 3-10. Full-wave rectification with a split secondary circuit (A) or a bridge circuit (B) improves the X-ray output. In the bridge circuit the anode is ungrounded.

Full-wave rectification can be obtained by using either two rectifier tubes, Fig. 3-10A, or four tubes in a *bridge circuit*, Fig. 3-10B. Each of these circuits passes both halves of the alternating cycle, but with proper polarity, to the X-ray tube. The two-tube *split secondary* circuit can be operated with the anode grounded, as shown. In the bridge circuit both filament and anode are at high potential with respect to ground but for many applications this is no disadvantage. In the bridge circuit the full voltage of the transformer is applied to the X-ray tube, whereas in the two-tube circuit the tube receives only one-half of the transformer voltage.

Transformer–rectifier units are commonly used at voltages up to 250 kV and higher voltage units have been constructed. Potentials greater than 500 kV can probably be best obtained by other means.

Transformers capable of supplying voltages of 250 kV or more are expensive and bulky. To reduce cost, several circuits have been devised for doubling the transformer voltage before it is applied to the X-ray tube. The Villard circuit and the Greinacher circuit are the most commonly used *doublers.*

Voltage multiplication can be carried beyond doubling in the Cockcroft–Walton circuit. A transformer of moderate voltage supplies a network of condensers and rectifiers. By a repetitive condenser charging, and charge sharing, the condensers are effectively put in series so that about 1 MV can be attained. This circuit can operate at either polarity and is used mostly in positive ion accelerators.

3.09 Constant Potential Circuits

All simple rectifier circuits have the disadvantage that the bremsstrahlung are generated at a highly variable, albeit DC potential. In any half-wave rectifier circuit the sinusoidal transformer voltage, Fig. 3-11A, will be passed on to the X-ray tube with a *waveform* similar to that shown in Fig. 3-11B. A half-cycle of the approximately sinusoidal wave is followed by a half-cycle when no radiation is generated. Full-wave rectifiers supply the missing half-cycles, Fig. 3-11C, but there are still undesirable low-voltage portions of the waveform.

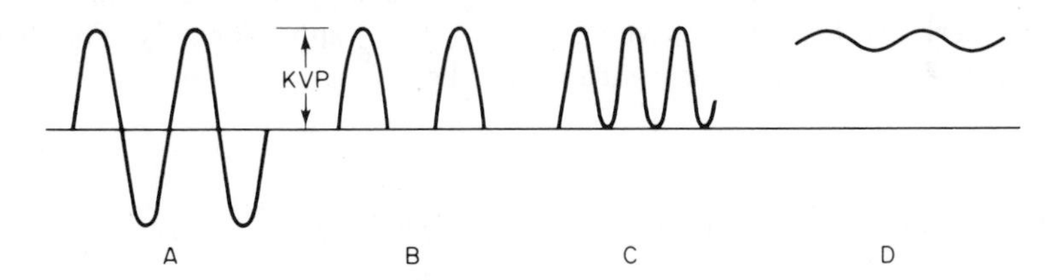

Figure 3-11. Voltage waveforms associated with X-ray tube circuits. (A) Sinusoidal supply voltage. (B) Output of a half-wave rectifier. (C) Output of a full-wave rectifier. (D) Fluctuations reduced to essentially CP by adding a condenser.

When the voltage across the X-ray tube is low, only soft radiations will be generated, Eq. (3-2). These will be useless if penetrating radiation is desired. In fact, soft radiation components may be highly undesirable, adding substantially to the radiation dose delivered to superficial layers but contributing almost no dose at depth.

It is customary to specify the hardness or quality of an X-ray beam by the *peak* value of the wave applied to the tube, Fig. 3-11B. This value is known as the peak kilovolts (kVP). Obviously peak kilovolts gives only a loose specification of the beam quality. Quality depends not only on the peak value of the voltage but also on the waveform since this determines the relative amounts of high- and low-energy radiations. A more precise specification of quality will be discussed in Sec. 8.12.

If a condenser C, dotted lines in Figs. 3-9 and 3-10, is added, the circuit is said to be *constant potential*, CP. The condenser will store energy during the high-voltage portions of the cycle and release energy during the low-voltage portions. This storing action reduces the voltage fluctuations to a *ripple*, which will be only a small percentage of the supply voltage amplitude. The voltage will then be specified as kilovolts, constant potential (kVCP) and will appear as in Fig. 3-11D.

The exact calculation of the amount of ripple obtained with a given circuit is complex but a simple approximation will suffice for most purposes. Let a capacitor of capacitance C be connected to an X-ray tube operating at V volts and I amperes, average values. Then the quantity of electricity stored in the condenser will be

$$Q = CV \tag{3-8}$$

coulombs if C is expressed in farads. During the low-voltage portion of the cycle the quantity of charge given up by the capacitor will be

$$q = It \tag{3-9}$$

where t is the time of current discharge, in seconds.

Practically all power in the United States is supplied at a frequency of 60 Hz (hertz) so for a half-wave circuit $t = \frac{1}{120}$ s, approximately. The drop in voltage due to the discharge current will be

$$v = \frac{q}{C} = \frac{It}{C} \tag{3-10}$$

and the fractional drop in voltage, or ripple, will be

$$\text{ripple} = \frac{v}{V} = \frac{It}{CV} \tag{3-11}$$

Illustrative Example

What is the percentage ripple in a half-wave 60-Hz circuit operating at 120 kV and 3 ma (milliamperes) with an 0.05 μf (microfarad) condenser?

$$\text{ripple} = \frac{It}{CV} = \frac{(3 \times 10^{-3}) \times \frac{1}{120}}{(0.05 \times 10^{-6}) \times (1.2 \times 10^{5})}$$
$$= 4.15 \times 10^{-3} = 0.415\%$$

Thus the circuit is essentially constant potential.

In a full-wave rectified circuit the discharge time t will be considerably shorter than the time of one-half a cycle and a given condenser will be more effective in reducing ripple.

3.10 Van de Graaff Electrostatic Generators

Many of the early X-ray workers were convinced, erroneously, that less injury was sustained by the X-ray worker if his tube was excited with a frictional electricity generator or "influence machine." Huge rotating-plate electrostatic generators were built but operation was erratic and they were supplanted by transformers.

In 1928, Van de Graaff announced a real advance in the design of electrostatic generators. The principle is illustrated in Fig. 3-12. A belt of some insulating material is driven at high speed by a motor. When a charging voltage, perhaps 20–40 kV, is applied across the belt from the needle points to inductor plates, a corona discharge will take place at the points and electric charge will be sprayed onto the belt. The belt carries the charge inside the top terminal, which has a smooth outer surface to reduce spark discharges. No electric field can be maintained inside a conductor such as the terminal and so a pointed metal comb set close to the belt will pick off the ascending charges and will deposit them on the terminal.

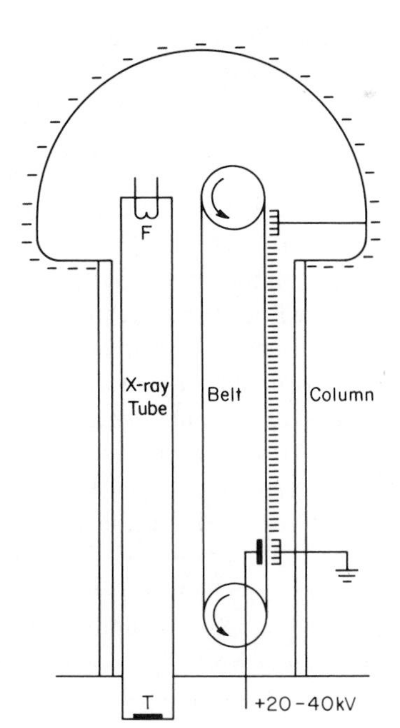

Figure 3-12. Schematic of a Van de Graaff generator. Both tube and column are sectionalized in order to obtain a uniform potential gradient. Output is CP.

High voltages can be built up readily by this scheme, the limit being set by the radius of the top terminal and the insulation of the supporting column. With the generator inside a steel tank filled to 200–300 psi with a dry insulating gas, voltages of 8–10 MeV can be realized. The outstanding feature of the Van de Graaff generator is its ability to produce truly CP voltages at very precisely set values.

3.11 Betatron

All the voltage sources so far described are essentially "brute-strength" arrangements in which the high potential is generated in a single step or in a relatively small number of steps. In such machines some parts must be insulated to withstand the full generator voltage; insulation becomes increasingly difficult as one goes above 1 MV.

In 1931, Lawrence and Livingston described a revolutionary method for accelerating charged particles by the repeated application of a relatively low voltage. The *cyclotron* is a resonance machine not suited for electron acceleration. Since the electron mass changes rapidly with energy, cyclotron resonance conditions cannot be maintained over a wide energy range. Several types of resonance accelerators were developed subsequent to the cyclotron and in 1941 Kerst described the *betatron*, a machine capable of accelerating electrons to very high energies. By directing these electrons onto a suitable target, X rays at previously unattainable energies were generated for use in deep therapy or for the inspection of thick metal parts.

The betatron can be considered as a special type of transformer with the usual secondary winding of many turns replaced by an evacuated tube or "doughnut." A pulse of electrons, accelerated to perhaps 50 kV is injected into the doughnut just as the current flowing in the primary magnetic coil passes through zero and starts to increase. As the magnetic flux through the center of the doughnut increases, each electron circulating around inside will gain energy but will continue to move in an orbit of constant radius. During acceleration the electrons complete many thousands of revolutions.

When peak primary current and magnetic flux are attained, the electrons will have acquired maximum energy. A current pulse is then sent through an auxiliary coil, altering the radius on which the electrons are orbiting. They will spiral out to strike a target and produce a pulse of X rays. The entire process will be repeated on the next magnetization cycle. Betatrons are usually designed to operate at voltages of 10 MV or more and since here the efficiency of X-ray production is high, target cooling is not required.

With a slight change in doughnut design and in pulsing arrangement the electron beam can be brought outside the doughnut for deep electron therapy, food processing, or sterilization.

In pulsed devices such as the betatron the fractional time of radiation emission is known as the *duty cycle*. Thus if a betatron emits f pulses per second, each having a duration of t seconds,

$$\text{duty cycle} = ft \tag{3-12}$$

A betatron may have a duty cycle of about 10^{-4}, which means that the radiation intensity in the pulses is 10^4 times as great as the average value. Because of the high output in the pulses, care must be taken in making measurements of betatron radiation to ensure proper instrument operation.

In the betatron, high-energy electrons are obtained without the presence of extremely high voltages. Problems connected with high-voltage insulation are avoided but are replaced by those associated with accurate timing of the various phases of pulsed operation.

3.12 Insulating Core Transformer

Although electrons can be accelerated to high energies by Van de Graaff generators or in betatrons, these devices have definite power limitations, with beam currents usually limited to the microampere range. This power limitation can be raised by at least an order of magnitude in the insulating core transformer, ICT. Invented by Van de Graaff, the ICT has been used as a source for the large-scale processing of foods and a wide variety of industrial products.

The usual transformer consists of a primary and secondary winding around a continuous iron core that carries the magnetic fields linking the two coils. The massive insulation required to securely isolate the high-voltage secondary coil places an upper limit on the voltage capability of the simple transformer. In the ICT, Fig. 3-13, the conventional iron core is broken up into a series of insulated segments. Each segment has its own secondary

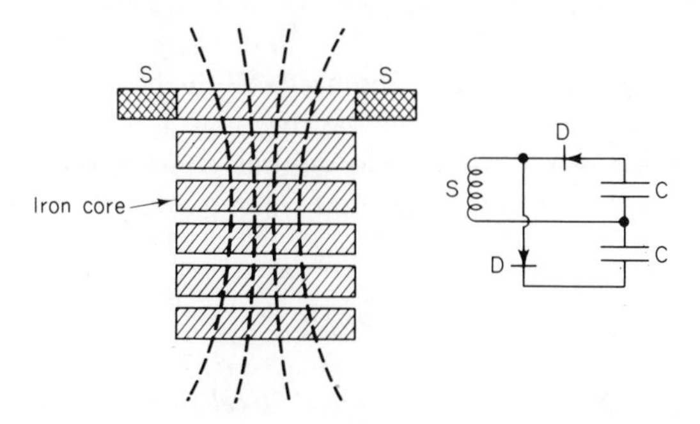

Figure 3-13. Schematic of an insulating core, or ICT, generator, showing one secondary voltage doubling circuit.

winding S which feeds a pair of condensers through solid-state diode rectifiers in a voltage doubler circuit. Each segment produces a DC voltage of perhaps 70 kV. Adjacent segments need be insulated only against this voltage but the segments connected in series produce a large summed voltage across the entire stack. The entire generator is enclosed in a steel tank filled with an insulating gas such as sulfur hexafluoride to a high pressure. ICT units can be operated at voltages of about 1.5 MV with currents of 25 ma or more.

3.13 Linacs

The most energetic bremsstrahlung produced by man has been obtained from linear electron accelerators or linacs. In the linac a series of evacuated conducting cavities are made of a size to be resonant with a powerful high-frequency oscillator, Fig. 3-14. When the cavity is excited by a short burst

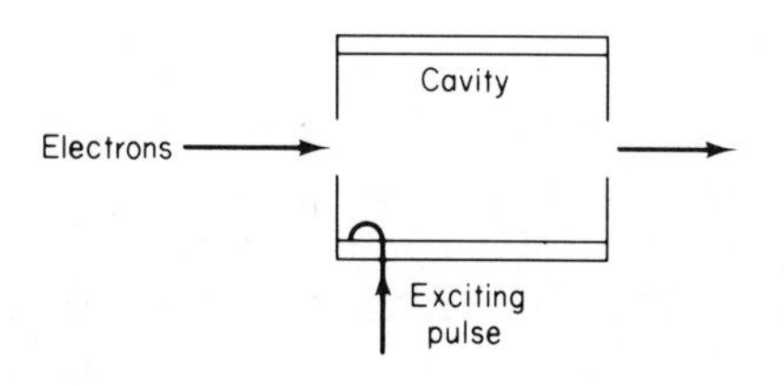

Figure 3-14. A single resonant cavity for electron acceleration in a linac.

of oscillations fed into a coupling coil, an electric wave will sweep down the cavity with nearly the velocity of light. An electron beam arriving at the cavity entrance at just the correct instant will be carried along on the wave as a surf rider is carried shoreward. As the bunch of electrons leaves one cavity, it picks up another wave sweeping down the next cavity, and so on. At the Stanford University accelerator facility, electron energies of 2 GeV have been attained in a linac 2 miles long.

In the more common sizes, 5–10 MV, the linac has become one of the most useful X-ray generators. Linacs are light and easily maneuverable since no heavy magnet or pressure tank is required. With the development of more powerful oscillator tubes the linac has become a reliable machine for a variety of radiation applications. The output consists of a series of short bursts, which must be taken into account when making beam measurements.

REFERENCES

BLEWETT, M. H., "*Characteristics of Typical Accelerators.*" *Ann. Rev. Nuc. Sci.*, **17**, 427, 1967.

BROWN, J. G., *X-Rays and their Applications.* Plenum Press, New York, N.Y., 1966.

KAPLAN, I. *Nuclear Physics*, 2nd ed. Addison-Wesley Pub. Co. Reading, Mass., 1962.

PATTEE, H. H., V. E. COSSLETT, AND A. ENGSTRÖM, ed., *X-Ray Optics and X-Ray Microanalysis*. Third International Symposium, Stanford University, August, 1962. Academic Press, New York, N.Y., 1963.

RICHTMYER, F. K., E. H. KENNARD, AND J. N. COOPER, *Introduction to Modern Physics*, 6th ed., International Series on Pure and Applied Physics. McGraw-Hill Book Co., New York, N.Y., 1969.

4

Stable Atomic Nuclei

4.01 Nuclear Constituents

The scattering experiments of Rutherford showed conclusively that an atomic nucleus contains an integral number Z times the elementary unit of positive electric charge and that Z is approximately one-half of the chemical atomic weight of the element. It was first thought that all the nuclear mass was derived from the number of protons, or hydrogen nuclei, in the nucleus.

Each nucleus was assumed to consist of enough protons to account for all the atomic mass. This number led to an excessive positive charge so a sufficient number of nuclear electrons was postulated to neutralize part of the charge and thus obtain the required Z value. Because of the small mass of the electrons, their addition would have little effect on the total nuclear mass. This model was used for some time to explain beta radioactivity, in which negative electrons are ejected from the nucleus.

There is now abundant evidence that the proton–electron model is incorrect. It is sufficient to state here that the "size" of the electron as given by the de Broglie relation, Eq. (1-20), is far greater than any nuclear diameter unless impossibly large values of velocity are assumed.

The correct structure became evident with the discovery of the neutron in 1932. The neutron is an uncharged particle with a mass slightly greater than that of the proton. A free neutron, outside a nucleus, undergoes radioactive decay into a proton and an electron. However, all the arguments against the existence of nuclear electrons apply equally well to the neutron, which, prior to decay, must be something other than a combination of a discrete proton and an electron.

For several years before the neutron was discovered, physicists had been measuring atomic masses using instruments known as mass spectrometers. The mass measurement depends on the fact that an ion of mass m and charge e, moving with a velocity v perpendicularly to a magnetic field B will be deflected into a circular trajectory of radius r, given by

$$r = \frac{m_0 v}{Be} \tag{4-1}$$

Equation (4-1) applies to ions moving at nonrelativistic velocities. Relativistic relations must be used at higher velocities. The most useful form relates the trajectory radius to the particle energy.

$$r = \frac{\sqrt{T^2 + 2m_0c^2T}}{Be} \tag{4-2}$$

In the mass spectrometer, ions are produced by any one of a variety of means, accelerated in an electric field, and then passed through an analyzing magnetic field. B and r are measured and since v (or T) and e are known, the mass of the particle can be calculated. Atomic masses are now known to such high precision that tabulated values are sometimes referred to as "exact masses." With values of Z and atomic masses in hand the number of each kind of nuclear constituent can be properly assigned.

Experimental evidence agrees with a nuclear model consisting of Z protons and N neutrons, making a total of $A = Z + N$ *nucleons* in the structure. With the proper combination both charge and mass can be accounted for without the need to invoke neutralizing electrons inside the nucleus. Z, N, and A are known, respectively, as *atomic number*, *neutron number*, and *mass number*. A nucleus can be specified by any two of the three numbers, usually Z and A. Structures are written ${}^{A}_{Z}$chemical symbol, as ${}^{1}_{1}H$, ${}^{235}_{92}U$. In a sense the atomic number specification is redundant since the chemical symbol is uniquely determined by it. As we shall see, the Z value is useful in balancing nuclear equations and it will be generally retained.

4.02 Nuclear Stability

Although the Z–N model of the nucleus fitted all the physical requirements, it was evident that atomic weight values, as measured on samples of naturally occurring elements, would require fractional values of N. This difficulty was resolved by the discovery that most chemical elements consist of a mixture of two or more species, each having the Z value appropriate to the element in question, but differing in N values.

Any nuclear structure which exists long enough to be identified is a *nuclide*. Nuclides with constant Z and varying N values are *isotopes*. Groups with constant N and varying Z are *isotones*. *Isobars* have constant A values

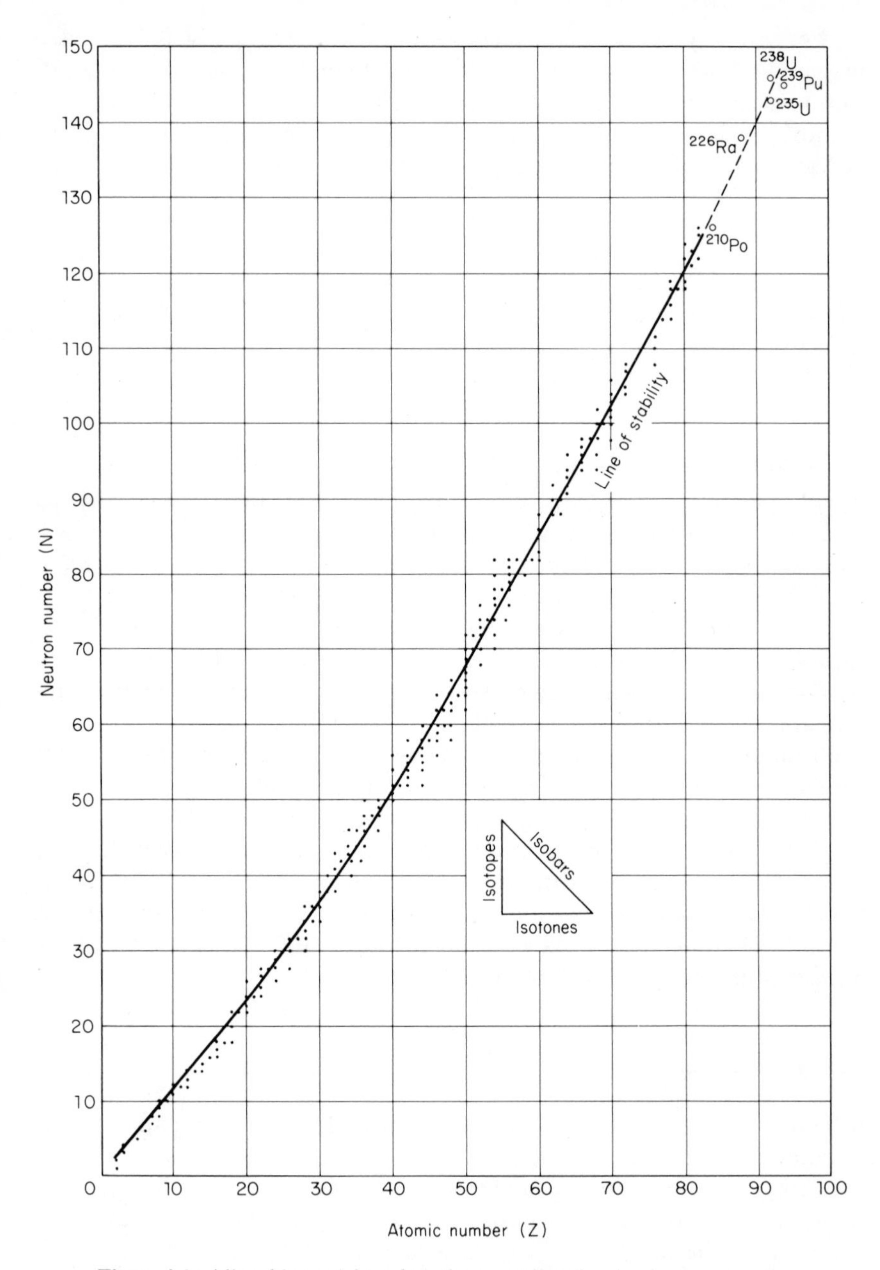

Figure 4-1. All stable nuclei tend to cluster rather closely along a smooth curve on a Z–N plot. Many nuclides of low abundance have been omitted from the plot and five heavy radioactive species have been included for comparison.

with varying Z–N combinations. It is sometimes convenient to classify nuclides in terms of neutron excess, or $(N - Z)$. Nuclides with this factor constant are *isodiapheres*.

Some nuclides appear to be absolutely stable, while others are known to undergo radioactive decay. There is no clear-cut distinction between the two groups. As our measuring ability increases, nuclides once thought to be stable are found to be undergoing decay at an extremely slow rate. It may well be that still more unstable structures will be found in the future.

When an N–Z plot of the stable nuclides is made, Fig. 4-1, the entries are found to fall fairly close to a smooth curve. Values of Z and N tend to be equal at small values and the stability line starts out with a slope of 45°. By $Z = 20$, neutrons tend to predominate until there is an N/Z ratio of 1.5 at the lead isotopes.

The five radioactive nuclides shown by open circles in Fig. 4-1 lie fairly close to an extension of the stability line. Other unstable species may lie well away from the line, on either side.

4.03 Natural Abundance

A census of stable nuclear species shows that structures with an even number of protons and an even number of neutrons predominate. Over 85% of the material in the earth's crust is composed of even–even nuclides, with $^{16}_{8}O$ and $^{32}_{16}Si$ accounting for over 75%. With a few cases of radioactive decay still uncertain the numerical species distribution is about

even Z–even N 156	even Z–odd N 55
odd Z–even N 50	odd Z–odd N 4

The four odd–odd structures, $^{2}_{1}H$, $^{6}_{3}Li$, $^{10}_{5}B$, and $^{14}_{7}N$, are all light elements with equal values of Z and N. The distribution suggests that nuclear stability is in some way connected with odd and even numbers.

Further evidence is obtained by considering the number of stable isotopes that are found for various values of Z. Odd-Z elements such as $_{9}F$, $_{37}Rb$, $_{55}Cs$, and $_{79}Au$ are observed to have one or at most two stable isotopes. In an odd-Z structure one Z–N combination appears to be stable and variations from this quickly lead to instability. In even-Z nuclei, stability can be maintained over a wider range of N values, as in $_{20}Ca$ with six isotopes and $_{50}Sn$ with nine. Note that the stability ranges tend to end at even values of N.

The distribution of a given element among its isotopes is denoted by the term *percent abundance*. Percent abundance of an isotope is the percentage of the total atoms in a composite sample which are of the particular species under consideration. Abundance values found in natural samples of an odd Z and an even Z element are listed in Table 4-1 and are plotted in Fig. 4-2.

TABLE 4-1
NATURAL ISOTOPIC ABUNDANCES

Nuclide		*Percent Abundance*
^{197}Au	$Z = 79$	100
^{198}Hg		10.0
^{199}Hg		16.8
^{200}Hg	$Z = 80$	23.1
^{201}Hg		13.1
^{202}Hg		29.8
^{204}Hg		6.9

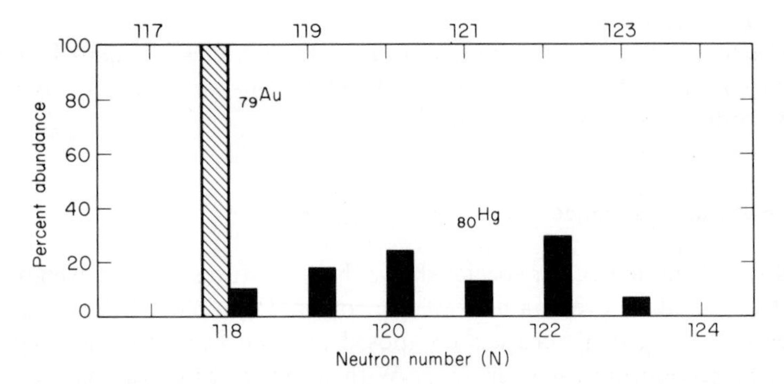

Figure 4-2. Typical isotopic abundances in an odd-Z and an even-Z element.

With a few exceptions the isotopic abundances in terrestrial samples are almost constant, independent of the origin of the sample. A notable exception is lead, $_{82}$Pb, where some of the stable nuclides may have been derived from the radioactive decay of heavier nuclei. Data from extraterrestrial samples are scarce but there appear to be significant differences from the abundance values found on earth.

4.04 Isotope Separation

Since the chemical behavior of an atom is determined by its orbital electron configuration, and hence by Z, reactions are essentially independent of N. Reaction rates may vary slightly because the heavier nuclei are somewhat less mobile than the lighter fractions but the basic reaction types will be identical. Reaction rate effects are detectable in the biological fractionation of hydrogen, where the stable isotopes ^{2}H and ^{1}H have a mass ratio of 2.

Chemical and biological processes do not appear to distinguish between the six isotopes of $_{20}Ca$, which is found with the same abundances in living tissues and in native minerals.

It is sometimes desirable to separate the individual isotopic constituents in a naturally occurring sample. A few chemical exchange reactions can be used where differences in rate constants will produce a small fractionation effect. In general, separation must be achieved through differences in physical rather than chemical properties. For example, gases diffuse through a porous barrier at a rate that is inversely proportional to the square root of the molecular weight. Although the degree of separation at a single barrier is small, many can be used to increase the overall fractionation. Gaseous diffusion is the method used most commonly in the large-scale separation of the two important isotopes of uranium, ^{235}U and ^{238}U. Centrifugation, thermal diffusion between hot and cold surfaces, and repeated electrolysis can be used in particular cases to obtain some isotope separation.

The highest degree of separation is achieved by the electromagnetic deflection of charged particles in specially designed mass spectrometers. Design emphasis is placed on obtaining large ion currents but at best the production of completely separated isotopes is small.

Illustrative Example

Calculate the time required to separate 1 g of copper, atomic weight, 63.54, into its constituent isotopes (^{63}Cu, 69.1%; ^{65}Cu, 30.9%) in a mass spectrometer which has an ion current of 1 ma.

Copper atoms in 1 g $6.02 \times 10^{23}/63.54 = 9.47 \times 10^{21}$
Assuming single ionizations, Cu^{+}
Total charge transport $(9.47 \times 10^{21}) \times (1.6 \times 10^{-19}) = 1520$ coulombs
Since 10^{-3} coulomb are transported per second:
Time required $1520/10^{-3} = 1.52 \times 10^{6}$ s or 17.5 days
The yield is $(63 \times 0.691)/63.54 = 0.684$ g of ^{63}Cu
$(65 \times 0.309)/63.54 = 0.316$ g of ^{65}Cu

4.05 Exact Nuclear Mass

Mass spectrometers and the accompanying techniques have been developed to a point which permits atomic and nuclear masses to be determined with extraordinary precision. Care must be taken to distinguish between the two masses. In a mass spectrometer the actual measurement is made on an ionized atom with one or more orbital electrons stripped off the neutral structure. Atomic mass is obtained by adding the masses of the stripped electrons to the measured mass. Nuclear mass is obtained by subtracting from the measured mass the total mass of the remaining electrons. Either of these high-precision values is sometimes called an exact mass, denoted by M_a or M, respectively.

The expected mass of an atom or a nucleus, W_a or W, can be calculated from the sum of the masses that go into each structure. Obviously

$$W_a = Zm_p + Nm_n + Zm_e$$

and

$$W = Zm_p + Nm_n$$

A comparison of W with the corresponding M value will always reveal a discrepancy. The case of ${}^4_2He$ is typical.

Illustrative Example

Compare the exact nuclear mass of ${}^4_2He$, 4.001502 u, with the total mass of its constituents. From the values given in Table 1-1,

$W = (2 \times 1.007277) + (2 \times 1.008665) = 4.031884$ u
$M =$ 4.001502
Loss in mass 0.030382 u
Energy equivalent $= 0.030382 \times 931.478 = 28.3$ MeV
(rounded off to three significant figures).

The energy equivalent of the lost mass, 28.3 MeV, has been radiated away during the assembly process to produce a structure that is more stable than the original state. Energy thus lost is known as *binding energy*, E_B. Every assemblage of neutrons and protons into a structure capable of more than a transient existence will have a positive value of the binding energy, Note that the greater the binding energy, the more stable is the assembled state of the system compared to its completely disassembled state.

Tabular values of exact masses are available from which binding energies can be calculated as in the example. When a value of binding energy rather than mass is desired, it is usually more convenient to work with MeV directly, using values that pertain either to the nucleus or the atom. A quantity, *mass excess*, Δ, is defined as the difference

$$\Delta = M_a - A(931.478) \text{ MeV} \tag{4-3}$$

where M_a must now be expressed in MeV. Tabulated values of Δ, which may be either positive or negative, are available.* M_a is calculated from the tabulated Δ value and is then subtracted from W_a to obtain the binding energy.

*Lederer, C. M., J. M. Hollander, and I. Perlman, *Table of Isotopes*, 6th ed. John Wiley & Sons, Inc., New York N.Y., 1967.

Illustrative Example

Calculate the binding energy of $^4_2$He from the mass excess.
From the referenced table, $\Delta = +2.4248$ MeV.

$$
\begin{aligned}
W_a &= (2 \times 938.256) + (2 \times 939.550) + (2 \times 0.511) &&= 3756.638 \text{ MeV} \\
M_a &= (4 \times 931.478) + 2.425 &&= \underline{3728.337} \\
E_B &= W_a - M_a &&= \quad 28.3 \quad \text{MeV}
\end{aligned}
$$

It is convenient, although not quite correct, to think of the binding energy as distributed uniformly among the total number of nucleons. For $^4_2$He the binding energy per nucleon, E_b, is $28.3/4 = 7.1$ MeV. This is the average energy required to remove a single nucleon from the complete structure. Total separation into the individual components will require the total binding energy of 28.3 MeV.

Binding energies can be calculated more directly from values of mass excess. For example, the synthesis of $^4_2$He from the separate particles is equivalent to a reaction

$$2(^1_0n) + 2(^1_1\text{H}) \longrightarrow {}^4_2\text{He} + Q \tag{4-4}$$

where Q represents any energy that may be gained by or lost from the system during the synthesis. Equation (4-4) is energetically equivalent to

$$2(M_n) + 2M_a(^1_1\text{H}) \longrightarrow M_a(^4_2\text{He}) + Q \tag{4-5}$$

where all quantities must be expressed in MeV. From Eq. (4-3),

$$
\begin{aligned}
&2[\Delta_n + (931.478 \times 1)] + 2[\Delta^1_1\text{H} + (931.478 \times 1)] \\
&\quad = \Delta^4_2\text{He} + (931.478 \times 4) + Q
\end{aligned}
$$

The terms involving nucleon numbers cancel as they must because we have the same number of nucleons on each side of the equation. We are then left with only the Δ values. Putting in tabulated values,

$$(2 \times 8.0714) + (2 \times 7.2890) = 2.4248 + Q$$

whence $Q = +28.3$ MeV as before. The positive value of Q indicates that energy has been lost from the system during the synthesis.

4.06 The Nuclear Force

Any loss in mass occurring during a nuclear synthesis results from the fact that the energy of assembly comes from the system itself. This energy loss is, in turn, the result of an attractive force between individual nucleons. The very fact that a group of nucleons exists as a stable structure requires such a force. An internucleon force of repulsion is readily recognized. Since each proton carries a positive electric charge, there will be an electostatic force of repulsion between them. This force varies as the inverse square of the proton–

proton separation and leads to a $1/r$ dependence of potential energy. Nuclear stability requires an attractive force at least as great as the p–p forces of repulsion.

The $1/r$ variation of potential energy creates a *potential barrier*, Fig. 4-3, which must be penetrated by any positively charged particle entering the nucleus. Kinetic energy sufficient to ensure penetration must be supplied by some external source to the incoming particle. At a critical distance from the center of the "target" structure, the incoming particle will suddenly experience an attractive force. This *nuclear force* has a zero value beyond the critical distance and an approximately constant value within. Thus the spatial dependence of energy is distinctly different from that arising from the more familiar inverse-square-law forces.

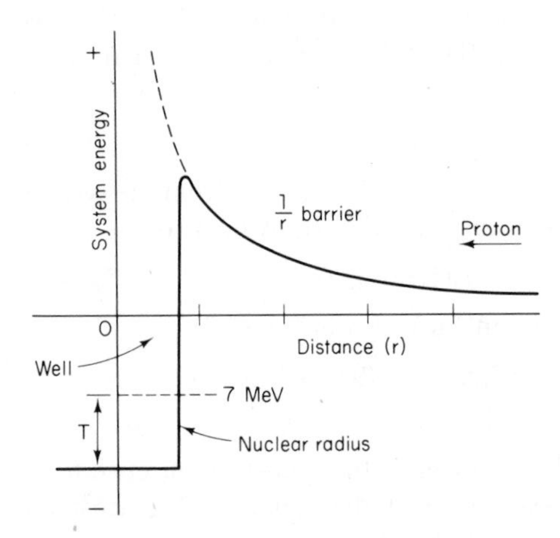

Figure 4-3. Nuclear protons create a potential barrier to incoming positively charged particles outside the potential well arising from the nuclear force.

Work will be done by the attractive nuclear force as it pulls the incoming particle into the nucleus and at this point the system radiates energy and loses mass. The original kinetic energy of the particle is recovered along with the binding energy of the particle into the final structure. When the binding energy is lost, the particle drops into a state of lower energy, or a *potential well*. As we have seen, the depth of this well is about 7.1 MeV in the case of $^4_2$He. The incoming particle, now a constituent of the new configuration, does not drop to the bottom of the potential well but comes into energy equilibrium with the other nucleons at a kinetic energy of several MeV.

Because of the wave nature of each nucleon the dimensions of nuclei are not sharply defined. One convenient definition of nuclear radius is the distance at which the nuclear force suddenly becomes effective. Measurements show this distance to be given by

$$r = r_0 A^{1/3} \tag{4-6}$$

where r_0 is a constant having a value of about 1.4×10^{-15} m or 1.4 fermis. Nuclei tend to be very nearly spherical and hence nuclear volume $V = \frac{4}{3}\pi r^3$ will vary as A, the total number of nucleons in the nucleus. Thus nuclei are formed with a nearly constant density. This density is enormous compared to the density of any ponderable objects known on earth but it may be attained in some stars.

Nucleons do not coalesce into infinitesimally small volumes as would be expected if only the attractive nuclear force were acting. It seems necessary to invoke a force of repulsion, effective only at very small distances, in order to obtain equilibrium values of the nuclear radii. Almost nothing is known about this repulsive force.

4.07 Nature of the Nuclear Force

Several features distinguish the nuclear force from the more familiar forces of electric or magnetic origin. The nuclear force is

1. Short-range. It has a nearly constant value out to about 1 fermi and then falls abruptly to zero.
2. Independent of electric charge. Forces between *n–n*, *n–p*, and *p–p* are nearly equal in magnitude.
3. A saturation force. Nuclear forces can act only between a few nucleons. ${}_{2}^{4}He$ is strongly bound because the nuclear forces are just saturated with the four particles. No stable nuclide with five nucleons exists because there is insufficient interaction between the fifth particle and the saturated four-nucleon structure to bind them together. For the same reason all eight-nucleon structures are unstable and are radioactive. The forces between the two tightly bound four-nucleon structures are too small to ensure absolute stability.
4. Always attractive. There is no evidence of polarity such as is seen with electric charges or magnetic poles.
5. A strong force. Gravitational attraction fails by many orders of magnitude (10^{35}) to account for the nuclear force. Electric and magnetic forces are ruled out for several reasons and are too small by a factor of about 10^6.

In 1935, the Japanese physicist Yukawa proposed a model to account for the nuclear force. Yukawa suggested that some sort of a subnuclear particle was exchanged or shared between two nucleons to create a force of attraction between them. Electron-sharing between atoms was known to produce an exchange force which binds the atoms into stable configurations. In the nuclear case a particle known as the *meson* was postulated as the exchange particle. Yukawa was able to calculate to a good approximation the mass of the then hypothetical particle. Later experiments confirmed the assumptions and identified the particle as the π meson, or pion, with a mass of 273 m_e.

According to the exchange model each nucleon will have one or more mesons associated with it. When there is an excess of one positive pion, the nucleon is a proton. One of the π^+ mesons may leave this nucleon to join with a neutral nucleon. The latter now becomes a proton, while the original, now uncharged, becomes a neutron. Thus a neutron and a proton may be only two different forms of a single particle, differing only by the net charge on the associated meson clouds. Proton–neutron interchanges can be equally well explained by the exchange of π^- mesons. Neutral pions must be invoked in order to account for *n–n* binding. Three types of pions are required and three types have been experimentally detected.

4.08 Tunneling Through the Barrier

The height of the potential barrier depends on the atomic number of the target nucleus and the charge carried by the incoming particle. Since electrostatic forces are relatively long-range, the electric field around a nucleus is due to the total charge Ze and so the barrier height will be proportional to Z. Uranium, $Z = 92$, presents a truly formidable barrier to the entrance of any charged particle.

The energy required to surmount a barrier will also be proportional to the charge carried by the incoming particle. An alpha particle, which is a doubly charged helium ion He^{2+}, will be confronted by a barrier height twice that presented to a proton, H^+. A neutron on the other hand will see no potential barrier at all. A neutron with only the kinetic energy of thermal agitation (0.025 eV at normal room temperature) will readily enter into the influence of the nuclear force of even the highest-Z nuclei.

According to classical mechanics an incoming charged particle must have initially a kinetic energy at least as great as the barrier height, E_1 in Fig. 4-4,

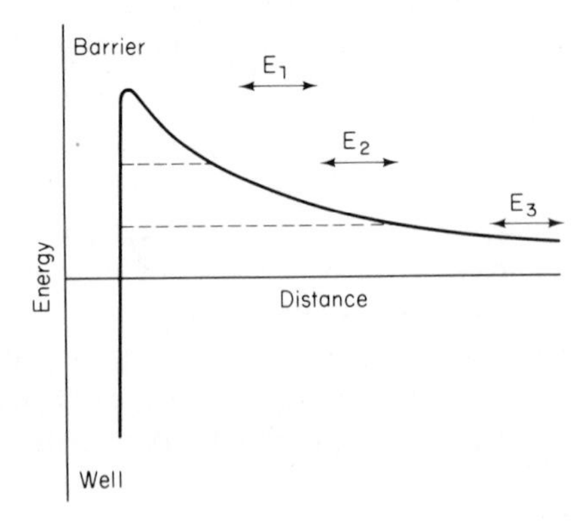

Figure 4-4. The probability of penetrating the potential barrier decreases rapidly as the energy of the incoming particle decreases because only successively thicker portions of the potential barrier can be reached.

if penetration is to be achieved. Similarly, a charged particle escaping from the nucleus would be required to have an energy equal to the sum of its binding energy and the barrier height. In fact, particles with energies much less than those required by the classical model are observed to pass through the barrier in either direction.

Barrier penetration by lower-energy particles, incomprehensible on classical grounds, is understandable in terms of the wave mechanics. Because of its wave nature each approaching particle has a finite chance of *tunneling* through the barrier rather than being absolutely required to pass over the top. As might be expected, the probability of tunneling decreases rapidly as the thickness of the barrier increases. Thus a particle with an energy E_2, Fig. 4-4, has a greater probability of tunneling than a particle with energy E_3 which can classically approach to only a thick portion of the barrier. At somewhat lower energies the penetration probability becomes so low as to be unobservable.

According to the wave mechanics there is no guarantee that an incident particle with an energy greater than the barrier height will penetrate and be captured by the nucleus. Energetic charged particles may undergo Rutherford scattering without capture, to make the overall probability of nuclear entry rather low.

4.09 Nuclear Models

In developing a model of the nucleus it was natural to lean heavily on the previous atomic experience. There are some similarities and some differences. Like the atom, the nucleus is quantized and the Pauli principle is operative. The nucleus lacks the central particle to play the role which it, as an entity, assumes in the atom. Internucleon forces can, therefore, be expected to lack some of the central symmetry seen in the atom.

A close analogy can be developed between the internucleon forces and those which hold a small liquid drop in a spherical shape. This analogy leads to the *liquid-drop* or *strong-interaction* model of the nucleus. Using the liquid-drop model a semiempirical equation can be developed for calculating the mass, and hence the binding energy, of any nucleus, using values of Z and A as inputs. The first two terms in Eq. (4-7) represent

$$M = 939.55(A - Z) + 938.26(Z) - 14.1(A) + 13.8(A^{2/3}) + 0.595(Z^2A^{-1/3}) + 19.3(A - 2Z)^2A^{-1} + 33.5A^{-3/4} \text{ MeV} \tag{4-7}$$

respectively, the contributions of the neutrons and the protons to the nuclear mass. This leaves only five adjustable constants to be determined from experimental data. When this is done, the predictions of the mass equation agree remarkably well with measurement when applied to the several hundred known nuclear species.

Measured values of nuclear mass are now known with high precision for almost all nuclides. When known, measured values will be used in preference to those calculated from the mass equation. This is most useful in establishing the applicability of the liquid-drop model and in demonstrating how nuclear properties vary as functions of Z and A.

We are particularly interested in the last term of Eq. (4-7), $33.5A^{-3/4}$, which is

negative for even Z–even N (even A)
zero for odd A
positive for odd Z–odd N (even A)

These relationships can be understood in terms of the pairings of nuclear spins.

We have seen that spin-pairing in an even-Z atom can lead to a zero angular momentum for the orbital electrons. In the nuclear case we must consider separately the pairing of neutrons and protons.

1. When Z is an even number and N is also even, A is necessarily even. There will be complete p-pairing and complete n-pairing to give a net angular momentum of zero. This corresponds to a relatively low energy state, a somewhat smaller value of nuclear mass, and we note that the energy term in Eq. (4-7) is now subtractive.
2. An odd value of A requires that either Z or A be even while the other is odd. One nucleon type will then be spin-paired while the other will not. Energy content and nuclear mass should be somewhat greater than with complete spin-pairing and the zero value of Eq. (4-7) is in accord with this.
3. An even value of A will be obtained whenever both Z and N are odd. There is now lack of complete spin-pairing in each set of nucleons. The still higher value of energy and nuclear mass requires the use of the positive sign for the spin-pairing term.

The relatively simple liquid-drop model provides a satisfactory explanation for many phenomena, including nuclear fission and radioactive decay. In other cases the model is inadequate to account for observed effects and must be supplemented.

4.10 The Shell Model

The liquid-drop model of the nucleus predicts properties which vary smoothly with Z and N except for the even–odd differences. In general, smooth variations are observed but in some regions there are discontinuities in a number of nuclear properties. Binding energy will serve as a case in point.

Figure 4-5 is a plot of the binding energy per nucleon, E_b, as a function

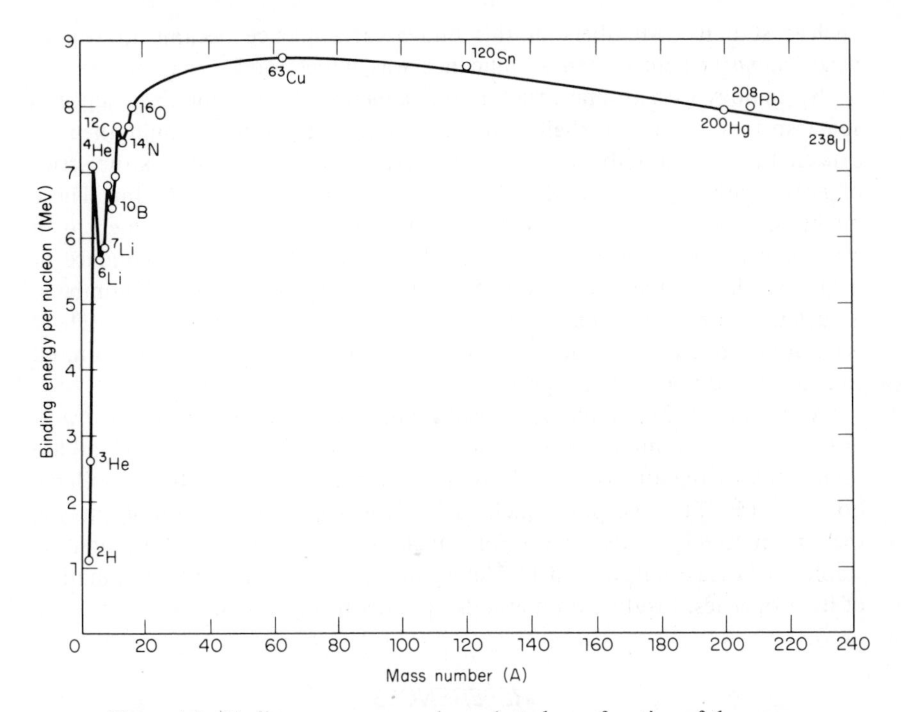

Figure 4-5. Binding energy per nucleon plotted as a function of the mass number. Only enough points have been plotted to establish the general features of the curve.

of mass number. The generally sharp rise in E_b at low values of A is explainable in terms of the liquid-drop model. From simple geometrical considerations a nucleon on the surface of a nucleus cannot take full advantage of the short-range exchange forces with adjacent nucleons. In the low-A region the volume–surface ratio is increasing rapidly and there is a relative increase in the utilization of the exchange forces which is reflected in an increase in E_b.

Forces of electrostatic repulsion account for the slow decrease in E_b above the broad maximum at $A = 63$. Because these are relatively long-range forces, each proton can interact with all others in the nucleus. This leads to an energy requirement for proton assembly that is proportional to Z^2/A. On the other hand total binding energy resulting from the nuclear force increases only as A so the disruptive effect becomes increasingly important as A (and Z) increases. The heaviest stable nuclide known is $^{208}_{82}$Pb.

Discontinuities not accounted for by the liquid-drop model can be seen in Fig. 4-5 at low-A values. Discontinuities in other nuclear properties occur at these same values of A and at higher values omitted from the plot. When either Z or N has one of the values 2, 8, 20, 28, 50, or 126, E_b has a greater value than at neighboring nuclides, which indicates a structure with a greater nuclear stability. Members of this numerical sequence became known as *magic numbers* before their significance was understood.

By analogy with atomic structure the sequence of magic numbers suggests some sort of a nuclear shell structure. Increased stability would then be expected at those numbers where shells were exactly filled, just as the series of noble gases appears at $Z = 2, 10, 18, \ldots,$ in the atomic realm. These might be thought of as atomic magic numbers. Theoretical treatment of a *shell* or *independent-particle* model of the nucleus leads to the sequence of magic numbers actually observed. Since there are now two sets of numbers, a nucleus can be *doubly magic,* as $^{208}_{82}Pb$, where $Z = 82$ and $N = 126$. There is no doubt that a shell structure exists in the nucleus and that this is required to explain many nuclear properties.

As technical skills have developed, man has produced a series of *trans-uranium* elements unknown in nature. Now identified up to $Z = 105$, all are radioactive, with half-lives that decrease rapidly as Z increases. It may well be, however, that at some point such as $^{252}_{126}X$, which would be doubly magic, there is an *island of stability.* If such a nucleus could be produced, it might be stable, or at least have a half-life sufficiently long to permit a determination of its properties. Production methods are now being sought.

REFERENCES

ELTON, L. R. B., *Introductory Nuclear Theory.* W. B. Saunders Co., Philadelphia, Pa., 1966.

FORD, K. W., *Basic Physics.* Blaisdell Publishing Co., Waltham, Mass., 1968.

KAPLAN, I., *Nuclear Physics,* 2nd ed. Addison-Wesley Pub. Co., Reading, Mass., 1962.

LEDERER, C. M., J. M. HOLLANDER, AND I. PERLMAN, *Table of Isotopes,* 6th ed. John Wiley & Sons, Inc., New York, N.Y., 1967.

LEIGHTON, R. B., *Principles of Modern Physics,* International Series on Pure and Applied Physics. McGraw-Hill Book Co., New York, N.Y., 1959.

PRESTON, M. A., *Physics of the Nucleus.* Addison-Wesley Pub. Co., Reading, Mass., 1962.

RICHTMYER, F. K., E. H. KENNARD, AND J. N. COOPER, *Introduction to Modern Physics,* 6th ed., International Series on Pure and Applied Physics. McGraw-Hill Book Co., New York, N.Y., 1969.

5

Radioactive Transformations

5.01 Discovery and Identification

Within a year following Roentgen's discovery of X rays, Becquerel observed somewhat similar radiations emitted spontaneously by uranium-containing minerals. This finding stimulated an intensive research effort that soon led to the discovery of several *radioactive* elements and two other types of radiation. Later investigations added to the list of active elements and identified the particular isotopes that were responsible for the radiations. Today the list of naturally occurring radioactive nuclides has been augmented by an even greater number of man-made products until more than 500 radioactive species are known. In each of these the *decay* of a *parent* nucleus into a *daughter* takes place spontaneously without the addition of any energy from the outside.

Three types of nuclear radiations were quickly recognized by the early workers:

1. Heavy particles, known originally as alpha rays and later as *alpha particles*, were identified by observing the characteristic spectral line emission of helium. At the moment of emission an alpha particle consists of a nucleus of $^{4}_{2}He$, or $^{4}He^{2+}$, which is just a tightly bound combination of two protons and two neutrons. As an alpha particle slows down in passing through matter, it acquires two orbital electrons to become finally a neutral atom of ^{4}He. Upon excitation this will emit the spectral lines of helium.
2. High-speed negative electrons were identified in some radioactive emis-

sions and were named *beta particles*. Much later positively charged electrons were detected in the emissions from some artificially produced radioactive nuclides. These particles were named *positrons* in contrast to the previously known *negatrons*. The two types of electrons have many properties in common in spite of the different charge polarities and so the term beta particle is used to denote either positron or negatron emission.

3. Gamma rays are photons of electromagnetic radiation with energies ranging from a few keV to perhaps 10 MeV. Except for the method of generation a gamma-ray photon is identical to an X-ray or bremsstrahlung photon of the same energy. Gamma-ray emission is the external manifestation of the process by which an excited nucleus goes to its ground, or most stable, state. It is the nuclear analog of the emission of photons with a few eV energy by an atom seeking its ground state after an excitation. In the nuclear case the excitation is usually, but not exclusively, the result of a particle emission in a radioactive decay. Thus gamma-ray emission usually follows the emission of an alpha or a beta particle and is therefore a characteristic of the daughter rather than the parent nucleus.

From either the shell or the liquid-drop model one might suspect that the radioactive emissions would consist of either ejected protons or neutrons. The latter would seem particularly likely because the uncharged particle would not be required to penetrate through the potential barrier. In fact, neither type of nucleon is emitted alone in radioactive decay. A single nucleon emission would result in a nuclear transformation of either $Z, N \longrightarrow Z, (N - 1)$ or $Z, N \longrightarrow (Z - 1), N$. When appropriate values of Z and N for parent and daughter are put into the nuclear mass equation, Sec. 4.09, each of the proposed transformations by nucleon emission is found to be energetically impossible.

Alpha-particle emission can be readily visualized because the component particles preexist in the parent nucleus, and the emission is energetically allowed from a good many nuclear configurations. Whenever a bound $2n$–$2p$ group is formed near the nuclear surface, it will be loosely bound to the other nucleons, Sec. 4.07, and it has a finite chance to escape. Energetically, heavier groups such as fully ionized ${}^{8}_{4}Be$ or ${}^{12}_{6}C$ could be emitted but these are never seen in spontaneous decay. Presumably the probability of formation of such a group is small and the potential barrier is impenetrable to the large charges.

Beta-particle emission, on the other hand, requires the creation of the particle at the moment of emission since there is abundant evidence against the presence of electrons as such in the nucleus. The mechanism of electron creation need not concern us here.

5.02 Rate of Radioactive Decay

The rate at which a radioactive species decays is a most important parameter of the transformation. Each individual decay is an event peculiar to the particular nucleus involved, completely independent of the state or behavior of any other nucleus. In any given species each nucleus is identical with all the others, except for the details of the internal fluctuations of energy among the constituent nucleons. It is the latter factor which determines the moment at which a nucleon will initiate a decay, and as a consequence each nuclear decay is completely independent of all others. At any instant each nucleus of a radioactive species has an equal probability of decaying.

Consider a sample consisting of N nuclei of a particular radioactive species. Since all decay probabilities are equal, the number of decays in a given short time will be proportional to the number of nuclei available for decay. That is,

$$\Delta N = -\lambda N \, \Delta t \tag{5-1}$$

where ΔN is the number decaying in a short time interval Δt. λ is a constant of proportionality known as the *decay constant* or the *disintegration constant*. Note that a minus sign is required in Eq. (5-1) because ΔN denotes a decrease in the number of nuclei available for decay. A rearrangement of Eq. (5-1)

$$\lambda = -\frac{\Delta N / N}{\Delta t} \tag{5-1a}$$

emphasizes that λ is numerically the fractional rate of decay over a time period so short that N can be considered to be constant.

Equation (5-1) can be put in differential form and integrated to give an expression for the value of N as a function of time.

$$N = N_0 e^{-\lambda t} \tag{5-2}$$

where N_0 = number of nuclei at time $t = 0$
e = base of natural logarithms

Equation (5-2) may be put in alternate forms that are sometimes more suitable for numerical calculation.

$$ln\, N = ln\, N_0 - \lambda t \tag{5-2a}$$

$$ln \frac{N_0}{N} = \lambda t \tag{5-2b}$$

According to Eq. (5-2) there will always be some parent nuclei remaining at any finite time, no matter how long this time may be. It is not possible, therefore, to specify the lifetime of any radioactive species. It is possible to specify *half-life T*, which is the time required for a sample to decay to one-half of its original population. From Eq. (5-2b),

$$ln \frac{N_0}{0.5N_0} = ln\, 2 = \lambda T$$

and

$$T = \frac{ln\, 2}{\lambda} = \frac{0.693}{\lambda} \qquad (5\text{-}3)$$

Half-life may be expressed in any convenient time unit but custom has settled on the use of the year, day, hour, minute, and second, abbreviated y, d, h, m, and s, respectively. λ will of course have the reciprocal units of y^{-1}, d^{-1}, h^{-1}, m^{-1}, and s^{-1}.

Known half-lives cover an enormous range, from more than 10^{10} y to less than 10^{-6} s, but in every case the time course of the decay follows Eq. (5-2) as long as there are enough undecayed nuclei to permit averaging individual decay probabilities over a reasonable number. Since λ is a nuclear and not an atomic property, it and hence T are unaffected by changes in state variables such as temperature, pressure, chemical combination, or state of aggregation.*

Average lifetime $\bar{T}$ is another term sometimes used in connection with radioactive decay. Average lifetime is calculated by summing all the individual lifetimes and then dividing this by the total number of nuclei involved. The result is

$$\bar{T} = \frac{1}{\lambda} = \frac{T}{0.693} = 1.44T \qquad (5\text{-}4)$$

Equation (5-2) describes the time course of the number of undecayed nuclei, or the amount of the radioactive substance that is present at any time after measurements have started. The corresponding activity A is the time rate at which the nuclei in the sample decay, $A = \Delta N/\Delta t$. From the definition of λ,

$$A = \lambda N = \lambda N_0 e^{-\lambda t} = A_0 e^{-\lambda t} \qquad (5\text{-}5)$$

Any convenient time unit can be used in equations describing radioactive decay but activity is customarily specified in either disintegrations per second (dps) or per minute (dpm). The commonly used unit of activity, the *curie* (Ci), is exactly equal to 3.7×10^{10} dps or 2.22×10^{12} dpm. Many samples are more conveniently specified in millicuries (mCi), or in microcuries (μCi). The reason for the apparently random choice of a number is explained after considering the decay of radium-226, Sec. 11.02.

5.03 Specific Activity

It is frequently desirable to express the concentration of a radioactive material in terms of its *specific activity*, which is the amount of activity per unit mass

*In some cases of transitions involving orbital electrons, small changes in λ can be observed under extreme changes in the state variables.

of the sample. Any convenient units of activity and mass may be used but custom has come to prefer mCi mg^{-1} or μCi mg^{-1}.

Illustrative Example

Calculate the specific activity of a 6.8-mg sample of ^{60}Co which is observed to decay at a rate of 8.5×10^8 dpm.

Activity $A = 8.5 \times 10^8/2.22 \times 10^{12} = 3.83 \times 10^{-4}$ Ci or 383 μCi
Specific activity $= 383/6.8 = 56.4$ μCi mg^{-1}

Each radioactive species has an *intrinsic* specific activity, which is the activity of a unit mass of the pure, undiluted material. Intrinsic specific activity depends only on the decay constant and the nuclear mass of the species in question.

Illustrative Example

Calculate the intrinsic specific activity of 5.3 y ^{60}Co.

Consider a 1-mg sample which will contain $[(6.03 \times 10^{23}) \times 10^{-3}]/60 = 1 \times 10^{19}$ nuclei of ^{60}Co.

$\lambda = 0.693/[5.3 \times (3.15 \times 10^7)] = 4.15 \times 10^{-9}\ s^{-1}$
Decay rate $= \lambda N = (4.15 \times 10^{-9}) \times (1 \times 10^{19}) = 4.15 \times 10^{10}$ dps
Activity $= 4.15 \times 10^{10}/3.7 \times 10^{10} = 1.12$ Ci
Intrinsic specific activity $= 1.12/1.0 = 1.12$ Ci mg^{-1}

A comparison of the two illustrative examples shows that the sample considered in the first was highly diluted. A sample such as that in the second, consisting only of the radioactive species, is said to be *carrier free*, denoted in tabulations by CF.

5.04 Alpha-Particle Emission

When an alpha particle is emitted, the parent nucleus ejects a group of four nucleons and is transformed into a daughter structure containing two less protons and two less neutrons. Symbolically,

$$^{A}_{Z}\text{X} \longrightarrow {}^{A-4}_{Z-2}\text{Y} + {}^{4}_{2}\alpha + \gamma \qquad Q_\alpha \tag{5-6}$$

where X and Y represent parent and daughter elements, respectively. The γ term is symbolic only, indicating the possibility of a photon emission. A particular decay may emit several, one, or no photons, depending on the microscopic details of the transformation. Q_α designates the energy available in the transformation. This energy, derived from differences in binding energy among parent, daughter, and that of the emitted alpha particle, divides among the kinetic energy of the alpha particle, kinetic energy of the recoiling nucleus, and gamma-ray emission, if any.

The type transition of Eq. (5-6) can be represented on either an A–Z or

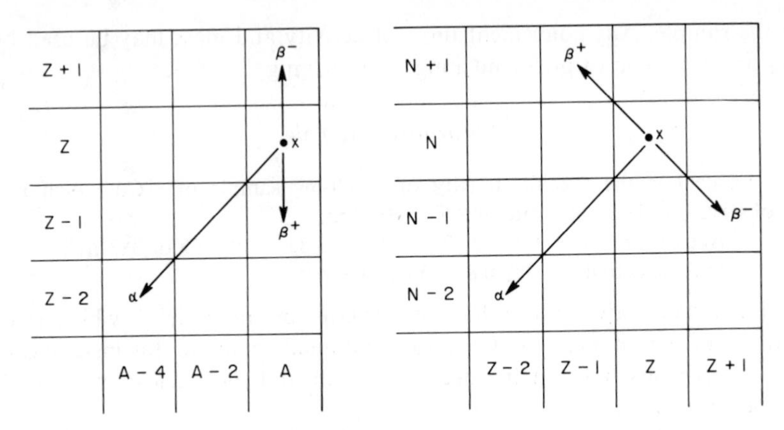

Figure 5-1. Alpha and beta transitions depicted on *Z–A* and *Z–N* plots.

a *Z–N* plot, Fig. 5-1. The pertinent parameters of a transition are displayed in a *decay scheme*, given in most compilations of decay characteristics. Figure 5-2 shows an abridged decay scheme for the alpha decay of ^{226}Ra.

$$^{226}_{88}\text{Ra} \longrightarrow {}^{222}_{86}\text{Rn} + \alpha + \gamma \qquad Q_\alpha = 4.869 \text{ MeV}$$

Following accepted convention positive-particle emission is depicted by an arrow drawn down and to the left. Although the decay scheme represents energy levels, they are shown only very roughly to scale, referred to the ground-state energy of the daughter. Complete decay schemes present much more information than is shown in the abridged versions here. Some of this information is discussed later; some is of interest only to the nuclear physicist.

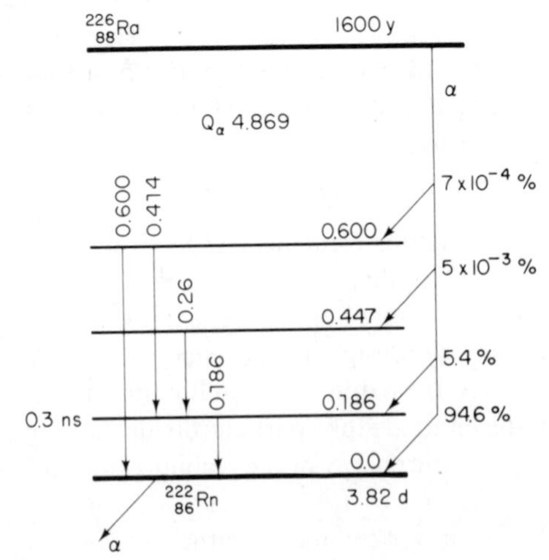

Figure 5-2. An abridged decay scheme for the alpha-particle decay of $^{226}_{88}$Ra.

According to Fig. 5-2, ^{226}Ra decays with a half-life of 1600 y, emitting alpha particles at four different energies. The most probable emission, 94.6%, goes to the ground state of the daughter ^{222}Rn. The three other particle transitions go to excited states of the daughter. Each of the excited states is relieved by gamma emissions until all transitions end at the ground state. Half-lives of the excited states are shown in the decay scheme unless they are very short.

Detailed studies of alpha-particle emission provide strong arguments for the existence of discrete energy levels in the nucleus. Each of the four alpha emissions shown in Fig. 5-2 is monoenergetic, and the sum of each alpha energy, the energy of recoil of the daughter nucleus, and the related gamma-ray energies will equal the energy available in the transition, Q_α.

5.05 Negative Beta-Particle Emission

When a negatron is emitted by a nucleus, the nuclear change is equivalent to

$$^1_0n \longrightarrow {}^1_1p + e^- \qquad (5\text{-}7)$$

The type equation for the decay is

$$^A_Z\text{X} \longrightarrow {}_{Z+1}^{\ A}\text{Y} + \beta^- + \nu + \gamma \qquad Q_\beta \qquad (5\text{-}8)$$

According to Eq. (5-7) the number of nucleons, and hence the mass number A, does not change in negatron emission. Parent and daughter nuclei are isobars. Z increases by unity because of the loss of one negative charge from the nucleus. As in alpha decay, γ symbolizes the possibility of photon emission and Q_β is the energy available in the transition. A new term, ν, representing the *neutrino* now appears in the equation.

Studies of beta-particle spectra show that in pure beta-particle emission such as that represented by Eq. (5-8) the electrons are emitted with a continuous energy distribution ranging from zero up to a maximum value T_m that is characteristic of each transition. Several classes of spectra came to be recognized, Fig. 5-3, but in every case there is a continuous energy distribution instead of the monoenergetic emissions seen with alpha particles. Details of the spectral differences will be considered in Sec. 6.06. It is sufficient here to recognize the continuous distribution and to note that this seems to argue against the existence of precise energy levels inside the nucleus.

In 1931, Pauli suggested that each beta decay is characterized by the simultaneous emission of two particles rather than one. The second particle would have to be electrically neutral and have an extremely small or even a zero mass because the charge and mass requirements of Eq. (5-8) are satisfied without a second particle. The second particle with these properties is extremely difficult to detect and several years passed before it was experimentally identified. Now the existence of the second particle, known as the

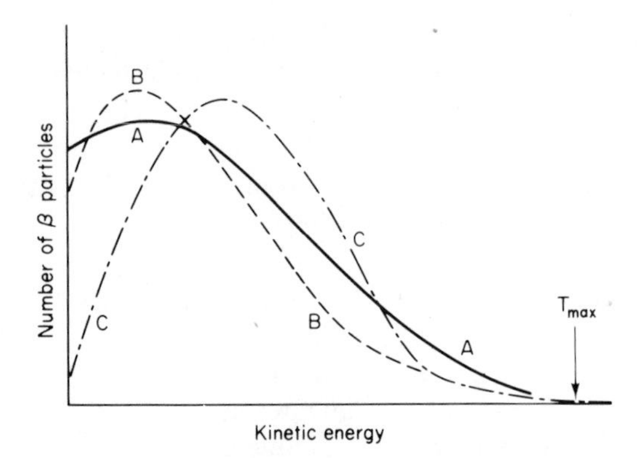

Figure 5-3. Typical energy distributions of beta particles emitted in radioactive decay. (A) An allowed negatron transition. (B) A forbidden negatron transition. (C) A positron transition.

neutrino, υ, is firmly established and its role in beta emission is understood. Strictly, the particle that accompanies negatron emission is an *antineutrino* but since neither type of neutrino has any biological significance, we shall not distinguish between them.

In every radioactive transition a fixed amount of energy is available for the negatron emission and in each individual decay this energy will divide between the two particles according to certain probability requirements. From the spectra shown in Fig. 5-3 we see that on the average the neutrino will carry away well over one-half of the available energy. This energy will be lost from the system under consideration because the neutral, massless neutrino interacts only very weakly with any matter through which it passes. For example, the radiation dose delivered to a patient or an animal from an ingested beta emitter will be determined by the average energy distribution of the β emission rather than by the maximum value of the transition energy.

Figure 5-4 shows the atomic decay scheme of the negatron emitter ${}^{41}_{18}Ar$. By convention, negative particle emission is depicted by an arrow drawn down and to the right.

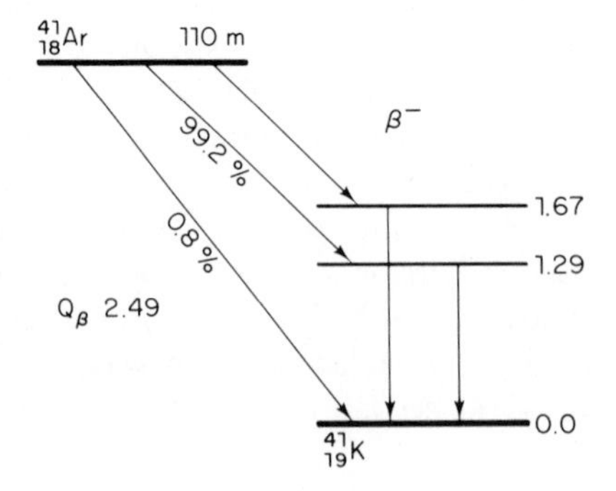

Figure 5-4. Atomic decay scheme for the negatron emitter ${}^{41}_{18}Ar$.

In the example shown, 99.2% of the transitions go to the 1.29-MeV excited state of ${}^{41}_{19}$K and consequently the beta-particle spectrum will consist predominantly of a 2.49 − 1.29 = 1.20 MeV distribution. With 0.8% of the decays going to the ground state there is a superposed distribution with a maximum energy of 2.49 MeV. A very few of the decays go to the 1.67-MeV state, leaving 0.82 MeV to be divided between each of these beta particles and its related neutrino.

In negatron decay no account need be taken of the mass of the electron ejected from the nucleus. Because the atomic number increases from Z to $Z + 1$, the daughter atom must acquire another orbital to regain electrical neutrality. The loss and the acquisition balance and so the self-energy of the beta particle does not appear in the diagram.

5.06 Positron Emission

A positron emission can be visualized as the result of a nucleon transformation:

$$ {}^{1}_{1}p \longrightarrow {}^{1}_{0}n + e^{+} \tag{5-9}$$

As in negatron decay the total number of nucleons remains constant, parent and daughter are isobaric, but the atomic number has decreased by one. The type equation becomes

$$ {}^{A}_{Z}\mathrm{X} \longrightarrow {}_{Z-1}^{\ \ A}\mathrm{Y} + \beta^{+} + \nu + \gamma \qquad Q_{EC} \tag{5-10}$$

Again a continuous beta-particle energy distribution is observed and simultaneous neutrino emissions are required.

Because of the decrease in atomic number the daughter atom must release one of its orbital electrons. In positron emission, orbital electron gain does not offset nuclear electron loss and so electron masses (energies) must appear in the decay schemes.

Figure 5-5 is the atomic representation of a typical positron decay. The

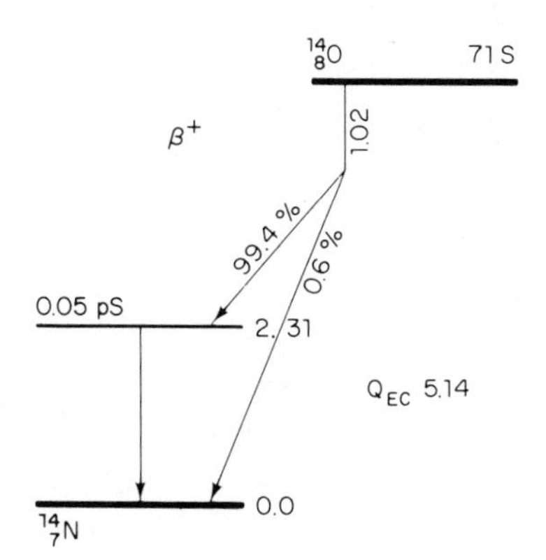

Figure 5-5. Atomic decay scheme for the positron emitter ${}^{14}_{8}$O.

loss of two electron masses is shown by the vertical arrow representing an energy loss of 1.02 MeV. Decay to the 2.31-MeV state of $^{14}_{7}N$ accounts for 99.4% of the transitions, with the remaining 0.6% going to the ground state. The respective beta-particle spectra will have maximum energies of $5.14 - 1.02 - 2.31 = 1.81$ MeV and $5.14 - 1.02 = 4.12$ MeV. The energy of a positron transition, 5.14 MeV, in this case, is designated by Q_{EC} for reasons which will appear presently.

It is evident that in positron decay the atomic mass of the daughter *must* be at least 1.02 MeV less than that of the parent because this mass is lost in the ejection of the two electrons. No such absolute requirement exists for negatron decay. With electron loss and gain canceling, negatron decay can take place when only a small transition energy is available. A case in point is $^{3}_{1}H$, which emits a negatron with a Q_{β} of only 18 keV. In the positron case at least 1.02 MeV must be available if the decay is energetically possible.

Positron emission will always be accompanied by the production of 0.511-MeV photons. As the positron loses energy in its passage through matter, it slows until it has a velocity that is comparable to that of the atomic orbitals. The probability of a positron–negatron reaction is then very high, according to

$$e^{+} + e^{-} \longrightarrow 2h\nu \qquad (5\text{-}11)$$

As the two charges neutralize, the two masses, each equivalent to 0.511 MeV, also disappear by conversion into two photons of equal energy. Momentum conservation requires the production of two oppositely directed photons which is known as *annihilation radiation.*

5.07 Electron Capture

A nuclide may have the type of instability that calls for an increase in the n/p ratio by Eq. (5-9) to improve its stability; yet it may have a daughter mass so large that positron emission is energetically impossible. The only course open to these nuclides is to achieve the equivalent of Eq. (5-9) by the capture of an orbital electron. Capture from the K shell is most probable since the K electrons lie closest to the nucleus; the process was originally known as K capture. L or M capture is possible and the more general term, electron capture, abbreviated EC, is now accepted. In electron capture the nuclear rearrangement is equivalent to

$$^{1}_{1}p + e^{-} \longrightarrow {}^{1}_{0}n \qquad (5\text{-}12)$$

which is equivalent to Eq. (5-9). The type equation becomes

$$^{A}_{Z}X + e^{-} \longrightarrow {}_{Z-1}^{A}Y + \nu + \gamma \qquad Q_{EC} \qquad (5\text{-}13)$$

Two versions of the decay scheme for $^{51}_{24}Cr$ are shown in Fig. 5-6. In the atom as a whole, Fig. 5-6A, there is no loss or gain of an electron, merely a

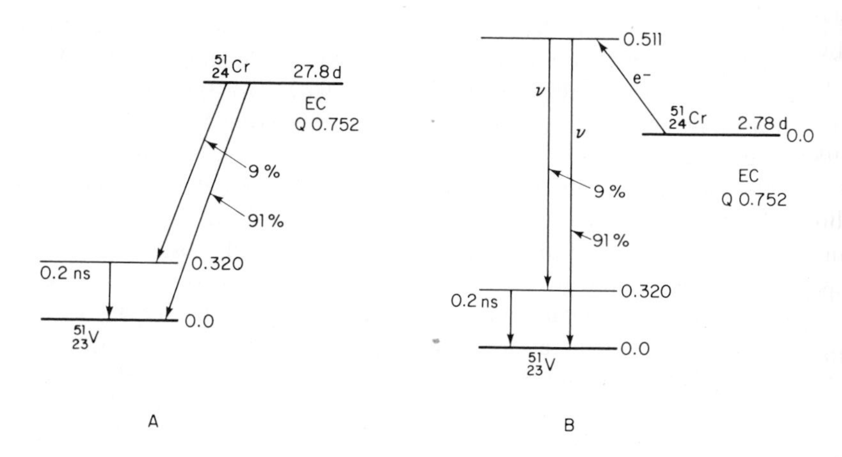

Figure 5-6. (A) Atomic, and (B) nuclear, decay schemes for the EC transition in $^{51}_{24}$Cr.

transfer from orbit to nucleus; hence there is no minimum energy requirement. Since electron capture is equivalent to the loss of one positive nuclear charge, EC transitions are shown by arrows down to the left. In the case shown, 9% of the transitions go to an excited state of $^{51}_{23}$V and 91% go to the ground state.

The details of an EC decay can be understood from a consideration of the nuclear decay scheme, Fig. 5-6B. For the nucleus alone the first step in the decay is the acquisition of 0.511 MeV from the captured orbital electron. Monoenergetic neutrino emissions then bring the nuclear energy levels down to the same states pictured in the atomic scheme. In the nuclear version of the decay scheme the neutrino emissions are shown specifically; in the atomic version they are implied by the EC arrows.

Just after electron capture the daughter atom has the proper total number of electrons in its orbitals but there is a vacancy in one of the lower shells. This vacancy will be filled promptly from an outer shell, this second vacancy will fill from a shell still farther out, and so on until the entire orbital structure has attained its lowest possible energy state. These orbital rearrangements will produce the X rays characteristic of the daughter element because the nucleus already has the Z value corresponding to it.

When an electron is captured by a nucleus, it is accelerated out of its stationary-state orbit and consequently bremsstrahlung will be produced during the movement. The resulting photons, known as *internal bremsstrahlung*, have energy distributions that range from zero up to the energy available to the transition.

The assay of a nuclide decaying by electron capture may present special problems because no charged particle, and perhaps no gamma ray, is available for detection. ${}^{51}_{24}Cr$, decaying according to Fig. 5-6, is a special case because 9% of the decays go to an excited state. The subsequent 320-keV gamma-ray emission provides a convenient radiation for the assay. If all the EC transitions go to the ground state, the characteristic X rays provide the only radiation available for detection. These photons will usually have quantum energies of only a few keV, Eq. (3-8), where detection techniques are difficult.

Electron capture can always be in competition with positron emission. Both attain the same end result and the EC mode is energetically possible in all cases where positrons can be emitted. Because of this, Q_{EC} is used to specify the transition energy for both decay modes. Reverse competition is not possible because many EC transitions take place at energies that are too low for β^+ emission.

5.08 Isomeric Transition

In the cases previously discussed, each gamma ray was emitted from an excited state that lasted for 10^{-6} s or less. Some excited states, known as *metastable states*, may exist for much longer periods of time and it is then profitable to consider the overall decay as two distinct processes.

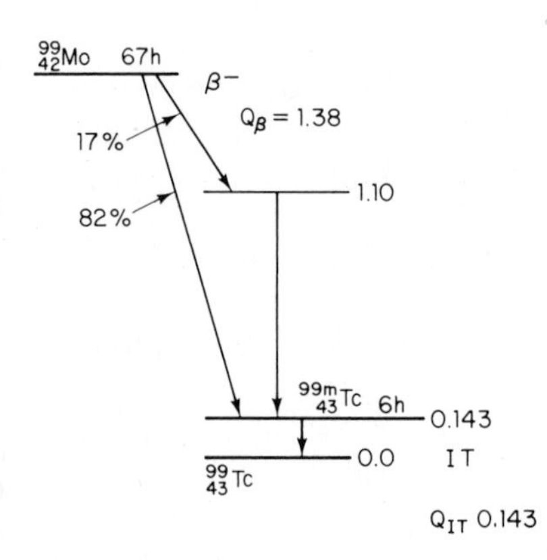

Figure 5-7. Beta decay of ${}^{99}_{42}Mo$ leads to a relatively long-lived metastable state of Tc, designated as ${}^{99m}_{43}Tc$.

Figure 5-7 shows the decay scheme for the beta decay of ${}^{99}_{42}Mo$. Two beta-particle groups go to excited states of the Tc daughter but one of these states quickly emits a gamma ray to bring all beta decays to the 0.143-MeV level. The remaining excitation is relieved by gamma-ray emission with a 6-hr half-life. This time is so long that the state is called metastable and designated ${}^{99m}_{43}Tc$. Isomeric transition, or IT, follows the type reaction

$$ {}^{A}_{Z}X^* \longrightarrow {}^{A}_{Z}X + \gamma \qquad Q \tag{5-14}$$

The only nuclear change is a loss of energy to the gamma ray and so the transition may be thought of as a case of nuclear isomerism.

The particular isotope used as an example in Fig. 5-7, ^{99m}Tc, has proved useful in diagnostic medicine. A ^{99}Mo generator with a half-life of 67 hr provides the ^{99m}Tc which is eluted from a separation column for use as needed in short-term studies.

5.09 Internal Conversion

We have seen how a nucleus may, in some cases, interact with an orbital electron to produce an EC decay. A nucleus in an excited state, expected to emit a gamma ray in dropping to the ground state, may also interact with the orbitals. Instead of emitting the gamma ray with energy E_γ, the nucleus may transfer this energy directly to an orbital electron. Interactions with K electrons are most probable but L and M interactions are also known. If the electron is bound to the nucleus with energy ϕ_e, the energy interchange will be

$$E_\gamma = T + \phi_e \qquad (5\text{-}15)$$

where T is the kinetic energy of the ejected electron. Equation (5-15) is identical in form with that governing the photoelectric effect, Eq. (1-17). Energywise it is as if the gamma ray was emitted and then interacted with the orbital electron in typical photoelectric fashion. In fact, the gamma ray is never emitted and the process of *internal conversion*, IC, is an energy conversion only.

From Eq. (5-15), conversion electrons will be monoenergetic and will not be accompanied by neutrinos. Values of ϕ_e can be obtained from tabulated values of the energies of X-ray absorption edges, Sec. 8.10 and Appendix 3.

Illustrative Example

Calculate the energy of the electrons emitted in the conversion of the 393-keV gamma ray from ^{113m}In, the nuclide formed by EC in ^{113}Sn.

From Appendix 3, the absorption edges in In are

$K = 27.9$ keV $\qquad L = 4.2$ keV

K-shell conversion: $393 - 27.9 = 365$ keV

L-shell conversion: $393 - 4.2 = 389$ keV

Figure 5-8 shows the conversion electron spectrum of ^{113m}In, with peaks occurring at the energies calculated in the illustrative example. In the figure the monoenergetic peaks are broadened by energy absorption in the sample being counted.

As Fig. 5-9 demonstrates, conversion electrons may be superposed upon the continuous energy distribution of beta particles. In the example shown,

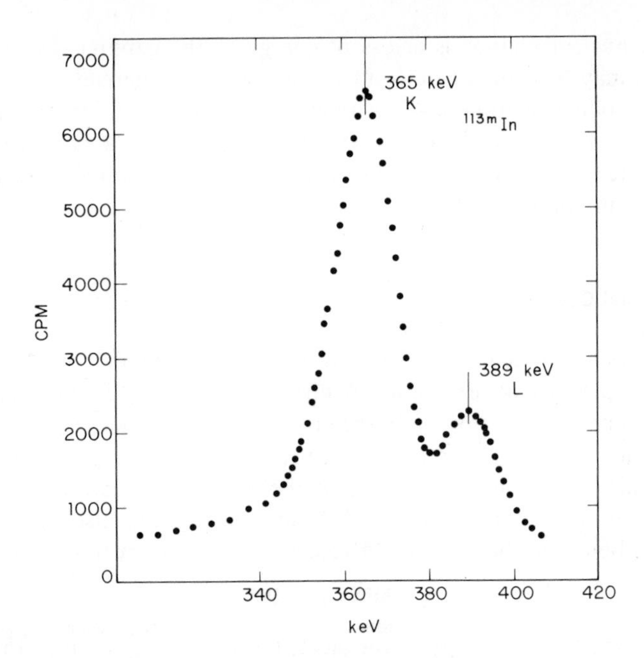

Figure 5-8. Energy spectrum of the electrons produced by the conversion of the 393-keV gamma ray in $^{113m}_{49}In$. The latter was produced by electron capture in $^{113}_{50}Sn$.

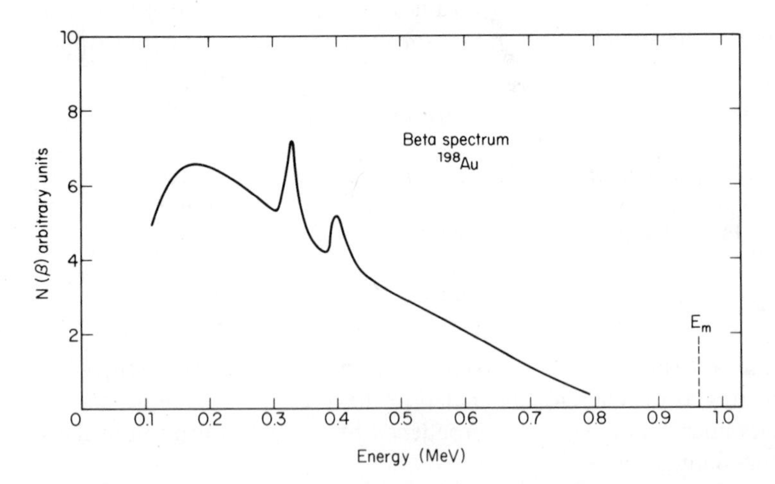

Figure 5-9. *K*- and *L*-conversion lines from the $^{198}_{80}Hg$ daughter are superposed on the continuous beta spectrum from the parent $^{198}_{79}Au$.

the beta spectrum of the parent ^{198}Au has superposed conversion lines from the daughter ^{198}Hg.

In most cases some of the excited states will decay by gamma-ray emission and some by internal conversion. The two competing modes lead to the definition of an *internal conversion coefficient* α as

$$\alpha = \frac{R_e}{R_\gamma} \tag{5-16}$$

where R_e is the rate of emission of conversion electrons and R_γ is the rate of gamma-ray emission by the transition in question. If desired, conversion coefficients can be calculated for conversion in each of the orbital shells, K, L, ...

5.10 Auger Electrons

Another type of internal conversion can follow an EC transition. We have seen that after electron capture the orbital structure is left with a vacancy in one of the lower shells, K, for example. Normally this vacancy would be filled by an $L \longrightarrow K$ transition with the emission of a K characteristic X ray. Alternatively the energy involved in the $L \longrightarrow K$ transition may be transferred to one of the outer, more loosely bound orbitals. In some cases the available energy may be divided between several outer electrons. The outer electrons receiving the energy will be ejected from their orbits with kinetic energies equal to the difference between the energy available and their binding energies. These are *Auger* electrons. Auger electrons have a line spectrum and no neutrino is involved. Auger electron energies are characteristic of the element that is involved in the decay and these electrons may be used as an analytic tool for element identification. After Auger electron emission, all the empty orbits will be filled to end the decay process.

5.11 Spontaneous Fission

Nuclear fission is usually thought of as the division into two roughly equal fragments of a nucleus that has been highly excited by the addition of energy from an external source. In some heavy elements, particularly those that lie beyond uranium, *spontaneous fission* takes place without the addition of external energy. Spontaneous fission is, therefore, a mode of radioactive decay, usually found in competition with alpha-particle emission. A typical reaction equation cannot be written because the relative masses of the two fission fragments are not constant from one nucleus to another.

Decay schemes denote spontaneous fission by SF and give the decay percentages of the competing modes. Thus the practically important nuclide

$^{252}_{98}Cf$ decays by α, 96.9%; SF, 3.1%. Alternatively the half-life for spontaneous fission alone is given along with the half-life for all decay modes combined. The entry would then be $^{252}_{98}Cf$, 2.6 y; SF, 85 y.

REFERENCES

EVANS, R. D., *The Atomic Nucleus*, International Series on Pure and Applied Physics. McGraw-Hill Book Co., New York, N.Y., 1955.

LAPP, R. E. AND H. L. ANDREWS, *Nuclear Radiation Physics*, 4th ed. Prentice-Hall, Inc., Englewood Cliffs, N.J., 1971.

RUTHERFORD, E., J. CHADWICK, AND C. D. ELLIS, *Radiations from Radioactive Substances*. Cambridge Univ. Press, Cambridge, mass., 1930.

SEGRÈ, E., ed., *Experimental Nuclear Physics*, **III**. John Wiley & Sons, Inc., New York, N.Y., 1959.

6

Charged Particle Emission in Radioactive Decay

A ALPHA PARTICLES

6.01 Theory of Alpha-Particle Emission

Each nucleon trapped inside the potential well will have an energy of several MeV and consequently will be in motion with velocities of the order of 10^9 cm s^{-1}. As a result of this motion, two-proton two-neutron groups will form occasionally near the nuclear surface. Because of the saturation nature of the nuclear force this group will be almost unbound to the remaining nucleons. There is then a finite probability of a group escape across the potential barrier even though it will present a height of perhaps 20 MeV to the doubly charged particle.

Classical theory requires the four-nucleon group to have an energy at least equal to the barrier height if it is to surmount the barrier and leave the residual structure. Any particle crossing the barrier will acquire kinetic energy as it moves away in the repelling field of the daughter nucleus. The classical model predicts then that alpha particles should be observed with kinetic energies of 20 MeV or so. In fact the observed energies lie in a rather narrow range of 4–10 MeV.

According to the wave mechanics, however, an alpha particle need not surmount the barrier but may tunnel through it, just as in the case of incoming charged particles. The two-way tunneling possibility is illustrated in Fig. 4-4. The probability of barrier penetration will increase with the energy of the particle since high-energy particles can tunnel through a relatively thin section of the barrier. Thus a particle with energy E_2, Fig. 4-4, has a greater probability of penetration than has a particle with energy E_3. The tunneling concept, applied to the $1/r$ shape of the barrier, provides a satisfactory explanation of the observed alpha-particle energies.

When an alpha-particle group inside the barrier has an energy greater than about 10 MeV, the probability of escape is very high and the corresponding half-life is very short. The existence of the parent nucleus may be too short to be identified. This would be the situation at or near E_1, Fig. 4-4. At lower energies, as E_2, the probability of escape decreases and we recognize half-lives ranging from microseconds up to 10^{10} y. At still lower energies, below 4 MeV, barrier tunneling becomes increasingly improbable and any half-lives are too long to measure by present techniques.

The wave-mechanical model predicts that the half-life of an alpha transition should increase as the particle energy decreases. This tendency can be seen by an examination of the entries in Table 11-1. The functional relationship is very strong, half-lives varying over a range of 10^{22} while energies change by less than a factor of two. Before considering this dependence in greater detail we must describe the variation of alpha-particle range with energy.

6.02 Range of Alpha Particles

Before energy measurements were made routinely, it was customary to express the initial energy of an ejected particle in terms of its range in air at standard conditions of temperature and pressure. Range measurements are usually simplified by maintaining a constant source–detector distance and varying the pressure of the intervening absorbing gas. Ranges thus determined can be easily converted to centimeters of the gas under the standard conditions. Alternatively the range may be expressed as an *equivalent range*, usually in units of mg cm^{-2}.

The results of a typical range measurement on a monoenergetic alpha-particle emission are shown in Fig. 6-1. Out to a distance of several centimeters from the source a constant number of particles will be counted per unit time. Each particle is losing energy along its path but none has been completely stopped. At point M, however, a few particles have lost so much energy that they can no longer actuate the detector. The count rate now drops rapidly with distance and tails off to zero at very low count rates.

This is just the behavior that would be expected from a group of particles, originally monoenergetic, that lose energy through a random series of collisions. The random nature of the collision energy losses results in a *range straggling* which is well characterized by the *normal distribution*, Sec. 12.09. Each ionization in air requires on the average an expenditure of about 34 eV. A typical alpha particle such as the 4.8-MeV emission from ^{226}Ra will then produce $(4.8 \times 10^6)/34 = 1.41 \times 10^5$ ionizations in coming to rest. With

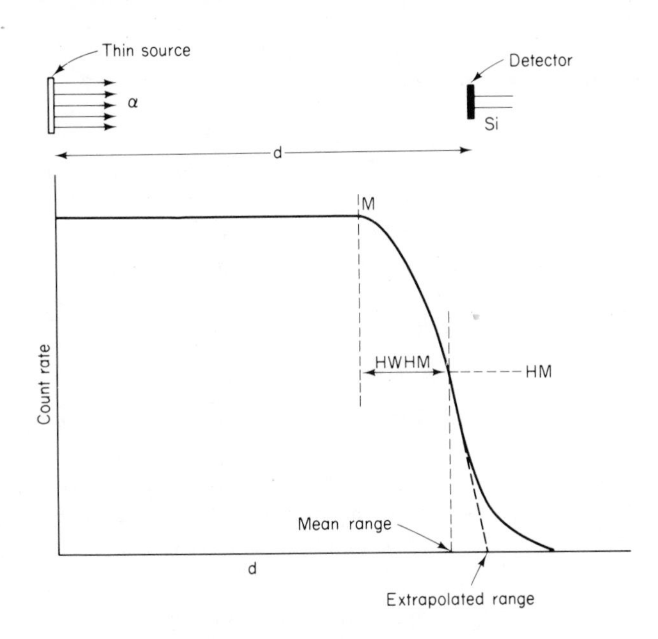

Figure 6-1. The range of alpha particles from a thin source as measured with a thin silicon or germanium detector. The portion of the curve showing straggling has been expanded for clarity.

a normal distribution the standard deviation of this number is $\sqrt{1.41 \times 10^5} = 376$, which is equivalent to a standard deviation in energy of $376 \times 34 =$ 12,800 eV. Thus there is about one chance in three that an alpha particle will arrive at the mean range with a residual energy greater than 12.8 keV or that it had been completely stopped before attaining the mean range.

Straggling is complicated by the fact that at energies of a few keV and below the doubly charged particle will acquire and lose one or more electrons several times before finally coming to rest in the neutral state. Range variations due to this process add to the statistical fluctuations in the ionizations to produce the observed straggling. Two ranges are customarily used to take straggling into account. The *mean range* R_m is the distance (or the equivalent range) at which the count rate is reduced to one-half of its maximum value. A tangent drawn to the range curve at this point, called the *half-maximum*, or HM, will intersect the range axis at the *extrapolated range* R_e. Only the high-energy half of the complete normal distribution curve will be seen and so it is customary to express the straggling as the *half-width* of the curve at the HM level. This factor is written as HWHM, Fig. 6-1. It can be shown that, with a normal distribution,

$$R_e = R_m + 1.06\,(\text{HWHM}) \tag{6-1}$$

6.03 Range–Energy Relationships

Bethe has derived expressions for the rate of energy loss of charged particles as they pass through absorbing media but some simple empirical relations are sufficiently accurate for many purposes. These expressions are particularly useful at the lower energies where the theoretical treatment is uncertain because of the repeated acquisition and loss of electrons by the slowing particle. Over the energy range encountered in radioactive decay, alpha-particle ranges in air may be given by

$$R = 0.325E^{3/2} \quad \text{or} \quad E = 2.12R^{2/3} \tag{6-2}$$

where R is the mean range in cm of standard air and E is the energy of the particle in MeV.

All the alpha particles emitted by radioactive nuclei have energies below 10 MeV, which permits the use of nonrelativistic velocity–energy relations. Putting the classical expression for kinetic energy into Eq. (6-2), we have for the initial particle velocity

$$v^3 = 1.03 \times 10^{27}R \quad (\text{cm s}^{-1})^3 \tag{6-3}$$

Equations (6-2) and (6-3) are forms of the *Geiger relation*, valid except at very low and very high velocities.

Alpha-particle ranges in substances other than air can be calculated approximately from the *Bragg–Kleeman* relation

$$R_s = \frac{3.2 \times 10^{-4} R_a \sqrt{A_s}}{\rho_s} \tag{6-4}$$

where the subscript s applies to the range, mass number, and density of the substance in question and R_a is the range of the particles in air. Figure 6-2 presents empirical range–energy relations for water (essentially equivalent to soft tissue) and for a typical photographic emulsion. The curve for water will give ranges in air to a good approximation if the abscissa values are read as millimeters instead of as micrometers.

6.04 Energy and Half-Life

The functional dependence between alpha-particle energy and decay constant of the transition was originally expressed in terms of the range by the *Geiger–Nuttall* relation

$$\log R = A + B \log \lambda \tag{6-5}$$

where R is the range in air and A and B are constants. From the range–energy relations it is evident that the energy–decay constant dependence has a similar logarithmic form with a different set of constants.

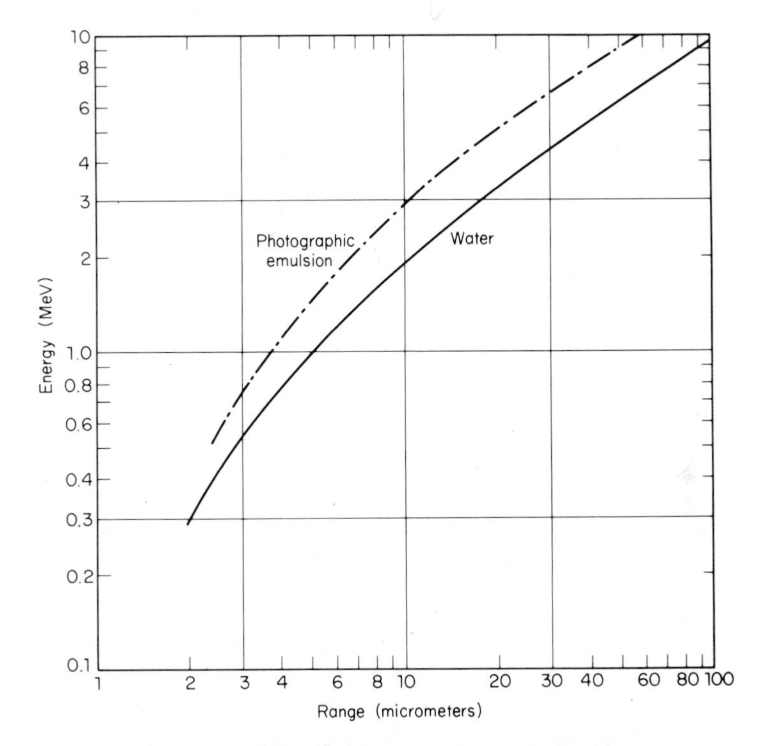

Figure 6-2. Alpha-particle range–energy relationships.

Measurements show that the Geiger–Nuttall relation is obeyed rather well by the transitions from even–even nuclei that go to the ground state of the daughter. Even–even transitions to excited states may have decay constants somewhat smaller than predicted and almost all even–odd or odd–even structures decay at very much slower rates. Transitions in odd–odd nuclei are complicated by the fact that in almost every case they are unstable against beta-particle emission as a competing mode of decay.

A decay channel showing an unexpectedly small decay constant is said to be *hindered.* The decay constant for a transition can be determined from a half-life measurement and this value compared to the decay constant calculated from Eq. (6-5) for an even–even ground-state transition of the same energy. The ratio of calculated/observed is the *degree of hindrance.* Figure 6-3 shows one way in which degrees of hindrance are displayed in compilations of decay schemes.* Note that all gamma-ray emissions have been omitted from the diagrams for simplicity. In this representation, degrees of hindrance

*Adapted from C. M. Lederer, J. M. Hollander, and I. Perlman, *Table of Isotopes*, 6th ed. John Wiley & Sons, Inc., N.Y., 1968.

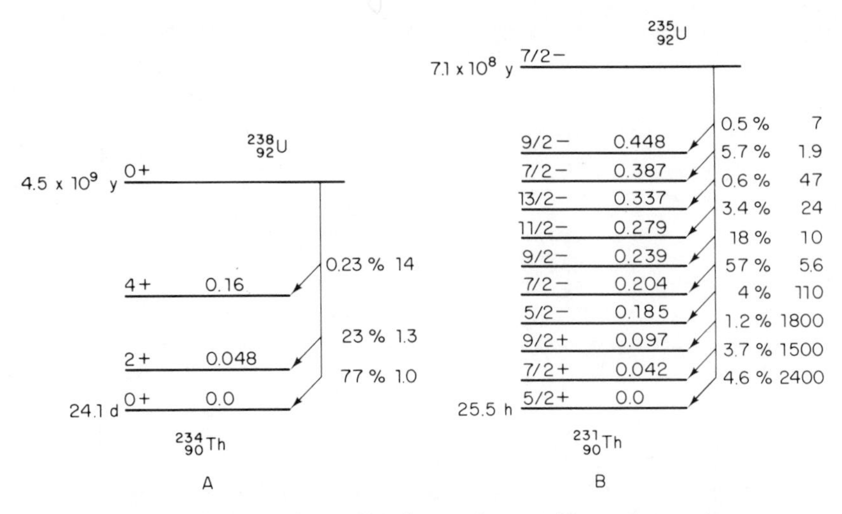

Figure 6-3. Typical alpha-particle decay schemes with angular momentum and parity designations.

are given in the right-hand column alongside each energy state. In even–even $^{238}_{92}$U the most probable transition to the ground state has a hindrance of unity, by definition. Alpha emissions going to the 0.160-MeV excited state of $^{234}_{90}$Th have less energy available for barrier penetration and these transitions would be expected to be less probable. In fact, they are even less probable than expected from simple energy considerations and a hindrance factor of 14 is noted in Fig. 6-3A.

Each energy state represented in Fig. 6-3 has a figure and an algebraic symbol near the left-hand end of the energy level. The figure gives the quantum number of the total angular momentum or spin of the state; the symbol designates the *parity* of the state. Parity is a parameter describing the symmetry class of the wave functions representing the state. Parity is without meaning in classical mechanics. We shall have little occasion to use either the spin or parity designations. It is sufficient to note here that the probability of a transition decreases as the change in spin between the states increases. The probability of decay is further reduced if there is a parity change.

The importance of spin and parity changes in determining decay probabilities can be seen by comparing Fig. 6-3A and B. In A, the most probable transition is to the ground state, which makes the most energy available for an assault on the potential barrier. Because of nucleon pairing the net spin is zero in both the initial and final states. In *B*, every state has some nonzero value of spin because complete pairing is impossible. The most probable decay mode (57%) is to a state 0.204 MeV above the ground state. Although this decay mode has less energy available, this lack is more than compensated for by the fact that in this transition there is no change in spin or parity. According to the decay scheme the hindrance factor in the ground-state transition is 2400 because a change in both spin and parity is involved.

The decay schemes shown in Fig. 6-3 are typical of the two classes of alpha decay. Even–even nuclei go predominantly to the ground state of the daughter with relatively few gamma-ray emissions. Odd–even or even–odd nuclei favor decay to an excited state and have relatively complex gamma-ray spectra.

An alpha particle has a mass that is appreciable with respect to even the heaviest nucleus from which it may issue. Momentum conservation requires a substantial recoil velocity, and hence energy, and this energy is not available to the alpha particle. Momentum and energy conservation laws lead to

$$E_\alpha = \frac{E(A-4)}{A} \tag{6-6}$$

where A is the mass number of the parent nucleus and E is the total kinetic energy of the transition. E will equal the reaction Q only in decays that go to the ground state; otherwise E will be less than Q by the energy of the gamma emissions.

6.05 Alpha-Particle Absorption

When an alpha particle passes through matter, its electric field interacts with the fields of the orbital electrons to produce a dense trail of atomic excitations and ionizations. On the average an energy transfer of about 34 eV will produce an ionization and even less is required for an excitation. Some electrons will be ejected with kinetic energies of 100 eV or so but even these energy transfers produce only a small reaction on the alpha particle. The particle slows as it loses energy but maintains an almost linear path to the end of its existence as an ion. Particle tracks can be made visible with a variety of techniques such as cloud chambers, bubble chambers, and photographic emulsions, and these clearly show the rectilinear nature of the trajectories.

The average ion density along a track is easily calculated. For example, we have already noted that a 4.8-MeV alpha particle produces about 1.41×10^5 ion pairs in coming to rest. From Fig. 6-2 this particle will have a range of 35 μm in tissue and from this the *average specific ionization* is calculated to be $(1.41 \times 10^5)/35 = 4030$ ion pairs μm^{-1}. In air the range is increased by a factor of about 10^3 and consequently the average specific ionization will decrease to about 4 μm^{-1} or 4×10^4 cm^{-1}.

The chemical and biological effects of the passage of a charged particle would be expected to depend on the value of the specific ionization along its track and this is indeed the case. More detailed considerations on the energy loss by an alpha particle are presented in Chapters 7 and 13.

B BETA PARTICLES

6.06 Nuclear Model of Beta Decay

Early measurements with the mass spectrometer identified the negative particles emitted in beta decay as electrons, previously made manifest in a variety of ways. Later, more sophisticated studies confirmed the identification and showed that the two particles are identical even to the most subtle quantum characteristic. Positrons are identical with negatrons in all respects except for the sign of the electric charge and are the corresponding particles in the *antisystem* of particles.

Two systems of particles appear to exist in nature. Each particle in one system, such as the negatron, appears to have an analog in the antisystem. In most cases as with the negatron–positron or the proton–antiproton pairs, the particles are identical except for the polarity of the charge. Neutral pairs such as the neutron–antineutron or the neutrino–antineutrino differ instead in symmetry characteristics comparable to left- and right-handedness.

Particles are created and annihilated in pairs. Thus an energetic photon may disappear with the formation of a negatron–positron pair, Sec. 8.06. In beta decay, negatron emission is accompanied by an antineutrino and positrons are accompanied by neutrinos. Conversely in our experience a free positron will quickly coalesce with a negatron with the creation of photons. If an antiparticle universe exists somewhere, a negatron would have a comparably short existence.

Several lines of evidence agree that electrons as such cannot exist in a nucleus prior to ejection. The size incompatibility has been mentioned in Sec. 4.01. Another argument is based upon the total number of particles comprising an odd Z–even A nucleus such as $^{14}_{7}N$. According to the proton–electron model this nucleus would consist of $14p + 7e$ for a total of 21 elementary particles. The proton–neutron model leads to $7p + 7n = 14$, an even number of nucleons. In the p–e model the total angular momentum or spin quantum number must be half-integral because there are an odd number of particles each with a spin of $\frac{1}{2}$. In the p–n model the even value of A leads to a total spin of either zero or some integer. Infrared spectroscopy can distinguish unequivocally between the two types of spin states and the evidence is entirely in favor of the proton–neutron structure.

Energy requirements for beta-particle decay are less stringent than are those for alpha decay. Electrons do not have to cross the nuclear potential barrier and consequently they do not have the energy limitations that are imposed on alpha particles. The well-known beta decay of ^{3}H has a Q_β of only 18.6 keV, which is far below the energy at which alpha emission is possible. Beta-emission patterns are more complex than alpha decays. With the energy constraint removed, many beta transitions can go to highly excited states to be followed by gamma-ray emissions as the excitations are relieved.

6.07 Range and Path Length

Although almost all beta particles are ejected with relativistic velocities, their masses are at most only a few times that of the orbital electrons with which they interact. Because of the comparable masses there is a substantial recoil velocity each time the beta particle interacts with an orbital. In contrast to an alpha-particle track the path of a beta particle is twisted and tortuous as it undergoes a closely spaced series of recoils. *Path length* refers to the total distance traveled by a beta particle in coming to rest. *Range* on the other hand is the linear distance from the point of emission to the end of the trajectory. The former distance is used in calculations of specific ionization and some other closely related terms. Range is the pertinent parameter used in determining beam penetration and in shielding calculations. As a very rough rule, average beta-particle ranges are about one-half the path length.

Beta particles undergo *collision energy loss* when they interact with orbital electrons to produce ionizations and excitations. In addition, bremsstrahlung will be produced at each inteaction because the recoil accelerations are large. As predicted by the Kramers relation, Eq. (3.5), *radiative energy loss* will increase with the initial energy of the particle and with the atomic number of the absorber. In the beta-particle energy range, and in most materials of biological interest, collision energy losses predominate. Then, to a good approximation, energy loss is relatively independent of atomic number of the absorber, depending rather on the electron density. Electron density in turn is nearly proportional to the ordinary mass density over quite a range of atomic numbers. Thus it is convenient to express beta-particle ranges in equivalent ranges, either in grams cm^{-2} or mg cm^{-2}, independent of the nature of the absorber.

Most beta-particle range measurements are made with aluminum absorbers. Over the energy range of most beta-particle emissions the aluminum values may be used for other substances in all but the most exact calculations. Several empirical relations between range and energy have been developed from experimental data. One satisfactory expression is

$$R = 0.542E - 0.133 \text{ g cm}^{-2} \qquad \text{for } E > 0.6 \text{ MeV} \tag{6-7}$$

$$E = 1.85R + 0.245 \text{ MeV} \qquad \text{for } R > 0.25 \text{ g cm}^{-2} \tag{6-8}$$

For the lower particle energies it is preferable to use plotted data such as that shown in Fig. 6-4.

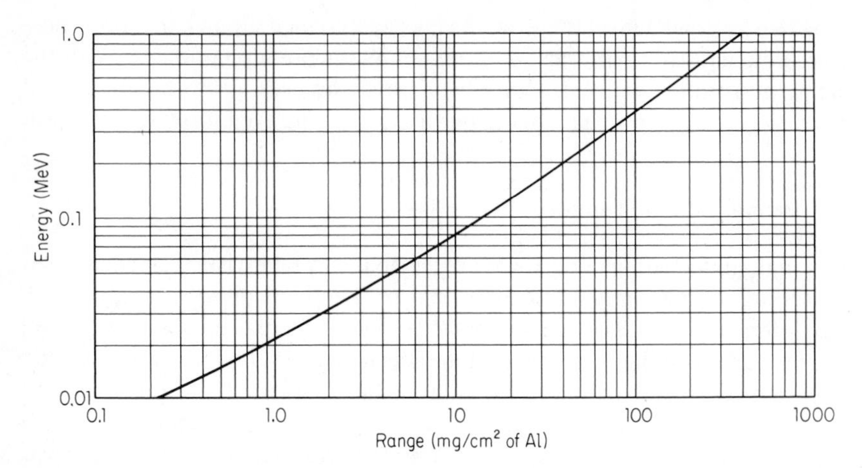

Figure 6-4. Equivalent range–energy relationship for low-energy electrons in aluminum.

6.08 Beta-Particle Ionization Patterns

As the relative range values indicate, beta particles interact much less strongly with orbital electrons than do alpha particles of equal energy. At any energy the beta particle has a much higher velocity than the alpha and so it spends much less time in the vicinity of each atomic electron. The rate of energy loss is further reduced because the beta-particle charge is only one-half that of the alpha.

Illustrative Example

Compare the average specific ionization of a 4.8-MeV beta particle with that previously calculated for an alpha particle of the same energy.

Assuming as before a required expenditure of 34 eV for each ion pair, we again have a total of 1.41×10^5 ion pairs.

From Eq. (6–7),

$$R = (0.542 \times 4.8) - 0.133 = 2.47 \text{ g cm}^{-2}$$

In normal air with a density of 0.001293 g cm^{-3} the linear range will be $2.47/(1.293 \times 10^{-3}) = 1.91 \times 10^3$ cm. The total distance traveled by the beta particle will be about twice this, or 3.8×10^3 cm. Then,

$$\text{average specific ionization (air)} = \frac{1.41 \times 10^5}{3.8 \times 10^3} = 37 \text{ ion pairs cm}^{-1}$$

In tissue of density 1.0 g cm^{-3}, the path length will be about 5.0 cm and the average specific ionization becomes

$$\frac{1.41 \times 10^5}{5} = 2.8 \times 10^4 \text{ cm}^{-1} \text{ or } 2.8\ \mu\text{m}^{-1}$$

With its relatively low value of specific ionization a beta particle has a good chance of passing through a living cell without any interaction. This chance is almost negligible with an alpha-particle passage. Differences in ion density and in the nature of the trajectories may lead to rather different biological effects, even though the basic ionizations and excitations are caused by ejected electrons in each case.

6.09 Half-Life and Beta-Particle Energy

Sargent was the first to show that a log E_m — log λ plot for many of the natural beta emitters resulted in a series of groupings along parallel straight lines. Clear-cut groupings are not observed for all of the hundreds of aritficially produced beta emitters that are now known but the log–log trend is still apparent. Figure 6-5 is a Sargent plot of only a few representative beta transitions.

A few light nuclei appear to have abnormally short half-lives for the energy that is available and these transitions are said to be *favored.* Most of the favored transitions involve *mirror nuclei* such as ${}^{3}_{1}H \longrightarrow {}^{3}_{2}He + \beta^-$ where the nuclear change results in an interchange of the Z and N values as a result

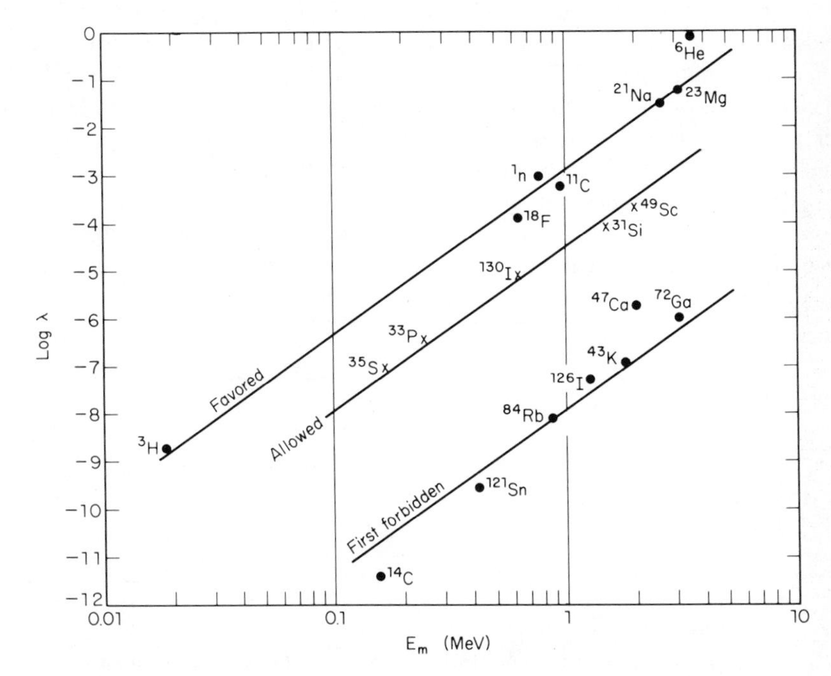

Figure 6-5. A Sargent plot of some typical beta emitters showing varying degrees of forbiddenness.

of the neutron–proton conversion. The parent nucleus will have a single unpaired nucleon with pairing of all the nucleons of the other type. For example, in $^3_1$H there is one single neutron pair and one unpaired proton. Since *n–n* and *p–p* forces are nearly equal, one of the paired nucleons can easily change its identity and the mirror nucleus is formed. There will be no change in nuclear shell occupancy since according to the Pauli principle there is room for nucleon pairs of each type in each energy shell.

Occupants of another, larger group on the Sargent diagram have "normal" half-lives related to the beta-particle energy and the transitions are said to be *allowed*. Other groups have still longer half-lives for a given energy and the transitions are said to be *first-*, *second-*, or *nth-forbidden*. Since the transitions do take place, albeit slowly, they would be better termed *hindered* rather than *forbidden* but the latter term is firmly established.

The classifications follow a set of *selection rules* which are based on the changes in angular momentum and parity. Allowed transitions are those in which the spin quantum number changes by only 0 or ± 1, with no change in parity. Hindered transitions are characterized by increasingly greater changes in angular momentum either with or without a change in parity. As more transitions are studied, it appears that some subgroups need to be introduced into the main forbiddenness classes.

Fermi developed a quantum mechanical theory of beta-particle emission whose validity has been amply confirmed by measurements. In the theory there is a mathematical function known as the Fermi integral function, f. The product of f and the half-life fT is a good measure of the forbiddenness of a transition and is frequently used to determine the parameters of a nuclear state from observed energy–half-life data.

The range of values is so great that log (fT) rather than fT itself is customarily listed in tabulations of nuclear properties. In one arrangement,* Fig. 6-6, the log (fT) values are placed to the right of the percent occurrence

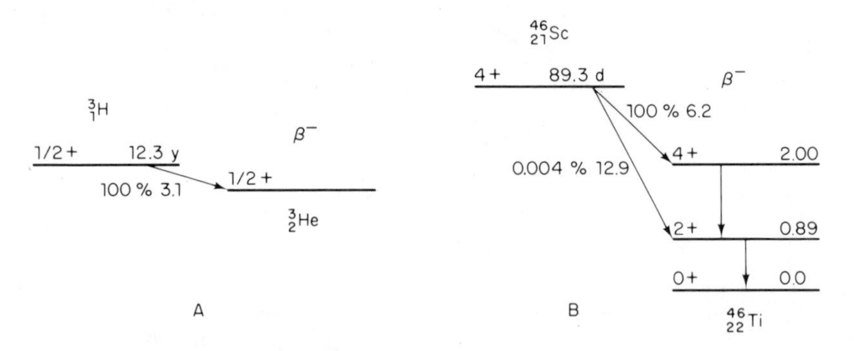

Figure 6-6. Beta-particle decay schemes with log (fT) values shown near each decay percentage.

*C. M. Lederer, J. M. Hollander, and I. Perlman, *Table of Isotopes*, 6*th ed.* John Wiley & Sons, Inc., New York, N.Y., 1968.

of each transition. Figure 6-6A shows the favored ^3H decay to have a log (fT) value of 3.1, with no change in spin or parity. In the decay of ^{46}Sc, Fig. 6-6B, an allowed transition to the second excited state of ^{46}Ti is characterized by no change in spin or parity and a log (fT) value of 6.2. Decays to the first excited state involve a spin change of 4/2 and log (fT) becomes 12.9, indicating a second forbidden transition. The spin change of 8/2 makes transitions to the ground state of ^{46}Ti highly forbidden, and they are not observed.

6.10 Average Beta-Particle Energy

Values of the average beta-particle energy $\bar{E}$ are important because these values rather than the maximum energies of the transition, E_m, enter into calculations of the radiation dose delivered in the immediate vicinity of the emitting nuclide. $\bar{E}/E_m$ ratios for the smoothly varying continuous beta-particle spectrum can be calculated from the Fermi theory. These ratios depend on the atomic number of the emitter, the transition energy, the degree of forbiddenness, and the details of the decay scheme. Note that in beta decay each E_m value may be taken as the Q_β of the decay, less any energy lost by gamma radiation. The mass disparity between the beta particle and the recoiling daughter nucleus is so great that the energy lost to recoil is negligible.

Figure 6-7 shows some of the $\bar{E}/E_m$ ratios evaluated for atomic number 30,

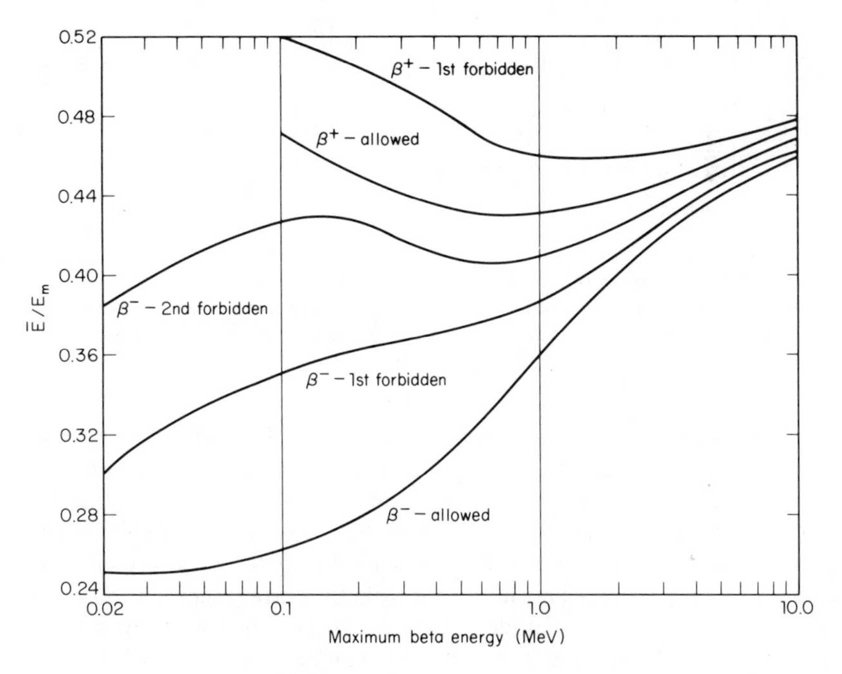

Figure 6-7. Values of $\bar{E}/E_m$ for $Z = 30$ evaluated from the Fermi distribution function. (Adapted from Dilman, *loc. cit.*)

with no allowance made for conversion electrons. Values of the energy ratio calculated from the Fermi distribution must be modified to take account of all conversion electrons. Tabular values of the energy ratio will include the effect of all beta-particle and electron emissions.*

6.11 Energy Determination

Because the spectral distribution of beta particles extends down to zero energy, there will be some decrease in count rate when even thin absorbers are put in the source–detector path. When the beta particles are unaccompanied by gamma rays the log (count rate) will be an approximately linear function of the absorber thickness, Fig. 6-8A. The exact shape of the absorption curve will depend on the degree of forbiddenness of the transition.

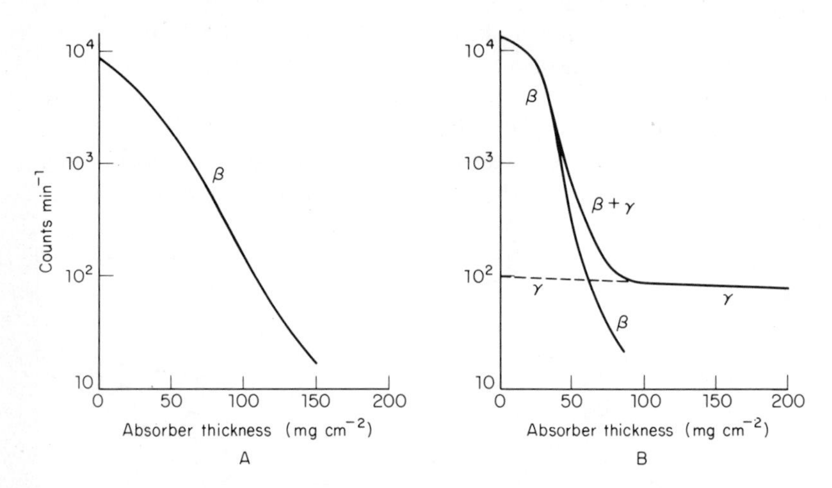

Figure 6-8. Absorption curves for (A) a beta emission with no accompanying gammas and (B) a beta plus gamma emission. The extrapolated gamma-ray contribution is subtracted from the total to obtain the beta absorption curve.

When there are accompanying gamma rays, the steep beta-particle absorption curve will merge into a linear portion with a gentle slope, Fig. 6-8B. This latter segment represents the contribution to the count rate of the penetrating gamma rays. To obtain the true beta absorption curve, the linear portion is extrapolated back to zero absorber thickness and a series of count-rate values on this curve are subtracted from the corresponding values on the composite curve. The remainders represent beta-particle contributions only and a true beta-particle absorption curve can now be drawn, Fig. 6-8B.

*L. T. Dillman, *Journal of Nuclear Medicine*, Supp. 2, **10**, March, 1969.

It is occasionally necessary to determine the maximum energy of a beta-particle emission from absorption data. Extrapolation of the count-rate–absorber-thickness data to zero count rate is unsatisfactory because of the gradual approach of the curve to the end point. An accurate end point can be obtained by *rectifying* the observed absorption curve to a linear form and then linearly extrapolating to zero count rate. Data rectification can be carried out by means of the Fermi function, which will be found tabulated in a publication of the National Bureau of Standards.* Details of the use of the method will be found along with the tables.

For most purpose beta energies can be determined by making comparative measurements on the absorption curve of the unknown and on one of known energy, preferably using transitions with equal degrees of forbiddenness. The most commonly used method of comparison is known as the *Feather analysis*. Details of the Feather method will be found in almost any textbook on nuclear physics.†

6.12 Isobaric Stability

All beta decays, whether by direct beta emission or by electron capture, are isobaric, for the mass number is unchanged by the transition. Beta decay is then to be expected whenever one member of an isobaric group, Z, has a greater mass than either of its neighbors, $Z - 1$ or $Z + 1$. In general, an A, Z nucleus will have only one decay channel either to A, $(Z - 1)$ or to A, $(Z + 1)$ as it attempts to approach the line of stability, Fig. 4-1. In some cases, however, both decay channels are available and the two decay modes are in competition. This duality can be explained in terms of the empirical equation for nuclear mass, Eq. (4-7).

Consider first the decay possibilities between isobars with an odd value of A. The last, or nucleon pairing, term of Eq. (4-7) will now be zero for all values of Z; with A constant, the remaining terms will show a parabolic relationship between M and Z, Fig. 6-9A. At or near some value of Z the nuclear mass will have a minimum value. A nucleus with a Z value less than that at the minimum will decay by one or by a series of negatron emissions until it reaches the minimum, where no other beta decays are possible. A nucleus with a Z value greater than that at the minimum will undergo positron or EC transitions until the minimum is reached. From the general shape of the parabola we would expect the transition energies to decrease as the minimum energy is approached. If selection rules do not interfere, corresponding

*National Bureau of Standards, *Tables for the Analysis of Beta Spectra*. Applied Mathematics Series 13. For sale by the Superintendent of Documents, U.S. Government Printing Office, Washington, D.C.

†R. E. Lapp and H. L. Andrews, *Nuclear Radiation Physics*, 4th ed. Prentice-Hall, Inc., Englewood Cliffs, N.J., 1972, p. 272.

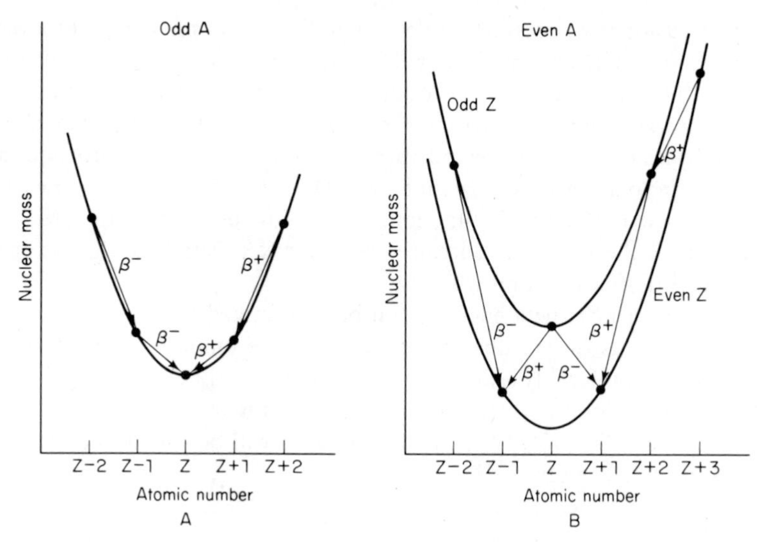

Figure 6-9. Nuclear mass (energy) –atomic number parabolas for isobars. (A) A single parabola accounts for transitions in all odd-*A* nuclei. (B) Two parabolas are required for even-*A* nuclei because of the changes in nuclear pairings with beta emission.

decreases in the decay constants should be observed. One example of these relations in a decay chain can be seen in Eq. (6-9).

$$^{139}_{53}\mathrm{I} \xrightarrow[2.7s]{} \beta^- + {}^{139}_{54}\mathrm{Xe} \xrightarrow[41s]{} \beta^- + {}^{139}_{55}\mathrm{Cs} \xrightarrow[9.5m]{} \beta^- + {}^{139}_{56}\mathrm{Ba} \xrightarrow[85m]{} \beta^- + {}^{139}_{57}\mathrm{La} \quad \text{(stable)} \tag{6-9}$$

When *A* is even, the final term in the nuclear mass equation can be either positive or negative and there will be two *M–Z* parabolas, Fig. 6-9B. Beta transitions will be toward the configuration of minimum mass as before but now each decay results in a shift from one parabola to the other. A nuclide located at the minimum energy point of the odd-*Z* parabola may have a mass greater than that of either of the two neighboring isobars and so two decay channels will be available. For example,

$$\begin{aligned} {}^{50}_{23}\mathrm{V} + e^- &\longrightarrow {}^{50}_{22}\mathrm{Ti} \quad (70\%) \\ &\searrow \beta^- + {}^{50}_{24}\mathrm{Cr} \quad (30\%) \end{aligned} \tag{6-10}$$

The ^{50}Cr daughter is more than 1 MeV heavier than the ^{50}Ti isobar but the transition between these two would require the emission of two positrons or a double electron capture. Either of these two-particle processes is highly improbable and the $^{50}\mathrm{Cr} \longrightarrow {}^{50}\mathrm{Ti}$ decay is not observed.

Figure 6-9B suggests that every odd–odd nucleus should be beta unstable. There are only four exceptions to this: $^{2}_{1}$H, $^{6}_{3}$Li, $^{10}_{5}$B, and $^{14}_{7}$N. All of these nuclei are located near the low end of the periodic table and have exact spin-pairing, with equal numbers of protons and neutrons. At the next odd–odd structure, coulomb repulsion requires an increase in the n/p ratio for stability, and $^{18}_{9}F$ is observed to be radioactive.

REFERENCES

DANIEL, H., "Shapes of Beta-Ray Spectra." *Rev. Modern Physics*, **40**, 659, 1968.

DILLMAN, L. T., "Radionuclide Decay Schemes and Nuclear Parameters for Use in Radiation-Dose Determinations." *J. Nuc. Med.*, Supp. 2, **10**, March, 1969, Part 2. Supp. 4, **11**, March, 1970.

RUTHERFORD, E., J. CHADWICK, AND C. D. ELLIS, *Radiations from Radioactive Substances*. The Macmillan Co., New York, N.Y., 1930.

TURKEVICH, A. L., ET AL., "Alpha Activity of the Lunar Surface at the Landing Sites of Surveyors 5, 6, and 7." *Science*, **167**, 1722, 1970.

WIDMAN, J. C., ET AL., "Average Energy of Beta Spectra." *Int. J. Applied Rad. and Isotopes*, **19**, 1, 1968.

7

Ionization Patterns of Charged Particles

7.01 Ion-Current Measurements

When positive ions and electrons are formed in a field-free space, they dissipate their energy in a series of interactions and then combine to return the irradiated region to electrical neutrality. In the presence of an electric field, as in a gas-filled volume fitted with electrodes, Fig. 7-1, ions can be

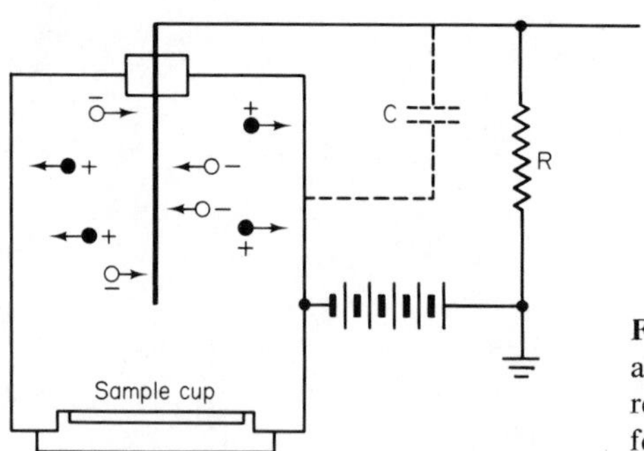

Figure 7-1. An ionization chamber with a polarizing battery and an external resistance R. The chamber electrodes form an electrical capacitance C.

collected and made manifest in an external circuit. Such a device is known as an *ionization chamber*. A polarizing voltage across the chamber will attract ions of opposite sign to each electrode. Electrons will flow through the circuit and will actuate indicating or recording meters.

Sensitive electroscopes and electrometers were in use when Roentgen and Becquerel made their pioneer discoveries, and the rapid early developments in the field were facilitated by the availability of these sensitive electrical instruments. Today vacuum tubes and solid-state electronic devices have largely, but not completely, replaced the electrometer. Pocket dosimeters used for personnel monitoring, and the thimble chamber–electrometer instruments serving as secondary standards of photon exposure, Sec. 10.05, utilize the deflections of small quartz fibers as indicators of charge collection.

As each type of ion moves toward the appropriate electrode, it will pass close to ions of opposite sign, moving in the opposite direction. These contacts open up the possibility of *recombination* or charge neutralization before the ions are collected at the electrodes. Ions lost by recombination do not affect the external circuit. The measured ion current is then submaximal and, in addition, may no longer be proportional to the ionizing power of the radiation incident on the chamber.

Recombination is proportional to the volume density of each type of ion and hence to the square of the radiation intensity, which produces ions of each charge. Recombination limits ion-current measurements to gas-filled chambers and to some specially designed solid-state detectors. Ion densities in liquids and solids are too high to ensure effective collection.

Recombination is also proportional to the time of flight of the ions to the electrodes. This time becomes smaller as the collecting voltage across the chamber is increased, Fig. 7-2. The condition of maximum current collection, known as *saturation*, should always obtain if meaningful measurements are to be made. In most cases radiation intensities are low enough to permit chamber saturation to be attained with only modest collecting voltages. Some accelerators, however, produce radiation in a series of pulses of only microseconds duration but of very high intensity. The average output of a pulsed

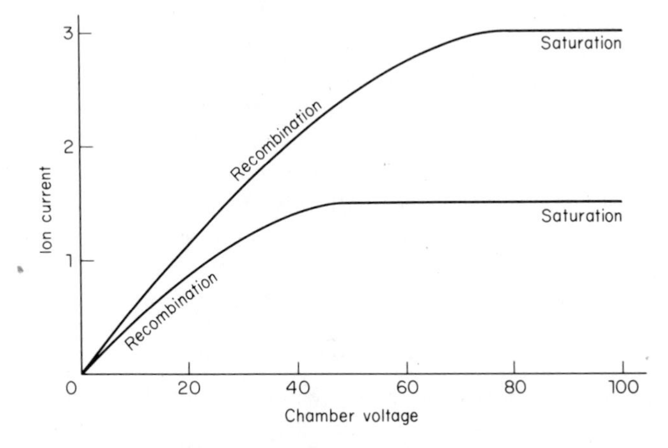

Figure 7-2. Current–voltage relationships in an ionization chamber for two values of radiation intensity.

source may not be high but care must be taken that the ion chamber does not saturate at the high-intensity pulses.

Under saturation conditions an ion chamber will give a response that is proportional to the total ionization within the sensitive volume. Alpha-particle ionization may be measured by introducing the source directly into the chamber or by admitting the particles through an ultrathin window. Under usual conditions beta particles will lose a substantial fraction of their energy in the chamber structures instead of in the gas, and the response may or may not be proportional to the radiation intensity. Special considerations must be applied to chambers designed for measuring the ionizing power of photons or neutrons, Sec. 10.05 and 10.11.

7.02 Linear Energy Transfer

The chemical and biological effects and effectiveness of a radiation depend on the spatial distribution of the ionizations and excitations produced by it. Sparsely spaced ionizations may be quite effective for one reaction; for another, yield may increase rapidly with ion density. Rough calculations have already been made of the average specific ionization along the track of a charged particle. These calculations must now be refined.

Zirkle introduced into radiation biology the term linear energy absorption, later changed to *linear energy transfer*, LET. The basic definition is LET $= (dE/dx)$ where dE is the energy transferred from a charged particle to a local volume of an absorbing medium over a path length dx. LET values are conveniently expressed in keV μm^{-1} but other units may be used. The specification to a local volume is elastic and may vary from one situation to another. In assessing radiation action on DNA, local volume may require restriction to a small portion of the molecule. For a gross effect on an organ a much larger volume might be considered local.

An energetic electron can transfer, in effect, no more than one-half of its energy in a single encounter with another electron. After a collision two *indistinguishable* electrons are in motion. The one with the greater energy will be counted as the primary, so the secondary can not have more than one-half of the original kinetic energy. Even with this restriction the secondary may have sufficient energy to expend a large portion outside the local volume.

A more precise definition of linear energy transfer has been developed by the ICRU* with LET values thus derived denoted by L, as

$$L_\Delta = (dE/dx)_\Delta \tag{7-1}$$

where Δ specifies a cutoff value of the energy acquired by the secondary

***Linear Energy Transfer*, Report 16, ICRU, Washington, D.C. 1970.

electron. Thus L_{100} means that LET is to be evaluated by including only those collisions in which 100 eV or less is acquired by the secondary electron. L_{∞} means that all collision processes are to be counted, regardless of the amount of the energy transfer; it does not mean that an infinite amount of energy is imparted.

When all collision energy transfers are included in the calculation, the L_{∞} value becomes equal to a quantity known as the *linear stopping power*, *S*. Like *L*, linear stopping power has the dimensions of energy per unit length. The ratio $(L_m/L_{air})_{\infty} = S_m$ is the *relative stopping power* of a medium *m* referred to air.

In most irradiation situations the absorber will be subjected to a distribution of *L* values because of the random nature of the interactions. In a very thin target single collisions may result in a relatively constant *L* value.

7.03 Energy Loss Per Ionization

The average energy loss per ion pair formed is an important parameter of radiation dosimetry. We have already used a value of 34 eV for this without qualification. In fact the energy loss in a gas per ion pair, *W*, depends somewhat on the type and energy of the charged particle and on the nature of the gas being ionized.

The ICRU has adopted a value of 33.73 eV for electrons in air. Variations of *W* with electron energy are negligible, at least for electrons with energies greater than 20 keV. The corresponding value for alpha particles is 34.98 eV and this value is applicable to the proton recoils from neutron collisions.

There is a tendency for somewhat smaller *W* values in some of the hydrocarbons that are used as quenching gases in pulse counters. Values in H_2, O_2, N_2, and air are nearly constant, and for present purposes a value of 34 eV has sufficient precision without further specification.

Values of *W* are calculated from a knowledge of the total amount of energy expended in a gas and the total number of ions formed by this expenditure. Much of the energy produces non-ionizing events; some is given to loosely bound electrons, and some goes to the more tightly bound orbitals in inner shells.

7.04 Clusters and Delta Rays

We have already noted that charged particles lose energy in a large series of discontinuous steps and that the average energy transferred is about 60 eV. When an electron receives 100 eV or so, it will produce a few ions on its own and these will be closely spaced because the kinetic energy is low. These small groups of ions are known as *clusters* from their appearance in cloud chamber photographs. The distribution of cluster sizes has been calculated from cloud chamber data obtained by Wilson,* Table 7-1.

TABLE 7-1
DISTRIBUTION OF CLUSTER SIZES

Ion pairs per cluster	1	2	3	4	>4
Fraction of clusters	0.45	0.22	0.11	0.10	0.12

Clusters will almost always be assessed as energy locally imparted and a value of Δ will be chosen to include them.

When an electron is capable of producing somewhat more than four ion pairs, it becomes a delta (δ) ray and it will leave a trail that is a distinguishable offshoot from the track of the original particle. If the original event took place in a small biological entity such as a cell, some delta rays will leave the cell to deposit at least a portion of their energy elsewhere. In assessing biological effects in terms of L_Δ, the value of Δ must be carefully chosen to fit the size of the volume of interest.

Protons, alpha particles, and other multiply charged ions leave linear trails with clusters and low-energy δ rays forming a roughly cylindrical region of ionization several micrometers in radius, Fig. 7-3. This type of

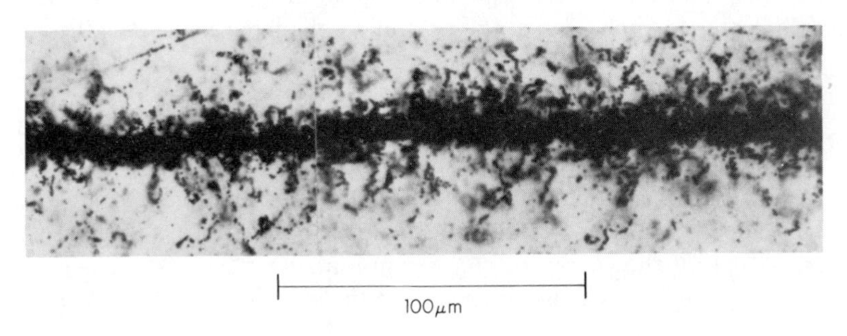

Figure 7-3. A portion of the column of ionization produced in a photographic emulsion by a multiply charged ion.

pattern is known as *columnar ionization*. Some care must be taken when an ionization chamber is used with columnar ionization. If the polarizing field in the chamber is at right angles to the column, it will quickly pull the ions out of the region of high ion density and high recombination probability.

*Wilson, C. T. R., *Proc. Roy. Soc. London*, **A**, 104, 192, 1923.

If the ionized column is parallel to the field, the ions will remain in the column during collection and the chance of recombination will be greatly increased, Fig. 7-4.

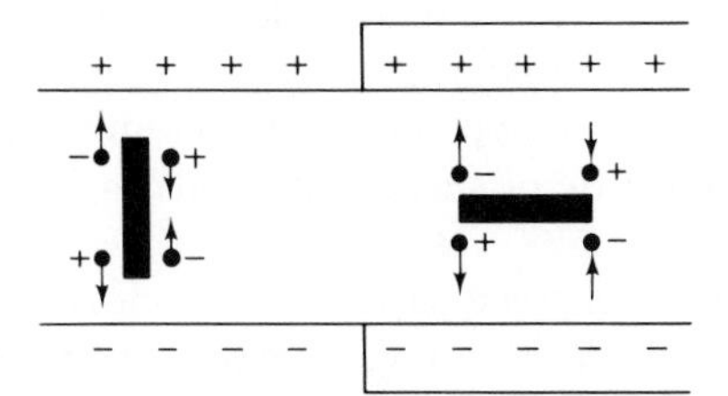

Figure 7-4. With columnar ionization the amount of recombination will depend on the orientation of the column.

7.05 Variability in Linear Energy Transfer

As an individual charged particle gives up its energy and slows, the value of L_Δ increases and reaches a maximum just before the particle loses its ability to ionize, Fig. 7-5. Each particle in an originally monoenergetic beam will behave similarly but fluctuations in the microscopic details of the energy transfers will broaden and reduce the maximum and will produce a range distribution due to straggling. The region of increased ionization seen near the end of a particle beam is known as the *Bragg peak.*

When charged particle beams are used in radiation therapy, advantage is taken of the Bragg peak whenever this is possible. The aim in therapy is to deliver the maximum possible radiation to the selected site while minimizing the energy deposited in healthy tissues. By a proper adjustment of particle

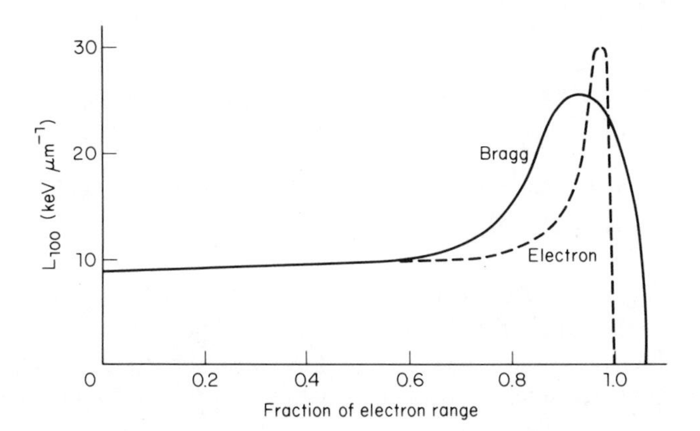

Figure 7-5. A single charged particle has a high L_{100} value just before it recombines. In a beam of many particles the maximum region is broadened to form the Bragg peak.

energy it is sometimes possible to put the Bragg peak at the depth of the lesion being treated while the superficial tissues are traversed by particles with lower L values.

When biological effects are to be interpreted in terms of the quality of the radiation causing them, fluctuations in values of L_Δ introduce serious complications. Biological effects depend ultimately on radiation-induced changes in cells or perhaps in individual molecules. It appears impossible to expose a population of small biological "targets" to exactly uniform radiation conditions.

Every charged particle in a monoenergetic beam may have identical values of L_Δ but the spatial distribution of the energy transfers will be far from uniform. One cell may experience no ionizations or excitations and thus be completely unaware of the presence of radiation; the adjoining cell may be subjected to many ionizing events when only one or a few are needed to alter the future life of the cell. Here a knowledge of the spatial distribution and the intimate details of each energy transfer is needed to evaluate the mechanism of action and the effectiveness of the radiation.

In coming to rest, a charged particle may distribute its energy over an L_Δ range of 100 : 1, Fig. 7-6. The curve shown is only schematic; the exact

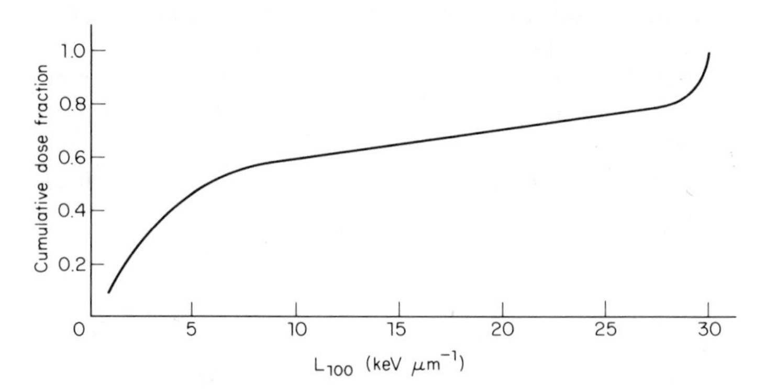

Figure 7-6. The cumulative dose fraction delivered by an electron with an initial energy of 200 keV as it passes through water.

behavior will depend on the original energy of the particle, the nature of the medium, and the value of Δ that is pertinent. In almost any radiation situation the biological system will be traversed by charged particles having a wide spread of energies. Values of L_Δ will be spread accordingly* and while aver-

*Baily, N. A., J. E. Steigerwalt, and J. W. Hilbert, "Changes in the Frequency Distribution of Energy Deposited in Short Pathlengths as a Function of Energy Degradation of the Primary Beam." *Rad. Res.*, **49**, 26–35, 1972.

age L_{Δ} values may be calculated, there is no assurance that these average values are universally applicable. Linear energy transfer is only one of the factors that enter into a determination of radiation effectiveness.

7.06 Energy Loss Relations

Bethe* has developed analytical expressions for the rate of energy loss from charged particles moving through an absorber. The expression applicable to protons and heavier particles is relatively simple. Except at extreme energies the particle velocities are not highly relativistic and each acceleration is so small that no account need be taken of radiative energy losses. For heavy particles the Bethe expression is

$$-\left(\frac{dE}{dx}\right)_c = \frac{4\pi z^2 e^4 NZ}{m_0 v^2}\left[\ln\frac{2m_0 v^2 E}{I(1-\beta^2)^2} - \beta^2\right] \text{ ergs cm}^{-1} \tag{7-2}$$

where z = net number of charges on the moving particle
e = elementary charge in esu
NZ = electron density in the absorber in cm^{-3}
m_0 = rest mass of the electron
v = particle velocity in cm s^{-1}
E = particle energy in ergs
I = average excitation energy in ergs
$\beta = v/c$

The corresponding relation for electrons is more complicated because of the greater recoil of the lighter ionizing particle.

$$-\left(\frac{dE}{dx}\right)_c = \frac{2\pi e^4 NZ}{m_0 v^2}\left[\ln\frac{m_0 v^2 E}{2I^2(1-\beta^2)} - \ln 2\,(2\sqrt{1-\beta^2} - 1 + \beta^2) + (1-\beta^2) + \frac{1}{8}(1-\sqrt{1-\beta^2})^2\right] \text{ ergs cm}^{-1} \tag{7-3}$$

Radiative energy losses may now become important particularly at high particle energies in high-Z absorbers.

$$-\left(\frac{dE}{dx}\right)_r = \frac{NEZ(Z+1)e^4}{137 m_0^2 c^4}\left(4\ln\frac{2E}{m_0 c^2} - \frac{4}{3}\right) \text{ ergs cm}^{-1} \tag{7-4}$$

According to Eq. (7-4), the rate of energy loss due to bremsstrahlung production is nearly proportional to NEZ^2. The total path length traversed by a particle in coming to rest is about proportional to its initial energy and

*H. A. Bethe and J. A. Ashkin, *Experimental Nuclear Physics*, E. Segrè, ed., Vol I. John Wiley & Sons, Inc. New York, N.Y., 1953.

inversely proportional to the electron density of the medium, NZ. Total bremsstrahlung production by a particle will then be about proportional to

$$NEZ^2 \times \frac{E}{NZ} = E^2Z \tag{7-5}$$

This is in accord with the results of the Kramers calculation, Eq. (3-4).

The Bethe equations are not applicable to low-energy particles, of perhaps 0.1 MeV for electrons and 1 MeV for alpha particles, because of uncertainties in interaction probabilities as the energy transfers approach the bonding energies of some of the orbital electrons. Each type of energy loss is proportional to z^2, which accounts for the strong ionizing capabilities of multiply charged ions. To a good approximation an alpha particle will have a value of linear energy transfer four times that of a proton of equal velocity.

7.07 Medium Polarization

In a condensed medium such as a liquid or a solid, many atoms lying close to the path of a charged particle will be neither excited nor ionized. All the nearby atoms, however, will come under the influence of the moving electric field which will temporarily distort or *polarize* the orbital electron structure as it sweeps past. Charges of like sign will be repelled by the moving field and unlike charges will be attracted, to create a sheath of transient dipoles, Fig. 7-7.

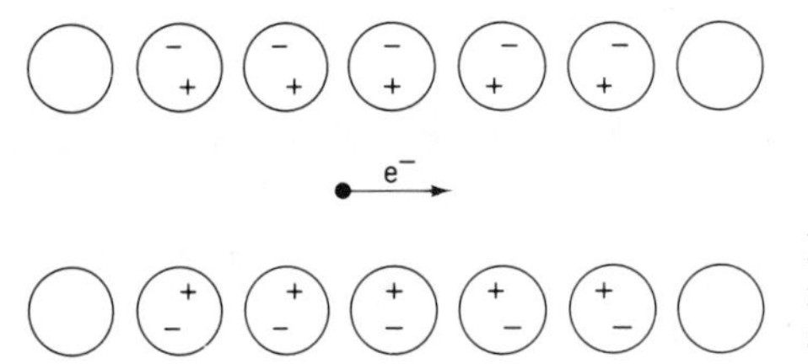

Figure 7-7. In a condensed medium a moving ion temporarily polarizes the atomic orbitals along its path.

Polarization effectively reduces the field strength at greater distances from the particle and hence reduces the number of atoms at risk of excitation or ionization. Values of L are reduced below those that would be expected from the simple theory. Corrections to L values due to polarization increase with NZ and with particle energy. In water the reduction in (dE/dx) may amount to about 10% at 10 MeV.

7.08 Cerenkov Radiation

The velocity of electromagnetic radiation in empty space, which is denoted by the symbol c, has a value of nearly 3×10^{10} cm s^{-1}. In ponderable, transparent media the velocity of these radiations will be substantially lower than c, perhaps $0.7c$ in some plastics and glasses. The velocity of visible light in a medium m will be given in terms of an index of refraction μ.

$$v_m = \frac{c}{\mu_m} \tag{7-6}$$

High-energy electrons may have velocities considerably greater than v_m and thus they will move in the medium with a velocity greater than that of the accompanying electric field. The lagging field will exert a decelerating force on the electron just as the water wave generated at the bow of a boat exerts a strong retarding force when the velocity exceeds a critical value. Visible and ultraviolet photons are produced as the electron is slowed by the retarding force of the slower moving field. Photons produced by this mechanism are known as *Cerenkov radiation* after their discoverer. The blue glow seen around the core of a swimming-pool type of nuclear reactor is primarily Cerenkov radiation.

7.09 Cavity Ionization

It is frequently necessary to determine the energy absorbed in a solid or a liquid where direct measurements of ionization are impossible. The *Bragg–Gray cavity ionization principle* can be invoked to infer the energy absorbed from the response of a small gas-filled chamber introduced into the medium. First proposed by W. H. Bragg, the cavity ionization principle was formalized and extended by L. H. Gray in 1936.

Consider a small gas-filled ionization chamber of volume V placed in the medium m whose ionization is to be measured, Fig. 7-8. The chamber volume must be so small that each electron traversing it will give up only a small

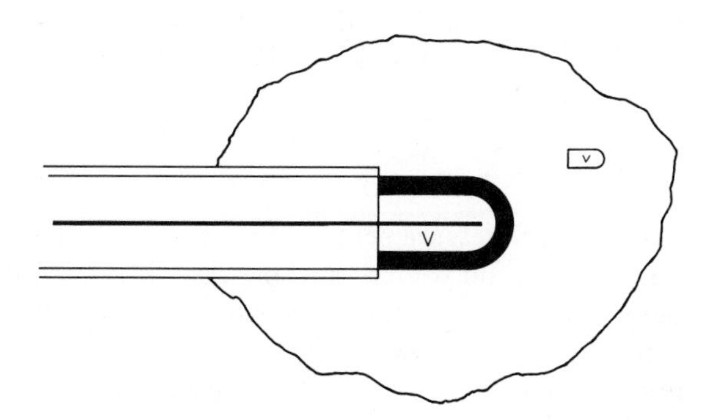

Figure 7-8. A gas-filled cavity V is inserted in a medium for measuring the energy imparted to a similar volume v. The chamber volume and wall thickness are exaggerated here for the purposes of illustration.

fraction of its energy to the gas. Then the presence of the chamber will have a negligible effect on the ion density throughout the medium. In particular the ion density in the chamber will have the same value that obtained in the volume of the medium that was displaced in introducing the chamber.

Let the relative stopping power of the medium relative to the gas in the chamber (usually air) be S_m and consider a small volume v in the medium. v is to be taken geometrically similar to V but with each dimension reduced by the factor $(1/S_m)$. Under these restrictions a particle crossing volume V will lose just as much energy as it will in crossing v. However, the cross section of V has an area S_m^2 times the corresponding area of v. The number of particles crossing V will be S_m^2 times the number that cross v and so the total energies deposited in the two volumes will be related by

$$E_a = S_m^2 E_m \tag{7-7}$$

But $V = S_m^3 v$ and so we have

$$\frac{E_a}{V} = \frac{S_m^2 E_m}{S_m^3 v} = \frac{E_m}{S_m v} \tag{7-8}$$

According to Eq. (7-8) the cavity ionization principle shows that the energy deposited in unit volume of the medium is just S_m times as great as the energy deposited per unit volume of the cavity.

An electrical measurement serves to determine J, the number of ion pairs per unit cavity volume, and since each ion pair requires on the average an expenditure of W eV,

$$E_a = JWV \tag{7-9}$$

and finally

$$\frac{E_m}{v} = JWS_m \tag{7-10}$$

S_m can be determined by an independent measurement; with this value and an electrical measurement of the ionization in a gas-filled cavity, it is possible to determine the energy imparted to a liquid or solid medium.

7.10 Pulse Counting

So far we have considered measurements designed to determine the ionizing capability of a given radiation source. In some cases it is sufficient, or even desirable, to count the number of ionizing events or *pulses* without regard to the ionization produced in each pulse.

An ion chamber is equivalent to an electrical capacitance C connected in series with an external resistance R, Fig. 7-1. When an essentially instantaneous pulse of ionization occurs in the chamber, current flow will quickly rise to a maximum value I_0 and will then decline according to

$$I = I_0 e^{-t/RC} \tag{7-11}$$

The product RC, which determines the speed of response of the system, is known as the *time constant*. When $t = RC$, the current flow will have decreases to $1/e = 0.37$ of the maximum value. With R in megohms and C in microfarads, t will be expressed in seconds.

If R and C are made small by circuit design, the time constant can be reduced to the point where perhaps 10^5 pulses per second can be detected and recorded. When R is small, the signal voltage developed across it will also be small but electronic amplifiers are capable of bringing most signals up to usable levels.

The use of amplifiers permits the detection and counting of some particular types of radiation selectively in the presence of others. Consider, for example, a beam of neutrons, which, in passing through a chamber, produce a number of proton recoils. At the same time several reactions, such as (n, γ), will produce a substantial photon fluence. Each proton recoil will lose almost all its energy in the chamber gas, producing perhaps 10^5 ion pairs. A photon absorption will produce an electron which may create perhaps 10^2 ion pairs in the gas before entering the chamber wall where it will lose the rest of its energy in producing uncountable ions.

The amplifier system designed for use with a fast chamber will have a sensitivity control or *discriminator* which can be set to reject all pulses below a desired size. With a proper adjustment of the discriminator a pulse-counting system can be made to reject all the small pulses arising from primary electron interactions and accept all the large pulses resulting from proton recoils. Thus neutron fluence can be measured in the presence of a strong photon radiation.

REFERENCES

Attix, F. H. and W. C. Roesch, ed., *Radiation Dosimetry*, Vol. I, II, and III. Academic Press, New York, N.Y., 1966.

Casarett, A. P., *Radiation Biology*. Prentice-Hall, Inc., Englewood Cliffs, N.J., 1968.

Kellerer, A. M., *Microdosimetry and the Theory of Straggling*. Report of a second panel on Biophysical Aspects of Radiation Quality, International Atomic Energy Agency, Vienna, 1967.

Linear Energy Transfer. ICRU Report No. 16, ICRU, Washington, D.C., 1970.

Platzman, R. L., *On the Primary Processes in Radiation Chemistry and Biology*, Symposium on Radiation Biology, Chap. 7, J. J. Nickson, ed. John Wiley & Sons, Inc., New York, N.Y., 1952.

Price, W. J. *Nuclear Radiation Detection*. McGraw-Hill Book Co., New York, N.Y., 1964.

Rossi, H. H., and W. Rosenzweig, "A Device for the Measurement of Dose as a Function of Specific Ionization." *Radiology*, **64**, 404–411, 1955.

8

Absorption of Ionizing Photons

8.01 Ionizing Radiations

Any particle bearing an electric charge and having kinetic energy sufficient to produce ions will interact or *couple* strongly with the orbital electrons in any medium through which it passes. Because of the strong coupling, energy losses to the orbitals will take place at closely spaced intervals even in the case of sparsely ionizing highly relativistic electrons. Charged particles having the requisite energy are said to be *directly ionizing*.

Directly ionizing particles lose energy to the orbitals in a large number of relatively small increments. An occasional encounter may transfer enough energy to permit the ejected electron to produce a recognizable ionization trail of its own. When the length of a secondary track exceeds some rather ill-defined limit, it is known as a *delta ray*. The average energy transfer, however, is less than 100 eV and the most probable transfer only about 20 eV.

Photons interact with orbital electrons through either their electric or magnetic field. Coupling between these fields and the coulomb field of an electron is very weak compared to the coupling between the coulomb fields of two charged particles. Even with this weak coupling, photons with quantum energies in the ultraviolet region, 2–10 eV, interact strongly through resonances with electronic energy levels, and consequently they have a very short range. Another region of strong absorption occurs at still lower energies where photons are in resonance with vibrational and rotational levels.

As photon energies increase above the electronic resonance region, the probability of a photon–electron interaction decreases and the radiation becomes more penetrating. High-energy photons may pass through many centimeters or meters of material without transferring any energy to it. Mere passage of a photon does not alter the material traversed; effects are to be expected only when there has been an energy transfer to the material.

When a high-energy photon does interact, the energy transfer will be relatively large. A photon may give up all its energy in a single encounter with one orbital electron which will then proceed to ionize in the usual way. Photons are one example of *indirectly ionizing* radiation.

Spatial distribution patterns of ionizations and excitations will depend on the type or *quality* of the radiation producing them. Any relation between biological or chemical effect and radiation quality must depend on the distribution patterns rather than inherent differences in the radiations themselves, for all are ultimately effective through ionizing electrons. The excitation and ionization pattern, and hence the biological effect, of an individual electron of given energy is independent of the nature of the radiation that produced it.

8.02 Geometry in Absorption Measurements

In some encounters a photon may give up only a portion of its energy to an orbital electron. The remaining energy goes into the creation of a new photon of lower energy than the original. The newly created photon may move off in any direction depending on the energy division among the three particles. Measurements of photon absorption are complicated by the presence of these new or scattered photons.

Consider an arrangement, Fig. 8-1, designed to measure the absorption characteristics of a *narrow beam* of photons. The photons emitted by source S are restricted or *collimated* to a narrow beam by the massive shield C. A second shield C' is placed in front of the detector D to delineate the area of its response. An absorber A is located close to the source. For the moment we restrict the discussion to photons whose energy does not exceed a few MeV.

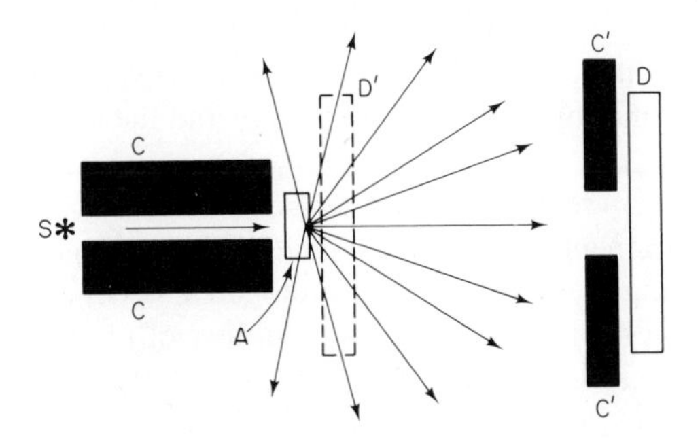

Figure 8-1. Source, absorber, and detector arrangements for measuring photon absorption.

Except for a small surface loss, all electrons liberated in A by photon absorption will be absorbed in A. Secondary photons originating in A by primary photon absorption will be emitted in all directions as shown. Few of these scattered photons will reach and activate the detector, which will intercept only those photons that pass through A without an interaction. This arrangement, said to have *good geometry*, measures *photon attenuation*, or the fraction of the incident photons removed by all absorptive processes in A.

When the detector is not collimated and is placed close to the absorber, as at D', it will respond to a large fraction of the scattered photons as well as those that have escaped all interactions. This arrangement, in *poor geometry*, serves to measure *energy transfer*, or the amount of energy actually deposited locally in the absorber.

The two concepts of attenuation and energy absorption have immediate practical applications. In considering the shielding of a photon source, one is concerned with the amount of the original radiation that is removed by absorbers. Narrow-beam conditions are seldom realized but modified narrow-beam attenuation constants can be applied. In calculating the energy delivered to some tissue or organism, one is interested in the amount of energy that is locally deposited rather than the amount that is removed from the beam.

8.03 Exponential Attenuation

Using the good geometry arrangement of Fig. 8-1, let the photon intensity at the detector be I_0 when there is no absorber in the beam. As used here, intensity specifies the rate at which photon energy crosses a unit cross section perpendicular to the beam direction. Intensity units might be ergs cm^{-2} s^{-1} or MeV cm^{-2} s^{-1}. If the beam is strictly monoenergetic, intensity can be expressed in photons cm^{-2} s^{-1}.

As absorbers are introduced, the intensity at the detector will decrease according to an exponential law.

$$I = I_0 e^{-\mu x} \tag{8-1}$$

where I is the beam intensity with absorber thickness x. μ is the *linear attenuation coefficient*, a function of the photon energy and the atomic number of the absorber.

Equation (8-1) is identical in form to Eq. (5-2), which describes radioactive decay. The exponential attenuation relationship arises for a reason analogous to that leading to an exponential radioactive decay; the probability of an absorbing event is proportional to the number of photons which pass the point under consideration. As before, the exponential form can be converted to the logarithmic

$$\begin{aligned} \ln I &= \ln I_0 - \mu x \\ \ln \frac{I_0}{I} &= \mu x \end{aligned} \tag{8-2}$$

There will be a certain absorber thickness which will reduce the beam intensity to just one-half of its original value. This is the *half-value thickness*, HVT, or the *half-value layer*, HVL, given by

$$\mathrm{HVL} = \frac{0.693}{\mu} \tag{8-3}$$

Another useful parameter is the *mean free path*, mfp, which is the average distance that a photon travels in the absorber before undergoing an interaction.

$$\mathrm{mfp} = \frac{1}{\mu} = 1.44\ \mathrm{HVL} \tag{8-4}$$

The probability of an interaction leading to photon absorption depends on the atomic composition of the absorber and not its state properties such as density or chemical form. Because of this independence from state variables, *mass attenuation coefficients* are more generally used than are the linear values. Mass attenuation coefficient in a material of density ρ is defined as

$$\mu_m = \frac{\mu}{\rho}\ \mathrm{cm^2\ g^{-1}} \tag{8-5}$$

which leads to the attenuation relation

$$I = I_0 e^{-\mu_m (x\rho)} \tag{8-6}$$

Photon attenuation may be expressed in terms of the number of absorbing atoms or electrons. If there are N atoms per cubic centimer of absorber, Eq. (8-6) can be written as

$$I = I_0 e^{-\sigma_a (xN)} \tag{8-7}$$

The coefficient per atom, σ_a in Eq. (8-7), has the dimensions of an area and so it is known as the atomic attenuation *cross section*. Cross sections may be expressed in $\mathrm{cm^2}$ but since most of them are comparable to the dimensions of the elementary particles, a special unit, the *barn*, b, is more frequently used. Both the barn, $1\ \mathrm{b} = 10^{-24}\ \mathrm{cm^2}$, and the millibarn, $1\ \mathrm{mb} = 10^{-27}\ \mathrm{cm^2}$, are commonly used in tabulating cross sections.

Photon absorption coefficients can also be expressed in terms of the number of orbital electrons per unit volume of absorber. In an absorber of atomic number Z, Eq. (8-7) becomes

$$I = I_0 e^{-\sigma_e (xNZ)} \tag{8-8}$$

where σ_e is the attenuation cross section per electron.

J. J. Thomson calculated the cross section per electron on the assumption that the electromagnetic field of the photon set the interacting electron into oscillation. The electron absorbed energy from the field and later radiated it away in degraded form. The Thomson model leads to

$$\sigma_0 = \frac{8\pi e^4}{3m_0 c} = 6.65 \times 10^{-25}\ \mathrm{cm^2\ per\ electron} \tag{8-9}$$

Thomson's model assumed only a single mode of photon–electron interaction and led to a cross section that is independent of atomic number and photon energy. This model is now known to be inadequate but the Thomson cross section appears as one factor in the more sophisticated treatments.

There are, in fact, three main modes of interaction, each with a unique dependence on Z and $h\nu$. Each attenuation coefficient will therefore consist of three components. Thus

$$\mu_m = \frac{\mu}{\rho} = \frac{\tau}{\rho} + \frac{\sigma}{\rho} + \frac{\kappa}{\rho} \quad \text{cm}^2\ \text{g}^{-1} \tag{8-10}$$

where τ, σ, and κ are the linear attenuation coefficients for photoelectric absorption, Compton scattering, and pair production, respectively.

8.04 Photoelectric Absorption

In photoelectric absorption the incident photon loses all its energy, with essentially all of it going to one of the orbital electrons, Fig. 8-2A. The electron involved will be ejected from the atom with an energy equal to that of

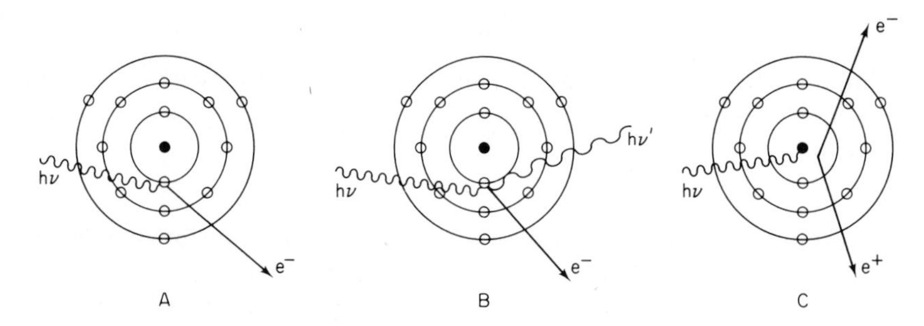

Figure 8-2. The three chief modes of photon–electron interaction. (A) Photoelectric absorption. (B) Compton scattering. (C) Pair production.

the photon, less the energy required to free the electron from its orbital position. The very small amount of energy imparted to the recoiling atom is customarily neglected. The energy relations are

$$T = h\nu - \phi \tag{8-11}$$

where ϕ is the energy binding the electron to its nucleus. Any orbital electron can participate in a photoelectric reaction provided that the photon energy exceeds the binding energy.

Photoelectric absorption is the most probable process at low photon energies but for the present we exclude discussion of photon energies only slightly greater than ϕ. When $h\nu$ is nearly equal to ϕ, resonance phenomena cause the absorption to change rapidly with energy; a separate discussion is required, Secs. 8.10 and 16.03.

Except at extremely high energies, the range of an electron is much less than the mean free path of a photon of equal energy. The ejected photoelectron will then expend its energy close to its place of origin, and therefore energy imparted to this electron is energy truly given up to the absorber.

After the photoelectron is ejected, the absorber atom will emit its characteristic X-ray spectrum as the orbital electron defect is removed. The process is exactly the same as in X-ray emission but now the secondary photons are called *fluorescent radiation* because they are excited by a primary photon instead of a bombarding electron.

In the case of a high atomic number absorber, ϕ is large (88 keV for lead) and the fluorescent radiation is quite penetrating. The fluorescent radiation gives up its energy at a great distance from its point of origin, which means that energy ϕ cannot be included in the local energy transfer. Living tissues on the other hand are for the most part composed of low-Z elements and here ϕ will be less than 1 keV. In this case the very soft fluorescent radiation will be absorbed very close to the point of emission and true energy transfer is $T + \phi$ instead of T alone. The very fact that ϕ is small in this case makes the distinction of little practical importance.

Neither theory nor experiment alone provides the proper values of the photoelectric cross sections over the entire range of photon energies and atomic numbers of practical interest. The coefficients are strong functions of both of these variables, varying, over a considerable range, about as

$$\tau \propto Z^3(h\nu)^{-4.5} \tag{8-12}$$

The general trend is a rapid increase in τ with atomic number and a rapid decrease as photon energy increases. Figure 8-3 shows τ_a values for two elements of biological importance. An increase in Z from 6 to 20 increases the photoelectric cross section by two orders of magnitude. Values for the two elements depicted have dropped by a factor of 10^{-6} at 1 MeV, where photoelectric absorption in the light elements is relatively insignificant. In heavy shielding materials such as lead, photoelectic absorption may be important at much higher energies because of the Z^3 factor.

The angular distribution of the ejected electrons is given by

$$dN = \frac{\sin^2 \theta}{(1 - \beta \cos \theta)^4} \, d\Omega \tag{8-13}$$

where dN = relative number of electrons in the small solid angle $d\Omega$ at an angle θ to the photon direction

$\beta = v/c$ for the electron

At low energies the distribution function has a maximum at $\theta = 90°$. As the photon energy increases, the electrons tend to be emitted in more sharply defined distributions tending toward the forward direction.

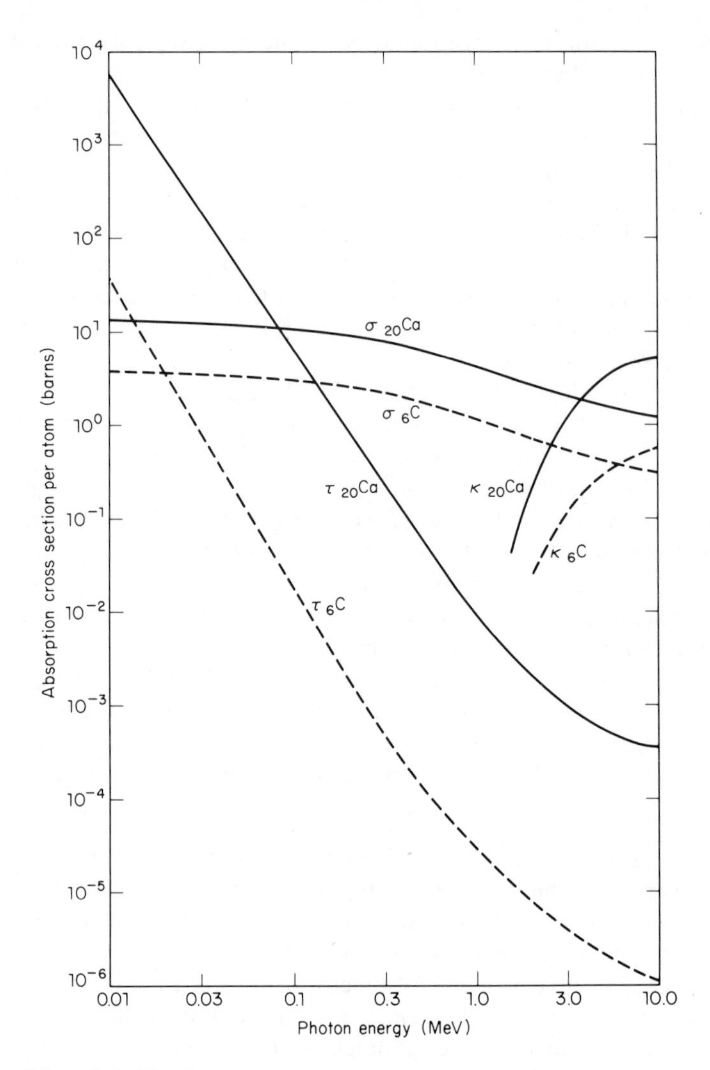

Figure 8-3. The three component cross sections for carbon and calcium as a function of photon energy.

8.05 Compton Scattering

Compton scattering, named after its discoverer, A. H. Compton, becomes important at photon energies where photoelectric absorption has fallen off because of the $(h\nu)^{-4.5}$ term. In Compton scattering, Fig. 8-2B, a primary photon is absorbed in interacting with an orbital electron. The latter is freed and a secondary, or scattered, photon of energy lower than the primary is created. Figure 8-4 shows the geometrical relations involved.

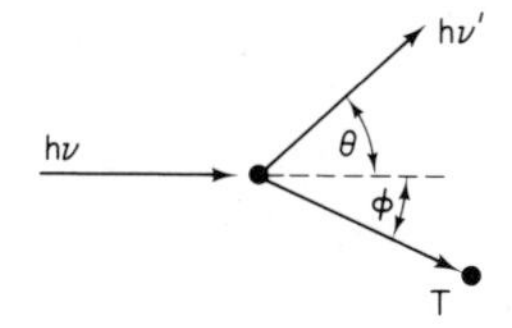

Figure 8-4. Geometrical relations in a Compton scattering event.

The Compton process can be analyzed by applying the relativistic forms of the laws of energy and momentum conservation to the collison. Each photon is considered to be a particle with energy $h\nu$ and momentum $h\nu/c$. Compton scattering is described most simply in terms of a wavelength shift between the primary and the scattered photons, $\Delta\lambda = \lambda' - \lambda$:

$$\Delta\lambda = \frac{h}{m_0 c}(1 - \cos\theta) \tag{8-14}$$

The constant factor $h/m_0c = 2.42 \times 10^{-10}$ cm or 0.0242 Å is known as the *Compton wavelength.*

Energy relations are usually expressed in terms of an energy parameter $\alpha = h\nu/m_0c^2$, which merely means that the photon energy is expressed in terms of the rest mass of the electron, 0.511 MeV. The energy expressions are

$$T = h\nu\frac{\alpha(1 - \cos\theta)}{1 + \alpha(1 - \cos\theta)} \tag{8-15}$$

$$h\nu' = \frac{m_0c^2}{1 + 1/\alpha - \cos\theta} = \frac{0.511}{1 + 1/\alpha - \cos\theta}\ \text{MeV} \tag{8-16}$$

Energy division between the Comption electron and the scattered photon depends strongly the original photon energy. At low energies the scattered photon carries away a large fraction of the available energy; at higher energies there is a large photon degradation for the same scattering angle.

Illustrative Example

Calculate the energy of a photon scattered at an angle of 90° for a primary photon of (a) 100 keV and (b) 1 MeV.

(a) From Eq. (8-14),

$$\Delta\lambda = 0.0242(1 - 0) = 0.0242\ \text{Å}$$

From Eq. (1-19),

$$\lambda = \frac{12.4}{100} = 0.124\ \text{Å}$$

$$\lambda' = 0.124 + 0.0242 = 0.148\ \text{Å}$$

$$E' = \frac{12.4}{0.148} = 83.6\ \text{keV}$$

which is 83.6% of the original.

(b) We use Eq. (8-16) directly, with $\alpha = 1.00/0.511$, $1/\alpha = 0.511$.

$$E' = \frac{0.511}{1.00 + 0.511 + 0} = 0.339 \text{ MeV}$$

which is 33.9% of the original.

The probability of a Compton interaction can be calculated from the Klein–Nishina equations*, one of the earliest and most completely verified results of the quantum mechanics. Compton processes take place with about equal probability at each orbital electron, regardless of the energy with which it is bound to its nucleus. Compton cross sections depend, therefore, only on the electron density in the absorber, independent of the Z values involved. Compton cross sections decrease with increasing photon energy but at a much slower rate than the corresponding photoelectric cross sections, Fig. 8-3. Compton scattering is effective at energies well beyond those at which photoelectric absorption has become negligible.

The Klein–Nishina equations also given the angular distribution of the scattered energy. Forward and back scattering (0° and 180°) are about equal up to around 50 keV, but from then on forward scattering predominates. At 20 MeV and above the scattered photons form a narrow beam along the direction of the original photons.

8.06 Pair Production

An entirely new absorption process takes place at photon energies greater than 1.02 MeV. Here, a photon can interact with an atomic nucleus and give up all its energy in the process of *creating* a negative electron–positive electron pair, Fig. 8-2C. The interaction actually creates the two electrons and therefore pair production must have a threshold equal to the energy equivalence of the two created masses, which is 1.02 MeV. Any photon energy above the threshold goes into kinetic energy of the electrons, with the positron receiving slightly more than the negatron. The created electrons tend to move off in the direction of the photon, a tendency which becomes more pronounced at high photon energies.

As the negatron proceeds, it loses energy to ionizations and excitations and is eventually captured to become an orbital electron. The position loses energy by the same processes but meets a quite different eventual fate. When most, or all, of the positron energy is lost, it will combine with a negative electron to produce two oppositely directed gamma rays, each having an energy of 0.51 MeV. These gamma rays are called *annihilation radiation* since the negatron–positron pair disappears in the process.

The theory of pair production is well understood but the results are not expressible in simple mathematical forms. Cross sections are low just above

*Evans, R. D. *The Atomic Nucleus*. McGraw-Hill Book Co., New York, N.Y., 1955, pp. 677–689.

1.02 MeV and increase to a maximum value at higher photon energies. Except at high atomic numbers and high photon energies the cross section varies nearly as Z^2, Fig. 8-3.

8.07 Other Attenuation Mechanisms

Small amounts of energy may be scattered from a photon beam by *Rayleigh* or *coherent* scattering. Each individual interaction of any of the types previously discussed is unrelated to any other absorptive process. In coherent scattering, photons scattered from adjacent atoms in a crystalline lattice arrangement interfere, either constructively or destructively. A diffraction pattern is produced as described in Sec. 3.03 and depicted in Fig. 3-3. The cross section for coherent scattering is very small and can be safely neglected in all calculations of shielding requirements or in radiation dosimetry.

High-energy photons can initiate *photonuclear* reactions in which a nucleon is ejected from the nucleus that absorbed the photon. With the exception of ^2H and ^9Be targets, all photonuclear reactions have a threshold of 8 MeV or more. Cross sections are so small that the process can be neglected in calculations of beam attenuation or of radiation dosage. Some of the product nuclei are radioactive and this fact must be kept in mind in utilizing high-energy photon beams. When photons below 8 MeV are used, any residual radioactivity will be below detectable limits.

8.08 Combined Attenuation

In general, all three of the main attenuation mechanisms will be operating simultaneously, with relative cross sections depending on photon energy and the atomic number of the absorber. Each process will follow an exponential law of the form of Eq. (8-1) and so the combined effect will also be an exponential, with an attenuation coefficient equal to the sum of the three constituent coefficients.

The mass attenuation coefficient is calculated as the sum of its three constituents according to Eq. (8-10). When one is interested in the local deposition of energy, the *mass energy transfer coefficient* (μ_κ/ρ) must be used instead of (μ/ρ). (μ_κ/ρ) has the same three components but each must be modified to take into account the energy removed from the primary photon beam but deposited at a distance by penetrating secondary photons. We now have

$$\frac{\mu_\kappa}{\rho} = \frac{\tau'}{\rho} + \frac{\sigma'}{\rho} + \frac{\kappa'}{\rho} \text{ cm}^2 \text{ g}^{-1} \tag{8-17}$$

in which the new coefficient can be related to the original attenuation values.

$\tau' = \tau(1 - \delta/h\nu)$. δ is the average energy emitted as fluorescent radiation when a photon of energy $h\nu$ undergoes photoelectric absorption. The fraction $\delta/h\nu$ is assumed to be lost from the immediate vicinity of the point of interaction.

$\sigma' = \sigma(E/h\nu)$. E is the average energy of the Compton electron per absorbed photon $h\nu$. $(E/h\nu)$ is the fraction of the original photon energy that is locally transferred. σ' can also be calculated from the Klein–Nishina equation for true absorption.

$\kappa' = \kappa(1 - 2m_0c^2/h\nu)$. $2m_0c^2$ is the energy lost in creating the two electron masses. When the positron annihilates, $2m_0c^2 = 1.02$ MeV is recovered as gamma radiation but this energy is not absorbed locally.

Figure 8-5 is a plot of the mass attenuation coefficients for lead, concrete, and sodium iodide. The first two substances are commonly used shielding materials; sodium iodide is a scintillating material used as a detector of photons. Each curve shows the strong dependence of photoelectric absorption on $h\nu$ and Z. At 1 MeV, Compton scattering predominates and the coe-

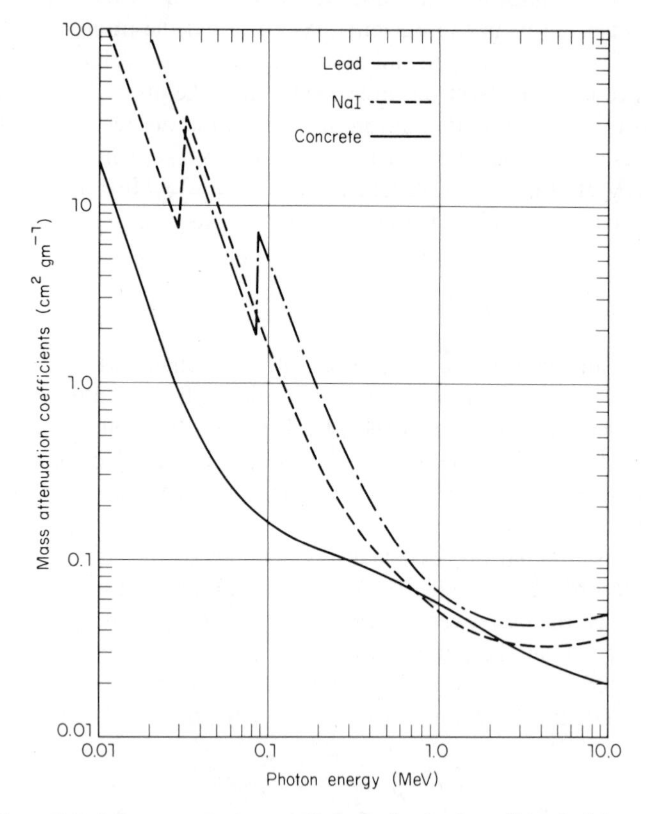

Figure 8-5. Mass attenuation coefficients for lead, sodium iodide, and concrete. Absorption discontinuities can be seen in the lead and the sodium iodide curves.

ficients are nearly equal but diverge again at higher energies because of differences in pair production. The discontinuities in the lead and sodium iodide curves occur at *absorption edges*, Sec. 8.10. Concrete consists primarily of low-Z materials and the absorption edges come at energies lower than those plotted.

Figure 8-6 is a plot of mass energy transfer coefficients for air, water, and calcium. Again the strong dependence on photon energy and atomic number is number is evident over the photoelectric region.

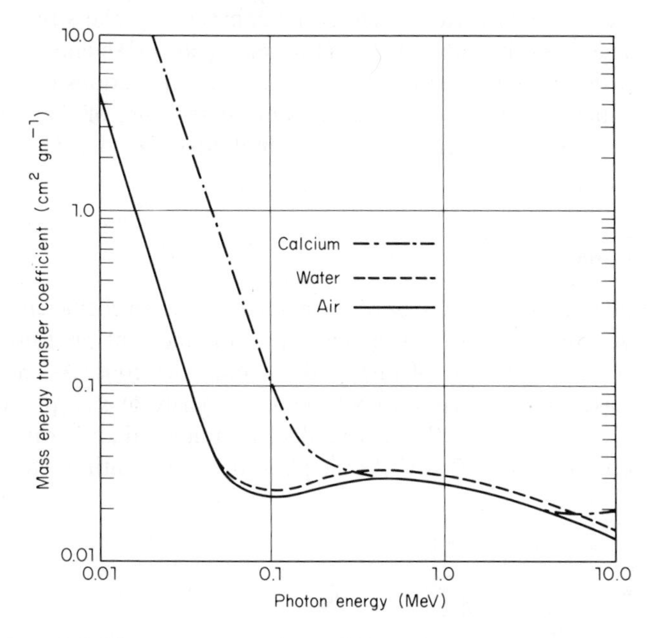

Figure 8-6. Mass–energy transfer coefficients for air, water, and calcium.

8.09 Broad-Beam Absorption

All the previous discussion applies strictly to narrow-beam geometry only, a condition seldom realized in practice. Under broad-beam conditions, Fig. 8-7, a detector even with good geometry will receive not only the direct, at-

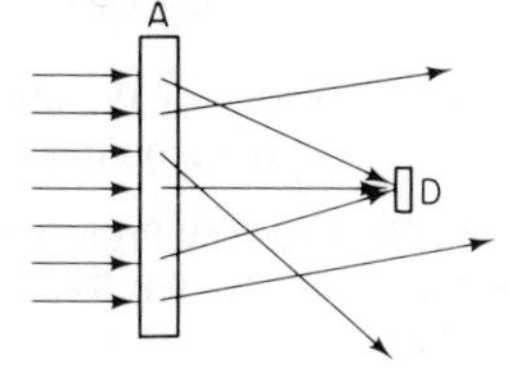

Figure 8-7. Under broad-beam conditions a detector receives photons scattered from all points of the absorber.

tenuated beam but also some photons scattered from all points in the absorber. With a broad beam the radiation intensity detected will be greater than that obtained under narrow-beam conditions. The use of narrow-beam attenuation constants will lead to an overestimate of the effectiveness of shielding.

A *buildup factor* B is defined as

$$B = \frac{\text{intensity under broad-beam geometry}}{\text{intensity calculated from narrow-beam constants}}$$

Buildup depends many factors such as the composition and size of the absorber, the energy of the radiation, and the general arrangement of surrounding material that might contribute to the scatter. Calculations can be made for some simple geometrical arrangements but best values of B are obtained from the actual operating conditions. Some buildup factors have been tabulated.*

8.10 Absorption Edges

Figure 8-8 illustrates the behavior of the mass attenuation coefficient in a lead absorber over the region where the energy of the incident photons may be equal to the binding energy of one of the orbital electrons. As the photon energy increases, the absorption coefficient, primarily due to photoelectric interactions, decreases smoothly until a discontinuity and an abrupt increase in absorption occurs at 13.0 keV. As the photon energy continues to increase,

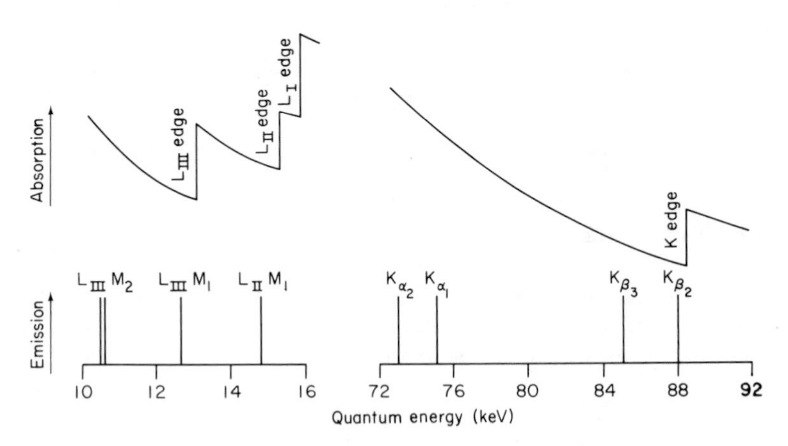

Figure 8-8. Relative energies of the absorption edges and characteristic emission lines of lead.

*"Medical X-Ray and Gamma-Ray Protection for Energies up to 10 MeV," *Structural Design and Evaluation Handbook*. NCRP Report No. 34. NCRP Publications, P.O. Box 3075, Washington D.C. 20014.

other sharp breaks appear at 15.2 and 15.9 keV. The absorption coefficient then continues to decrease smoothly until a final discontinuity is seen at 88 keV.

The abrupt increases are due to resonant absorption when the incoming photon has just the energy required to free a particular orbital electron from its nucleus. A photon with less energy is unable to transfer any energy to that particular electron. The energies where absorption is enhanced are known as *absorption edges* because of their abrupt appearance. The three low-energy edges seen in Fig. 8-8 are due to the resonance ejection of electrons from various sublevels in the L shell. The single edge at 88 keV comes from the removal of a K electron. M, N, and still higher-order edges exist but they occur at very low energies where experimental detection is difficult.

Characteristic X-ray emission energies involving any particular electron are always less than the absorption edge energy for the same electron. An absorption edge corresponds to the complete removal of an electron from the atom. This requires the energy ϕ in the photoelectric relation, Eq. (8-11). A characteristic X-ray photon, on the other hand, is emitted when an electron drops from one bound state to another. In Fig. 8-8 the edge at 88 keV represents a transition from the K shell to infinity. The K_α emission lines at 73–75 keV are the result of $L \longrightarrow K$ transitions, the K_β lines are from $M \longrightarrow K$ transitions, and so on.

An element cannot absorb its own emission lines by resonance because the emission lines will always lie on the low-energy side of the absorption edge. Thus the K_α lines of $_{82}Pb$ will be most effectively absorbed by $_{75}Re$, where the K edge occurs at 76.7 keV.

Any orbital vacancy created by photon absorption will be subsequently filled from outer orbits, just as in the case of the generation of characteristic X rays following electron impact. The fluorescent radiation that follows photon absorption forms a more nearly monoenergetic source of X rays than can be obtained by electron bombardment of a target. In the latter case a substantial bremsstrahlung production accompanies the emission of the characteristic wavelengths.

8.11 Heterogeneous Photon Beams

In practice, photon beams are seldom monoenergetic and even if they are initially so, they rapidly become heteroenergetic by Compton scattering and the production of fluorescent radiation. Heteroenergetic beams are inevitable from X-ray tubes since the bremsstrahlung spectrum has a continuous distribution from zero up to the maximum energy of the exciting electrons.

Absorption processes tend to selectively remove the low-energy components and thus *harden* the beam. In general the soft components of an

X-ray beam are undesirable and some absorbing *filters* will be put into the beam to remove them. In X-ray therapy, for example, a hard, penetrating beam may be desired in order to deliver an effective radiation does to some deep tissue. Any soft components in the beam will be strongly absorbed by the skin and other superficial tissues, to produce undesirable radiation effects outside the treatment volume. Bremsstrahlung required for irradiating superficial tissues can be generated by applying a low voltage to the X-ray tube, which will usually have a thin beryllium window to minimize losses in the tube itself. Doses to deeper tissues will be low because of the strongly attenuated beam.

Any external X-ray beam will be subject to some *inherent filtration* as it passes through the tube window and perhaps other materials such as cooling oil. Inherent filtration can be reduced by making the window as thin as possible and by using a material of low atomic number to reduce photoelectric absorption. Beryllium appears to be the material of choice for low-absorption windows.

The amount and kind of *added filtration* introduced for beam hardening depends on the nature of the primary beam and the amount of hardening desired. In a beam generated at about 200 kVP the added filtration might consist of 0.2–0.5 mm of copper followed by 0.2–0.5 mm of aluminum and perhaps 1–3 mm of carbon in the form of Bakelite or another plastic. The copper filter will harden the beam by its strong photoelectric absorption of low-energy photons. At the same time a strong copper fluorescent radiation will be produced by the absorption of photons at energies above the copper absorption edges. The copper fluorescent emissions will be absorbed in the aluminum, whose still lower-energy fluorescent emissions will be absorbed by the carbon. Carbon fluorescence occurs at such a low energy that it will be strongly absorbed even in air. Note that the absorber sequence Cu–Al–C is important. Any reversal of the filter order will reduce the effectiveness of the combination.

8.12 Effective Beam Energy

The spectral distribution or *quality* of the radiation produced by an X-ray tube is not easy to determine with precision. Quality is usually specified by giving the peak voltage at which the radiation was generated, kVP, and the half-value thickness, HVL, of the useful beam after it has passed through any added filtration. The HVL is obtained from an absorption curve, Fig. 8-9, using absorbers appropriate to the beam energy that is involved. To reduce anomalous absorption, the K edge of the absorber should not be at more than 0.2–0.5 of the peak potential. Three absorbing materials are commonly used for HVL determinations in accordance with the schedule of Table 8-1.

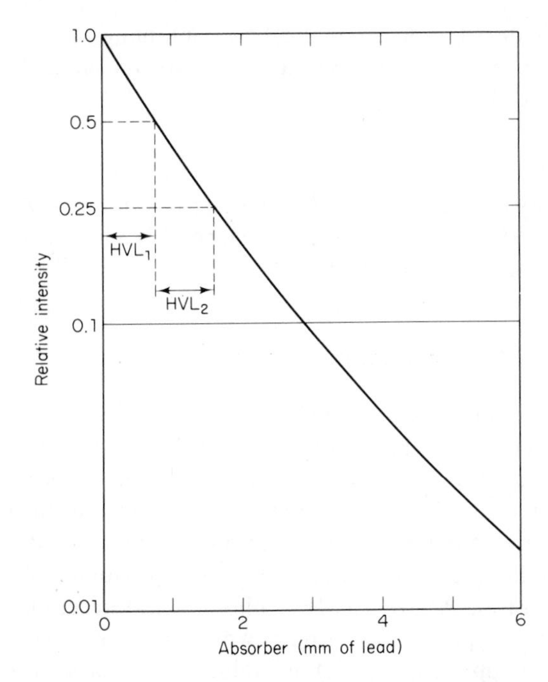

Figure 8-9. Absorption curve for the determination of the homogeneity coefficient.

TABLE 8-1
RECOMMENDED FILTERS FOR HVL DETERMINATIONS

Element	*K edge, keV*	*Useful Range, keV*
Aluminum	1.6	50–150
Copper	8.9	200–400
Lead	88.0	400–upward

The *effective energy* of an X-ray beam can be determined by comparing the measured HVL value with values obtained with monoenergetic beams. Although effective energy is a useful specification of beam quality, its limitations must be clearly understood. Two beams with equal values of HVL are equal only in terms of the absorption measurements used in making the HVL determinations. Two heteroenergetic beams with identical values of HVL may have quite different spectral distributions and may be unequally effective in a particular application.

The addition of absorbers for making an HVL determination will harden the beam by an amount which depends on its spectral distribution. A useful measure of this distribution is obtained by continuing the absorption measurements and comparing the second HVL with the first. A *homogeneity coefficient*, h, is defined as

$$h = \frac{(\text{HVL})_1}{(\text{HVL})_2} \tag{8-18}$$

The homogeneity coefficient for the beam described in Fig. 8-9 has a value of $h = 0.80/0.90 = 0.88$. This indicates that the beam is relatively homogeneous, with a small low-energy component.

8.13 High-Energy Photon Absorption

As technical developments pushed the energy limits of particle accelerators higher and higher, photon beams of corresponding energy became available. Electron beams of more than 1 GeV are now in use and still higher energies are to be anticipated. At the higher energies the clear distinction between energy locally deposited and that transferred to a distance tends to disappear.

As energies increase above 1.02 MeV, the probability of electron-pair production increases. Above 10 MeV, photonuclear reactions become appreciable. Various types of meson pairs will be produced when photon energies exceed the appropriate threshold value. All of these mechanisms act to increase the absorption probability or, conversely, to decrease the mean

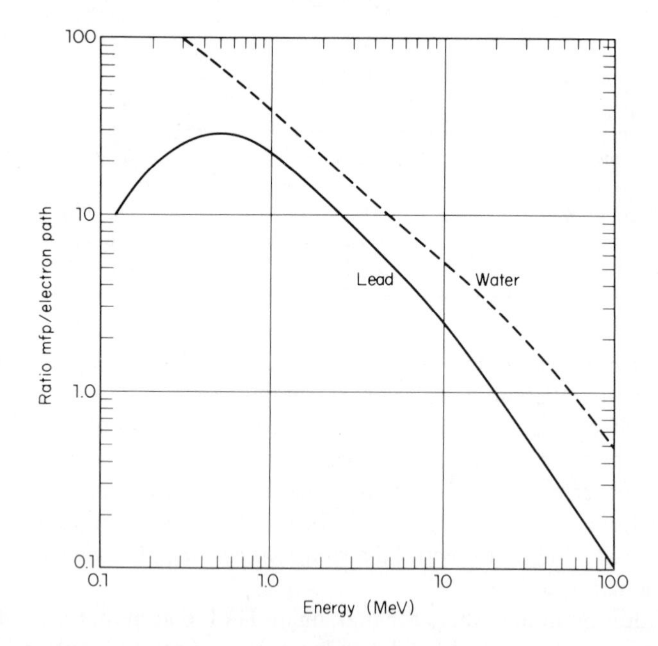

Figure 8-10. Ratio of the photon mean free path to the maximum path length of an electron of the same energy.

free path of high-energy photons. Absorption minima can be seen in the lead and sodium iodide absorption curves, Fig. 8-5. Elements of lower atomic number show minima at higher energies.

Electrons produced by photon absorption have energy spectra extending up to the maximum photon energy. The range of these electrons increases steadily with energy until they equal or even exceed the mean free part of the photons that released them. The ratio of photon mfp/electron path length is shown plotted in Fig. 8-10 as a function of energy. In lead the two lengths are equal at about 20 MeV; in water equality is only reached at 60 MeV.

Above 2–3 MeV there is a decreasing lack of distinction between local and distant energy transfers. As we shall see, this introduces complications into the determination of the radiation dose delivered by high-energy photon beams.

8.14 Nuclear Photon Absorption

Photon–nucleon coupling is so weak that absorption by nuclei is only a minor component of the total attenuation of a photon beam. Other results of nuclear absorption are, however, of some significance. Consider a typical nuclear reaction

$$^{12}_{6}\mathrm{C} + h\nu \longrightarrow {}^{11}_{6}\mathrm{C} + {}^{1}_{0}n + Q \tag{8-19}$$

The ejected neutron in Eq. (8-19) was originally bound to the nucleus with some 7.5 MeV; and this energy must be supplied by the photon if the reaction is to go. When exact mass values are inserted in this type of reaction, Q is generally found to be negative by about 8 MeV. Photon-induced reactions will have, therefore, a threshold energy of about 8 MeV, with cross sections that are appreciable only at energies well above this.

In many cases, as in Eq. (8-19), the reaction product is radioactive so high-energy photon irradiations may leave an absorber with an appreciable residual activity. The threshould value plays an improtant role in the sterilization of foods and other materials by photon irradiation. High-energy photons are desirable in order to obtain deep beam penetration but, above the reaction threshold energy, account must be taken of any radioactivity that may have been induced. Large radiation doses are known to be carcinogenic in living tissues and a strict, nonquantitative extrapolation would lead to the conclusion that high-energy photon irradiation of food has added a carcinogenic characteristic to it.

8.15 Recoilless Nuclear Resonance Absorption

Discrete energy levels exist in the nucleus, roughly analogous to the levels seen at much lower energies in the atomic orbital electrons. Photon absorption by a nucleus can raise it to an excited state without initiating a nuclear reaction typified by Eq. (8-19). Because of the high energies involved, there is a complication not seen in atomic absorption where only a few eV are transferred. As in the atomic case, nuclear absorption of a photon will be followed by photon emission as the nucleus returns to its ground state. Let us first consider the emission process.

When a photon of energy $h\nu$ leaves a nucleus, it carries away a momentum $p = h\nu/c$ and consequently the emitting nucleus will recoil with an equal but oppositely directed momentum. There will be an energy of recoil, $T_r = p^2/2m$, and hence all the energy of the transition, a very precisely fixed quantity, will not be transferred to the photon. When a photon of energy $h\nu$ is absorbed by a nucleus, it will impart momentum $p = h\nu/c$ and hence energy $p^2/2m$ to the nucleus and this energy will not be available for nuclear excitation.

Illustrative Example

Calculate the energy lost to recoil when a nucleus $A = 60$ emits or absorbs a 50-keV photon.

In this calculation mechanical energy units rather than MeV should be used.

$$h\nu = 50 \times (1.6 \times 10^{-9}) = 8 \times 10^{-8} \text{ erg}$$

$$p = \frac{h\nu}{c} = \frac{8 \times 10^{-8}}{3 \times 10^{10}} = 2.67 \times 10^{-18} \text{ g cm s}^{-1}$$

$$T_r = \frac{(2.67 \times 10^{-18})^2}{2 \times 60 \times (1.67 \times 10^{-24})}$$

$$= 3.56 \times 10^{-14} \text{ erg}$$

$$= 0.0222 \text{ eV}$$

The recoil energy, comparable to the energy of thermal agitation at room temperature, is a small faction of the photon energy. The recoil energy is, however, large compared to the width of the nuclear energy levels. This precludes the resonance absorption by a nucleus of its own emitted radiation. Resonance absorption is well-known in the optical spectrum where emission lines can be readily absorbed by the same atomic species that emits them.

Mössbauer was the first to point out that nuclear resonance absorption can be achieved under proper conditions, and his name has become attached to the process of recoilless nuclear absorption. In the first place Mössbauer incorporated the emitter and the absorber into a crystal lattice structure. Held to adjoining atoms by interatomic bonds, the recoiling nucleus shares its momentum and energy with many neighbors and thus greatly increases the effective value of m in the momentum–energy relation. In the second place the emitting source was moved relative to the absorber with a velocity just sufficient to bring the two energies into exact resonance through a Doppler frequency shift. Resonance is made manifest by a sharp increase in the number of photons absorbed.

The Mössbauer effect has become a very powerful technique for the determination of nuclear energy levels with high precision and for studying changes in interatomic bonding energies.

REFERENCES

ATTIX, F. H. AND W. C. ROESCH, ed., *Radiation Dosimetry*, Vol. I, II, and III. Academic Press, New York, N.Y., 1966.

COMPTON, A. H. AND S. K. ALLISON, *X-rays in Theory and Experiment*. D. Van Nostrand Co., Inc., Princeton, N.J., 1935.

HUBBELL, J. H., *Photon Cross Sections, Attenuation Coefficients, and Energy Absorption Coefficients From* 10 *keV to* 100 *GeV*. National Bureau of Standards, No. 29, August, 1969.

MÖSSBAUER, R. L., "Recoilless Nuclear Resonance Absorption." *Ann. Rev. Nuc. Sci.*, **12**, 123, 1962.

Physical Aspects of Irradiation, Report 10b of the International Commission on Radiation Units and Measurements. Issued as National Bureau of Standards Handbook 85, March, 1964.

Radiation Quantities and Units. ICRU Report 19. International Commission on Radiation Units and Measurements, 7910 Woodmont Ave., Washington, D.C. 20014.

9

Neutrons

9.01 Free and Bound Neutrons

Inside a nucleus a bound neutron is maintained by the very rapid meson exchanges with neighboring nucleons. Outside, a free neutron has at best a fleeting existence as it is removed either by nuclear capture or radioactive decay. Free neutrons exist for only a fraction of a second after active production of them has ceased.

A few high-energy positive-ion accelerators are capable of producing neutrons with energies of a few hundreds of MeV, and some neutrons in the cosmic-ray flux are observed with energies of many GeV. Most neutrons generated on earth will have initial energies of 20 MeV or less. Neutrons are divided into energy categories, Table 9-1, but the class boundaries are ill-defined because there are no unusual properties that change abruptly at particular energies. The following classification is somewhat arbitrary.

TABLE 9-1
NEUTRON CLASSIFICATIONS

High energy	> 10 MeV	Slow	0.03–100 eV
Fast	10 keV–10 MeV	Epithermal	1 eV
Intermediate	100 eV–10 keV	Thermal	0.025 eV

Neutrons with an average kinetic energy of 0.025 eV are in thermal equilibrium with their surroundings at a temperature of 293°K. At this energy the mean neutron velocity is 2200 m s^{-1} and hence thermal neutrons are sometimes referred to as "2200 m s^{-1} neutrons."

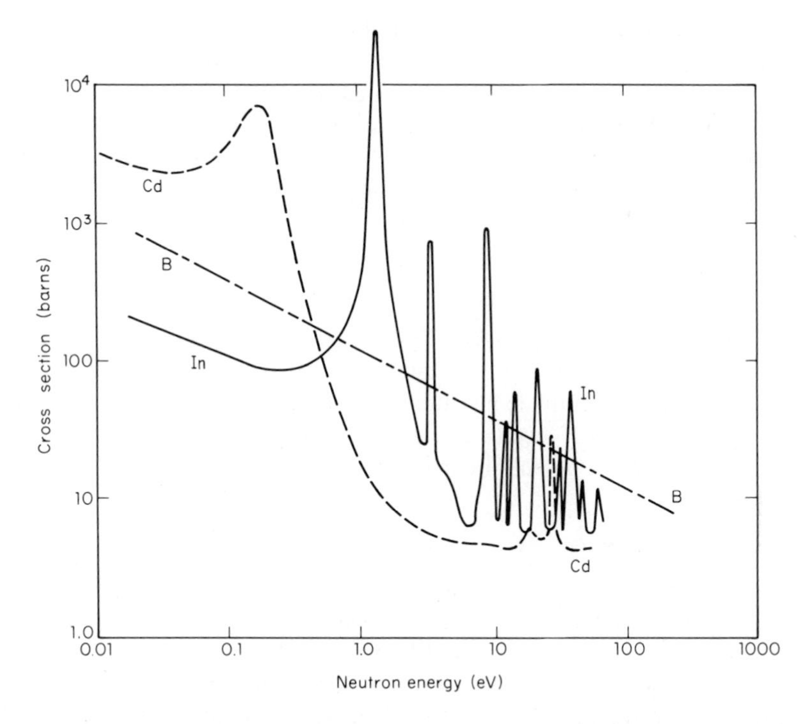

Figure 9-1. Some mass attenuation cross sections for low-energy neutrons. Boron shows an almost exact 1/v dependence. Cadmium has a broad, strong resonance at 0.1–0.5 eV. Indium has many sharp resonances which are superposed upon a general 1/v trend.

The element cadmium has a strong resonance absorption for neutrons at about 0.2 eV, Fig. 9-1. A cadmium filter will absorb nearly all neutrons of 0.5 eV or less, the exact value depending the thickness of the filter. Those neutrons of greater energy, transmitted by the cadmium, are sometimes known as "cadmium neutrons."

Lacking a net electric charge, high-energy neutrons undergo straightforward mechanical collisions with the atomic nuclei which they encounter in flight. These collisions rapidly degrade the energy toward the equilibrium value that characterizes the temperature of the neutron environment. Unhindered by the potential barrier a neutron of any energy can enter an atomic nucleus. Once inside, capture probability is high because nucleon–nucleon coupling is strong. The probability of neutron capture would be expected to increase as velocities decrease because then the particle spends more time in the vicinity of each nucleus which it encounters, and this is generally observed.

By chance, or in the absence of atomic nuclei, as in empty space, a neutron may escape nuclear capture. It is then doomed by radioactive decay. Neutrons are isobaric with protons, $A = 1$, and since they are heavier by 782 keV, a beta decay $n \longrightarrow p + \beta^-$ is to be expected. This radioactive decay is observed with a half-life of 12 m. Prior to decay the neutron is not a bound combination of a proton and an electron. All the arguments against the existence of nuclear electrons apply to the neutron as well as to heavier structures, and these preclude the existence of a p, e combination.

9.02 Neutron Production

Free neutrons were first produced by nuclear reactions initiated by alpha particles and this is still the method used in designing many calibration sources. Consider an intimate mixture of an alpha emitter such as ^{210}Po with a finely powdered low-Z metal such as beryllium. The alpha particles readily pass the relatively low potential barrier and enter the Be nucleus to initiate a reaction given by

$$^{9}_{4}\text{Be} + ^{4}_{2}\text{He} \longrightarrow ^{13}_{6}\text{C} \longrightarrow ^{12}_{6}\text{C} + ^{1}_{0}n + Q \tag{9-1}$$

This is one example of an (α, n) reaction which is usually written without intermediate product, known as the *compound nucleus*, in this case $^{13}_{6}$C. In shorthand notation, Eq. (9-1) would be written ^{9}Be $(\alpha, n)^{12}$C.

The total energy available in the reaction, Q, is divided as kinetic energy between the neutron and the recoiling nucleus, ^{12}C. With bombarding energies of only a few MeV all velocities will be nonrelativistic and the energy division between the neutron and the recoil will be described by the laws of classical mechanics.

Illustrative Example

Calculate the energy of a neutron produced according to Eq. (9-1) by a 5.3-MeV alpha particle from ^{210}Po.

Taking atomic mass values from Appendix 4,

$$8394.6 + 3728.3 = 11177.7 + 939.5 + Q$$

$Q = 5.7$ MeV and the total energy available is $5.7 + 5.3 = 11.0$.

Momentum conservation requires that the momenta of the neutron and the recoiling nucleus be equal and oppositely directed. That is,

$(mv)_n = (mv)_r$ whence $T_n m_n = T_r m_r$, and since $T_n + T_r = 11.0$ MeV,

$$T_n = 11.0(12/13) = 10.2 \text{ MeV}$$

A spectrum of neutron energies will be produced in an actual source, Fig. 9-2, because the bombarding particles will lose varying amounts of kinetic energy by collisions before they are captured by a beryllium nucleus. Some variations in the neutron energy spectrum can be achieved by a proper choice

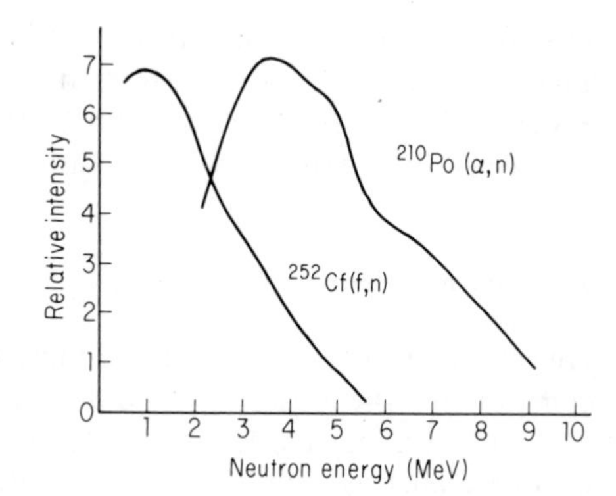

Figure 9-2. Energy distributions for neutrons produced in a commonly used α, n reaction and by the spontaneous fission of ^{252}Cf.

of alpha-particle sources and the target nuclei but practical choices are limited. Neutron sources lack the fine energy control that can be obtained by accelerated charged particles.

Four alpha-particle emitters have been found useful for constructing neutron sources. ^{210}Po has almost no gamma-ray emissions but with its half-life of only 138 d the neutron output declines rapidly. ^{226}Ra has the advantage of a long half-life but the strong gamma-ray emission of the Ra daughters adds a usually undesirable photon radiation to the neutron output. The 24,000-y half-life of ^{239}Pu provides an almost constant neutron output and the source is relatively free from gamma-ray contamination. Physically large sources are required by the low intrinsic specific activity of ^{239}Pu. About 16 g of this isotope is required to obtain 1 Ci of alpha activity. A much smaller source of equal activity can be made with ^{238}Pu, whose half-life of 86 y is conveniently long. Neutron output of each of these (α, n) sources is low, about 10^6 to 10^7 neutrons per second per curie of alpha activity.

Positive-ion accelerators can be used to produce neutrons by utilizing the energetic ions to induce nuclear reactions. One of the most useful reactions involves the bombardment of a tritium target by a beam of deuterons. The reaction is

$$^{3}_{1}\mathrm{H} + ^{2}_{1}\mathrm{H} \longrightarrow ^{4}_{2}\mathrm{He} + ^{1}_{0}n + 17.6\ \mathrm{MeV} \tag{9-2}$$

About 14 MeV of the Q of the reaction appears as the kinetic energy of the neutron. Because the potential barrier of the tritium target is low and the projectile is also singly charged, the reaction goes with good yield at deuteron energies of only 200–300 keV. Relatively simple accelerators can be used to produce energetic neutrons with this reaction.

Californium-252 is an artificially produced radioactive nuclide which undergoes spontaneous fission in about 3% of its decays. The half-life of $^{252}_{98}$Cf, 2.6 y, is relatively short and the corresponding high intrinsic specific activity permits the construction of sources that are physically much smaller than those utilizing reactions of the type of Eq. (9-1). ^{252}Cf holds so much promise for investigative and clinical medicine that the United States Atomic Energy Commission has constructed extensive facilities for its production. Figure 9-2 compares the spectrum of the ^{252}Cf fission neutrons with that obtained from a typical (α, n) reaction.

9.03 Thermal Neutron Reactors

Nuclear reactors, operating on the induced fission of ^{235}U, constitute the most common and the most intense sources of neutrons now available. Their importance warrants a separate discussion.

Every nucleus with an A value greater than about 200 can be made to fission, or split, into two large fragments when sufficient excitation energy is added. The most common way of adding the required excitation energy is to bombard the target with neutrons, which can enter a nucleus without encountering the potential barrier that is presented to a charged particle. When a neutron comes under the influence of the nuclear force, the binding energy of the added particle becomes available for excitation and in heavy nuclei the response may be fission rather than the ejection of a single nucleon that is more common in lighter structures. Some nuclei such as ^{235}U have a large cross section for fission by thermal neutrons and are said to be *fissile;* others like ^{238}U require fast neutrons to produce fission and are *fissionable.*

A single reaction covering the fission process cannot be written because the details of the split into the two fragments,usually unsymmetrical, vary from one fission to another. One example of ^{235}U fission is

$$^{235}_{92}\text{U} + {}^{1}_{0}n \longrightarrow {}^{89}_{35}\text{Br} + {}^{144}_{57}\text{La} + 3{}^{1}_{0}n \tag{9-3}$$

Both atomic numbers and mass numbers balance in Eq. (9-3), which represents the situation at the instant of fission. Almost all fission products are radioactive, emitting negatrons until the initial n/p ratio is reduced to a value compatible with nuclear stability. Thus each initial fission product forms the head of a *fission chain.* In the example cited the final result of the fission product decays is

$$^{235}_{92}\text{U} + {}^{1}_{0}n \longrightarrow {}^{89}_{39}\text{Y} + {}^{144}_{60}\text{Nd} + 3{}^{1}_{0}n + 7{}^{0}_{-1}e \tag{9-4}$$

With the proper geometrical arrangement, and if the total amount of ^{235}U exceeds a certain *critical mass,* enough of the neutrons emitted in a fission can be caused to initiate other fissions to sustain a *chain reaction.* Figure 9-3 is a schematic arrangement of a nuclear reactor or pile. Fission takes place in one of the uranium fuel rods and the emitted fast neutrons travel outward, losing energy through collisions with the atoms in a *moderator.* The graphite or other moderator material is chosen for maximum neutron energy absorption with a minimum of loss by capture. When a moderated

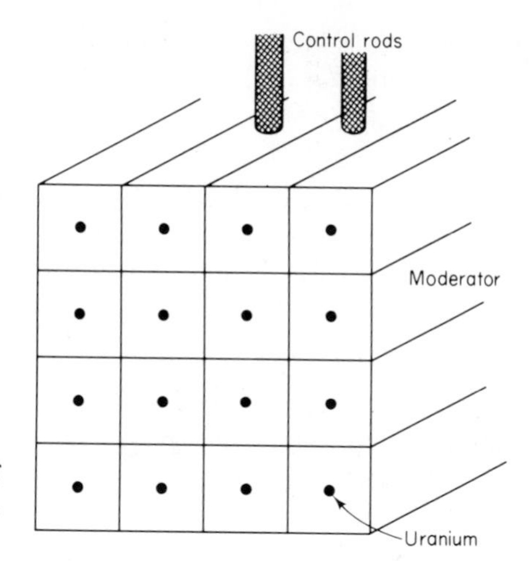

Figure 9-3. A schematic arrangement of fuel, moderator, and control rods in a nuclear reactor.

neutron reaches a second fuel element, it is at thermal energy and capable of producing a fission in another ^{235}U nucleus.

Steady-state operation of a reactor is obtained by adjusting a series of control rods which contain strong neutron absorbers such as boron or cadmium. By withdrawing the rods, some of the excess reactivity can be used to bombard samples placed in the thermal neutron flux.

The fission process itself and the decaying fission products in the fuel elements are intense sources of gamma radiation. All nuclear reactors are operated behind heavy shielding with all adjustments made by remote control.

9.04 Pertinent Definitions

Neutrons are another example of indirectly ionizing radiation. Lacking an electric field, they give up their energy in a series of relatively widely spaced collisions, each involving a substantial transfer of energy. Almost all the energy transferred goes into the release of charged particles. In the second step of the energy transfer process, the charged particles interact strongly with the charges in the matter through which they pass and leave a trail of closely spaced ionized and excited molecules. A few basic definitions applicable to neutron beams are in order before considering in detail the mechanisms of interactions between neutrons and an absorber.

Before they are absorbed, neutrons undergo a series of either elastic or inelastic collisions; hence neutrons will be found moving in all possible directions in any medium, even though they were originally in a well-directed beam. The total number of neutrons that enter a sphere per unit of cross-sectional area is known as the *particle fluence* or simply *fluence*. For a sphere of cross section a,

$$\Phi = \frac{\Delta N}{\Delta a} \tag{9-5}$$

where the Δs indicate that the variables N and a, while small, are still large enough to permit meaningful averages.

Flux density ϕ is the time rate of particle entry per unit area or

$$\phi = \frac{\Delta \Phi}{\Delta t} \tag{9-6}$$

Without further specification, Φ and ϕ apply to the total number of neutrons without regard to their energy distribution. The two basic definitions can of course be restricted and specified to apply to any desired energy or energy range.

Fluence and flux density can be expressed in terms of the *number density* n which is just the total number of neutrons of all energies in unit *volume*. Consider for simplicity a monoenergetic beam of neutrons with a velocity v cm s^{-1}. Then in 1 second the neutrons contained in a volume of v cm^3 will flow across a unit area and in time t seconds

$$\Phi = nvt \tag{9-7}$$

and flux density becomes

$$\phi = nv \tag{9-8}$$

When a neutron fluence is not monoenergetic, the simple relations of Eqs. (9-7) and (9-8) must be replaced by integrals of n over the range of velocities desired.

A beam of monoenergetic neutrons traversing an absorber will be attenuated according to

$$\Phi = \Phi_0 e^{-\mu/\rho(\rho x)} = \Phi_0 e^{-\mu_m(\rho x)} \tag{9-9}$$

As in Eq. (8-6) for photons, μ_m is the mass attenuation coefficient. Here, μ_m is the fraction of the incident neutron fluence that undergoes a collision in unit equivalent density (ρx). By analogy with Eq. (5-1a),

$$\mu_m = \frac{\Delta \Phi / \Phi}{\Delta(\rho x)} \tag{9-10}$$

A mass energy transfer coefficient (μ_K/ρ) can be defined as the product of μ_m (the fraction colliding) and the average fraction of the neutron energy that is given up to charged particles in one collision. Then for neutrons of energy E

$$\Delta E_K = \Phi E \frac{\mu_K}{\rho} \Delta m \tag{9-11}$$

where ΔE_K is the kinetic energy imparted to charged particles in a small mass of absorber Δm. The value of Δm is chosen large enough to permit meaningful averaging but so small that ΦE may be considered to be constant throughout it. Rearranging Eq. (9-11),

$$\frac{\Delta E_K}{\Delta m} = \Phi E \frac{\mu_K}{\rho} = K \tag{9-12}$$

From Eq. (9-12) the *kerma*, K, is defined as the sum of the initial kinetic energies of all the charged particles released by indirectly ionizing radiation in a unit mass of an absorber. Kerma will have the dimensions of an energy per unit mass, ergs g^{-1} or joules g^{-1}.

Any neutron beam, even though originally monoenergetic, will rapidly become heterogeneous because of random collision energy losses. To take care of the actual heterogeneous conditions, kerma is defined generally in terms of an integral of $\Phi E(\mu_K/\rho)$ over the energy range of interest.

9.05 Attenuation Mechanisms

Neutrons with energies greater than thermal are slowed by a series of elastic and inelastic collisions with atomic nuclei. In the first type, both total kinetic energy and momentum are conserved; in an inelastic collision, some kinetic energy is lost to nuclear excitation. A neutron that has escaped capture and radioactive decay will continue to degrade in energy until it equilibrates with the thermal agitation energy of its environment. Collisions will then produce energy fluctuations above and below the mean until the neutron disappears through capture or decay.

When the neutron energy has decreased to 100 eV or so, the probability of capture becomes appreciable. For many nuclei the probability of capture is inversely propertional to the neutron velocity, giving rise to the so-called $1/v$ law. In general, neutron capture becomes more probable as thermal equilibrium is approached.

A number of possibilities follow neutron capture. The compound nucleus will be in an excited state because it has suddenly acquired both the binding energy and the kinetic energy of the absorbed particle. Photon emission, (n, γ), neutron emission, (n, n), charged particle emission, (n, p) or (n, α), and fission, (n, f) are the most probable modes of nuclear de-excitation. Each will have its own cross section and these will combine to form a total cross section for capture, $\sigma_c = \sigma(n, \gamma) + \cdots + \sigma(n, f)$. The total cross section for beam attenuation will include a term for neutron removal by scattering, $\sigma_T = \sigma_c + \sigma_s$. Cross sections are usually expressed in *barns* or in millibarns, Sec. 8.03.

Each cross section is a function of the particle energy and the detailed structure of the absorbing nucleus. There are many resonance levels and the attenuation coefficients are more complicated than are their photon counterparts. The latter vary smoothly and slowly with Z and are independent of the nuclear neutron number N. In $_{48}$Cd the values of σ_c for thermal neutrons are $^{112}Cd = 0.03$ b, $^{113}Cd = 20{,}000$ b, $^{114}Cd = 1.1$ b. The neighboring element $^{113}_{49}In$ has a capture cross section of less than 10 b. Boron enriched ^{10}B is used in neutron detectors and as an absorber because $\sigma_{th}(n, \alpha) = 3840$ b while in the more abundant ^{11}B the total capture cross section is only 5 mb.

Neutron diffraction is a powerful supplement to X-ray diffraction in crystal structure determinations because of the large scattering cross sections of hydrogen and some other low-Z elements, which are almost transparent to photons. Neutron radiography* has become an increasingly popular analytical tool because of its ability to reveal hydrogenous material and isotopic differences which are undetectable with photon beams. At neutron energies of 10 MeV or more, nuclear capture has a low probability but it is not zero. When capture does take place at these energies, the capturing nucleus is raised to such a high level of excitation that it may respond by *spallation*, (n, s). In spallation the nucleus undergoes a complete breakup into a few large groups of nucleons, some small aggregates, and a few individual nucleons. Spallation is to be distinguished from fission where two massive fragments and a few neutrons are formed from the excited nucleus.

In our upper atmosphere there is an appreciable flux density of high-energy neutrons formed by the absorption of protons and other constituents of the cosmic radiation. Some of these neutrons have energies greater than the total binding energy of even the heaviest nucleus and are capable of initiating a complete nuclear disassembly. Photographic emulsions and cloud chamber photographs show many ionization tracks originating at the point of disintegration. These tracks range from the sparsely ionized trails of high-speed mesons to the dense ionizations of multiply charged heavy ions. These cataclysmic events are known as *stars* from the appearance of the ion tracks. Even at high altitudes the rate of star formation is low but it is frequent enough to be considered in planning high-altitude manned missions. It is conceivable that the high ion density in a star might produce serious functional damage to some small, critical biological structure.

9.06 Collision Energy Loss

In an elastic collision the kinetic energy transferred to a stationary nucleus will be a function of the nuclear mass and hence of mass number A. Maximum energy transfer will take place in a head-on collision where the recoil takes place along the original line of flight. In a head-on collision the fraction of the original energy transferred, E_f, is

$$E_f = \frac{4A}{(1 + A)^2} \tag{9-13}$$

E_f has a maximum value of 1.00 when $A = 1$ for H and decreases to 0.27 for

*Berger, H., "Neutron Radiography." *Ann. Rev. Nuc. Sci.*, 21, 335, 1971.

^{12}C and 0.22 for ^{16}O. If we set $A = \frac{1}{1836}$, we find that the maximum energy transferred to an electron is only 0.002, or 0.2%, of the original neutron energy.

Even at neutron energies of only 100 eV or so, enough energy can be transferred to a hydrogen atom to break all chemical bonds. Neutron collisions are, therefore, characterized primarily by the ejection of *recoil protons*, with lesser numbers of heavier recoiling ions.

Energy transfers in neutron collisions are large compared to the amounts involved in charged particle interactions. Relatively few interactions are required to degrade a neutron to thermal energy and so it is classed as an indirectly ionizing particle even though trails of ions are left behind its passage.

In an inelastic collision some kinetic energy is lost as the struck nucleus is raised to an excited state which will be relieved later by gamma-ray emission.

$$^{A}_{Z}\text{X} \longrightarrow {}^{A}_{Z}\text{X}^{*} \longrightarrow {}^{A}_{Z}\text{X} + \gamma \qquad (9\text{-}14)$$

An inelastic collision may be thought of as an example of neutron capture and reemission at a lower energy.

$$^{A}_{Z}\text{X} + {}^{1}_{0}n \longrightarrow {}^{A+1}_{Z}\text{X}^{*} \longrightarrow {}^{A}_{Z}\text{X} + {}^{1}_{0}n + \gamma \qquad (9\text{-}15)$$

9.07 Radiative Capture

Some neutron capture reactions require the addition of several MeV of kinetic energy but others are most probable at low or thermal energies. A good many nuclei obey the $1/v$ law, as seen from the attenuation curves in Fig. 9-1. Boron obeys the law quite precisely at least up to the energy limit of the plot. Indium has a number of strong resonances but in spite of these the general $1/v$ trend is apparent. The broad deep resonance in cadmium obscures all but a small segment of $1/v$ absorption.

A $1/v$ dependence seems reasonable on simple principles. The time a neutron spends in the vicinity of an atomic nucleus is proportional to $1/v$ and the capture probability might well be expected to be proportional to this time.

Radiative capture is the most probable neutron capture reaction. Typically

$$^{31}_{15}\text{P} + {}^{1}_{0}n \longrightarrow {}^{32}_{15}\text{P} + \gamma + Q$$

or, in short,

$$^{31}\text{P}(n, \gamma)^{32}\text{P} \qquad (9\text{-}16)$$

where Q is symbolic of the overall energy requirements of the reaction. This type of capture is usually written, for short, $^{31}\text{P}(n, \gamma)$ and is called an *n–gamma* reaction. In an n, γ reaction, neither the incident nor the ejected particle experiences any hindrance from the nuclear potential barrier. The neutron need have only the kinetic energy of thermal equilibrium but it will still contribute its binding energy to the compound nucleus. Stability is then achieved by the emission of one or more gamma rays.

The stability following gamma-ray emission may be only relative because the product, as in Eq. (9-16), may be radioactive.

$$^{32}_{15}P \longrightarrow ^{32}_{16}S + \beta^- + Q \tag{9-17}$$

The product is an example of an *artificially* produced radioactive nuclide.

9.08 Other Capture Reactions

Somewhat more energy is usually required when proton emission follows neutron capture in an n, p reaction.

$$^{32}_{16}S + ^{1}_{0}n \longrightarrow ^{32}_{15}P + ^{1}_{1}p + Q$$

or, in short,

$$^{32}S(n, p)^{32}P \tag{9-18}$$

In an n, p reaction the proton must escape across the nuclear potential barrier. As we have seen before, escape will be favored as more energy becomes available to the escaping particle and, hence, from the initiating neutron. The product, ^{32}P in this case, will have identical properties whatever the mode of production.

A definite threshold exists for n, α reactions such as

$$^{19}_{9}F + ^{1}_{0}n \longrightarrow ^{16}_{7}N + ^{4}_{2}He + Q \tag{9-19}$$

Now the binding energy of four nucleons is lost while only one is supplied and, in addition, a double charge must cross the potential barrier.

Other capture reactions such as n, $2n$ and n, $3n$ are known. The first of these must have an energy threshold of the order of 8 MeV, a representative binding energy, because two particles leave the nucleus and only one enters. Similarly n, $3n$ reactions will have a threshold of about 16 MeV.

In a few heavy nuclei even a thermal neutron can induce a fission, or n, f reaction. In other nuclei, fission can only be induced by fast neutrons. For elements below lead, fission cross sections are so small that they may be neglected.

9.09 Radioisotope Production by n, γ

The cross section for the radiative capture of thermal neutrons, denoted variously as σ_a (activation), σ_r, or σ_{th}, is the most important component of the total capture cross section. Radiative capture cross sections are usually much larger than those for other neutron-induced reactions. These large values are well suited to utilize the high neutron flux densities found in thermal reactors, sometimes in *thermal columns* specially designed to thermalize neutrons with a minimum of loss by higher energy capture.

Consider a sample consisting of N atoms of a particular nuclear species having a characteristic value of σ_a. Let the sample be small enough to ensure that all portions of the target be exposed to a uniform neutron flux density ϕ cm^{-2} s^{-1}. From the basic definition of a cross section the number of reactions induced in a short time dt will be

$$dR = \phi N \sigma_a \, dt \tag{9-20}$$

If a stable reaction product is formed, it will accumulate at a nearly constant rate until N becomes appreciably depleted.

In general the product of an n, γ reaction is radioactive and in this case the product will begin to decay as soon as it is formed, at a rate proportional to the amount that is present through the decay constant λ. The situation is analogous to the production of a radioactive daughter by an extremely long-lived parent, Eq. (11-8). At the end of an irradiation time t there will be

$$R = \frac{\phi N \sigma_a}{\lambda}(1 - e^{-\lambda t}) \tag{9-21}$$

atoms of the radioactive product.

If the sample is removed from the radiation at time t, production will cease but decay will continue. At any later time t', the activity A, which is usually the quantity of interest, will be

$$A = R\lambda = \phi N \sigma_a (1 - e^{-\lambda t}) e^{-\lambda t'} \text{ dps} \tag{9-22}$$

Some care must be taken in using Eq. (9-22). ϕ will customarily be in units of cm^{-2} s^{-1}, so for consistency all times must be taken in seconds. N must correspond to the number of atoms of the particular species to which σ_a is applicable. Tables will give σ_a in barns, which must be converted to cm^2 before use in Eq. (9-22). With all of these precautions taken, the activity calculated will be in disintegrations per second, dps.

Illustrative Example

A 500-mg sample of chromium is irradiated for 10 days in a nuclear reactor whose flux density is 6×10^{12} thermal neutrons cm^{-2} s^{-1}. Calculate the activity of ^{51}Cr formed by ^{50}Cr(n, γ) at the time the irradiation is terminated. The following information is obtained from various tables of atomic and nuclear properties. Atomic weight of Cr, 52.0; natural abundance of ^{50}Cr, 4.3%; σ_a, 17 b; ^{51}Cr, $T_{1/2}$, 27.8 d.

$$N = (6.02 \times 10^{23}) \times 0.5 \times \frac{0.43}{52} = 2.5 \times 10^{20} \text{ nuclei of } ^{50}\text{Cr}$$

$$\lambda = \frac{0.693}{27.8 \times (8.64 \times 10^4)} = 2.89 \times 10^{-7} \text{ s}^{-1}$$

$$t = 10 \times (8.64 \times 10^4) = 8.64 \times 10^5 \text{ s}$$

$$A = (6 \times 10^{12}) \times (2.5 \times 10^{20}) \times (17 \times 10^{-24}) \times (1 - e^{-0.249})$$

$$= 5.6 \times 10^8 \text{ dps} = \frac{5.6 \times 10^8}{3.7 \times 10^7} = 15.1 \text{ mCi}$$

The total number of ^{51}Cr nuclei produced, obtained precisely by an integration, will be slightly overestimated by considering N in Eq. (9-20) constant, and taking dt as the total irradiation time. Then $R = (6 \times 10^{12}) \times (2.5 \times 10^{20}) \times (17 \times 10^{-24}) \times (8.64 \times 10^{5}) = 2.2 \times 10^{16}$ nuclei. Less than 10^{-4} of the available target nuclei reacted and so depletion was properly neglected in the activity calculation.

9.10 Specific Activity

At the end of an irradiation involving an n, γ reaction, the radioactive product will be mixed with a relatively large amount of unreacted target atoms. Product and target are isotopes of the same element so separation can only be achieved by the slow, expensive method of electromagnetic deflections. The dilute product may be useless for many biological purposes because normal physiological processes may be disturbed by an excess amount of the element in question when enough activity is administered for study.

Illustrative Example

Calculate the amount of chromium that would be administered if 25 μCi of the ^{51}Cr from the previous example was required for a tracer study.

$$\text{specific activity} = \frac{15.1}{0.5} = 30.2 \text{ mCi g}^{-1} \text{ of Cr}$$

$$\text{administered chromium} = \frac{0.025}{30.2} = 8.3 \times 10^{-4} \text{ g}$$

The biologist rather than the physicist must determine whether or not the administration of 0.83 mg of chromium is acceptable.

If the specific activity is too low for use, it may be possible to produce the radioactive material in a reactor with a higher flux density and to irradiate it for a longer period of time. If material prepared with these conditions is still unsatisfactory, recourse must be had to production by a charged particle reaction. In the example above, ^{51}Cr could be made by the proton bombardment of vanadium in a suitable accelerator.

$$^{51}_{23}V + ^{1}_{1}p \longrightarrow ^{51}_{24}Cr + ^{1}_{0}n \tag{9-23}$$

The product is now not isotopic with the target and chemical separation is possible. High specific activity material, sometimes completely carrier free, can be produced with charged particle reactions.

In a few special cases the *Szilard–Chalmers reaction* may be used to obtain high specific activity material from an n, γ reaction. A classic example is the production of ^{128}I by $^{127}I(n, \gamma)$. An organic compound such as ethyl iodide is irradiated with thermal neutrons. When the postcapture gamma ray is emitted, the recoil energy imparted to the ^{128}I nucleus is sufficient to break the chemical bonds and liberate free ^{128}I from the organic molecule. The reaction is

$$C_2H_5\ {}^{127}I + n \longrightarrow C_2H_5 + {}^{128}I \tag{9-24}$$

Chemical methods are available for separating the free I from the organic target molecules.

The Szilard–Chalmers reaction is one example of the fact that in a nuclear reaction involving particle or photon ejection the recoiling nucleus will have sufficient energy to break intermolecular bonds. Because of this it is not possible, in general, to simply activate one atom in a molecule and at the same time retain the original molecular configuration. The radioactive product must be separated from the target and then be synthesized into the desired chemical structure.

9.11 Medical Uses of Neutrons

As each new type of radiation became available, it was considered for possible advantages over those previously in use. Neutrons appeared to offer some advantages over photons for the destruction of malignant tissues because of the dense ionizations produced by the recoil protons. Early attempts at neutron therapy were not encouraging because of the extensive damage produced in normal tissues. New techniques and improved dosimetry may lead to more favorable results from neutron irradiation.

Attempts have been made to enhance the effects of neutrons and to localize them in a desired tissue by the preradiation administration of elements with particular nuclear properties. Boron is an element which has been studied in detail. Natural boron contains ^{10}B in an abundance of nearly 20% and this nucleus has a σ_c for (n, α) of over 3800 b. If a boron-containing compound could be found that would concentrate in a malignancy, subsequent neutron irradiation would induce

$${}^{10}_{5}B + {}^{1}_{0}n \longrightarrow {}^{4}_{2}He + {}^{7}_{3}Li + Q \tag{9-25}$$

in good yield. Over 7 MeV of energy is released in this reaction and this will be dissipated in the immediate vicinity by the heavily ionizing alpha particle and the recoiling ^{7}Li. Unfortunately sufficiently high boron concentrations have not been attained and boron-n therapy has not had the success originally hoped for.

Some diagnostic uses of neutrons have been made by the activation of radioactive isotopes in living tissues. Under proper circumstances some elements can be made sufficiently active for radioactive assay without excessive doses of radiation to the patient.

REFERENCES

Auxier, J. A., W. S. Snyder, and T. D. Jones, "Neutron Interactions and Penetrations in Tissue," *Radiation Dosimetry*, 2nd ed., Vol. I. F. H. Attix and W. C. Roesch, ed. Academic Press, New York, N.Y., 1968.

Halpern, I., "Nuclear Fission." *Ann. Rev. Nuc. Sci.*, **9**, 245, 1959.

Lane, J. A., "Economics of Nuclear Power." *Ann. Rev. Nuc. Sci.*, **16**, 345, 1966.

Measurement of Neutron Flux and Spectra for Physical and Biological Applications. NCRP Report No. 23, National Council on Radiation Protection and Measurements, Washington, D.C., 1960.

Neutron Fluence, Neutron Spectra and Kerma. Report No. 13 of ICRU, Washington, D.C., Sept. 15, 1969.

Protection Against Neutron Radiation. NCRP Report No. 38, National Council on Radiation Protection and Measurements, Washington, D.C., Jan. 4, 1971.

10

Dosimetry of Ionizing Radiations

10.01 Erythema Dose

X rays were used in diagnostic and therapeutic medicine for several years before a satisfactory unit of radiation dosage was established and instruments were developed to measure it. Prior to that time a number of poorly understood effects had been used for dose estimation. A variety of chemical compounds showed color changes upon exposure and several of these chemical systems were adapted to radiation dosimetry. Radiation effects on photographic emulsions were known but the need for carefully controlled processing was not fully appreciated.

Physicians noted early the delayed reddening of the human skin at the point of entry of an X ray beam and this effect was utilized as a biological dosimeter. *Skin erythema dose* or SED was defined as that amount of radiation (usually X rays or gamma rays from ^{226}Ra) that would just produce a detectable reddening of the human skin.

Maximum erythema does not develop for a week or more unless the exposure is excessive, an obvious disadvantage in a basic unit of dosage. Humans vary tremendously in their response to the ultraviolet components of summer sunlight and similar variations occur in the response to more energetic photons. The skin response is primarily a function of the amount of soft, easily absorbed radiation with a lesser dependence on the more penetrating components which expend their energy over greater volumes of tissue. Equal erythematous responses may be accompanied by quite different doses at depth, the relative values depending on the quality of the radiation.

10.02 The Roentgen

In 1928, the International Commission on Radiological Protection developed and recommended the use of a unit named the roentgen, originally abbreviated lowercase r but now denoted by the capital R. The roentgen unit was defined in terms of the ionization produced in a standard volume of air and this definition has been retained. Refinements in concepts and nomenclature have taken place and today there is a sharp distinction between the terms *exposure* and *dose*, previously used interchangeably. The roentgen is a unit of exposure produced by the absorption of photons in air. Other units must be used for other radiations and for other absorbing substances.

The ICRU defines exposure X as the quotient

$$X = \frac{\Delta Q}{\Delta m} \tag{10-1}$$

where ΔQ is the sum of all the electric charges of either sign produced in air when all the electrons released by photons from an air mass Δm are absorbed in air. The Δ's signify that m is chosen large enough to permit meaningful averaging and yet so small that only a negligible fraction of the photon fluence is absorbed in it. The unit of exposure is the *roentgen*, R.

$$1 \text{ R} = 2.58 \times 10^{-4} \text{ coulomb kg}^{-1} \text{ released} \tag{10-2}$$

This unit is numerically identical with an earlier definition in which 1 R = 1 esu of charge per 0.001293 of air (1 cm^3 under standard conditions).

Consider a small volume of air of mass Δm in the path of a photon beam, Fig. 10-1A. Photons passing through this volume will give up some of their energy to the air, the fraction deposited being determined by the value of the mass energy transfer coefficient $(\mu_\kappa/\rho)_{\text{air}}$. If the energy flow or fluence is ψ cm^{-2}, the energy imparted to the mass Δm will be

$$\Delta\epsilon = \psi \left(\frac{\mu_\kappa}{\rho}\right)_{\text{air}} \times \Delta m \tag{10-3}$$

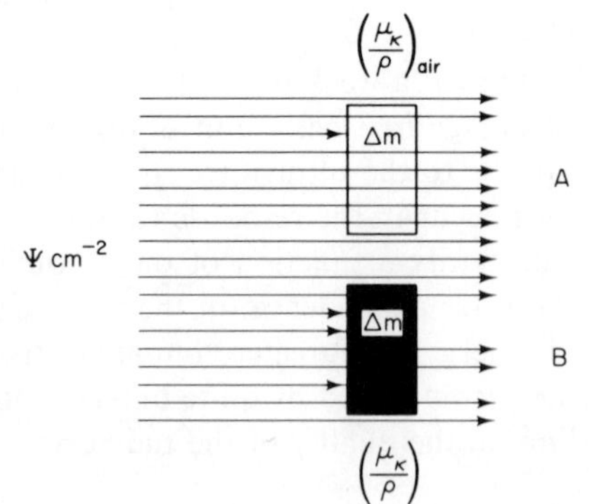

Figure 10-1. (A) A mass of air and (B) some other substance exposed to a uniform photon fluence.

The number of ions produced will be proportional to ΔQ so an electrical measurement will serve to determine it.

Some of the electrons released in Δm may leave the volume but if this is surrounded by enough air, each of these electrons will lose all its energy before striking any other material. Under this condition the energy lost from the volume will be just compensated for by electrons that originate outside it and deposit energy as they pass through.

A measurement made under these conditions will characterize the ability of a photon beam to ionize a standard substance, air. This is the significance of *exposure*, the only quantity that is expressed in roentgens. The approximate energy transfer that is involved in an exposure of 1 R is readily calculated.

Illustrative Example

Calculate the energy transferred to 1 g of air by an exposure of 1 R.

From the definition, there will be 2.58×10^{-7} coulomb of charge released in 1 g.

$$\text{number of ion pairs } N = \frac{2.58 \times 10^{-7}}{1.6 \times 10^{-19}} = 1.61 \times 10^{12}\ \text{g}^{-1}$$

$$\begin{aligned}\text{energy absorbed} &= WN = (1.61 \times 10^{-12}) \times 33.73 = 5.44 \times 10^{13}\ \text{eV g}^{-1}\\ &= (5.44 \times 10^{13}) \times (1.6 \times 10^{-12}) = 87.0\ \text{ergs g}^{-1}\end{aligned}$$

Note that the value calculated depends on the experimentally determined value of W and hence it cannot be considered to be a definition of the roentgen unit.

10.03 Energy Limitations

The mass energy transfer coefficient for air is a function of photon energy and the energy fluence required to produce an exposure of 1 R will vary accordingly, Fig. 10-2. At each photon energy, a *number fluence* Φ can be related to the energy fluence by $\Phi \times h\nu = \psi$. Figure 10-3 is a plot of the number fluence required to produce an exposure of 1 R.

Note that the basic definition of the roentgen does not specify any time over which the fluence must be received. Time rates are of considerable practical importance and extensions to dR/dt, $d\psi/dt$, and $d\Phi/dt$ are readily made, using time units convenient to the problem at hand. These quantities are known as exposure rate, energy fluence or energy flux density rate, and fluence rate or flux density, respectively.

Practical difficulties of measurement intervene to limit the energy range over which the definition of the roentgen can be fulfilled. Measurements are made with a *standard air chamber* which must be very carefully constructed

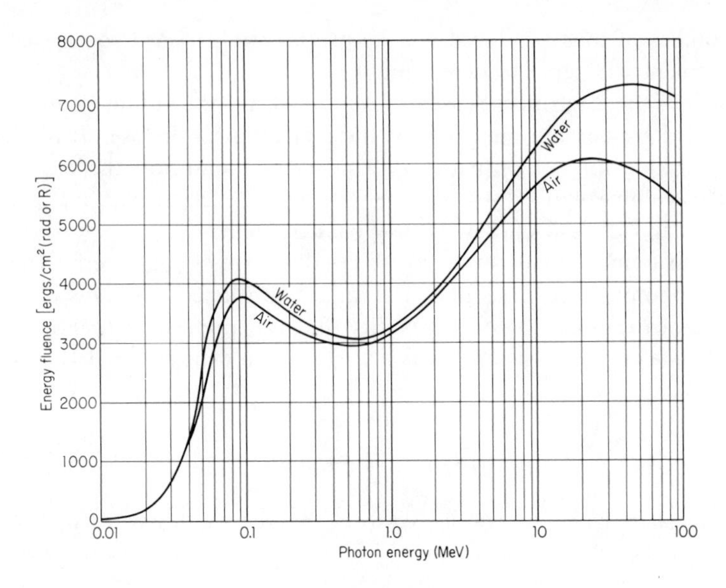

Figure 10-2. Photon energy fluence required to produce 1 R (air) or 1 rad (water).

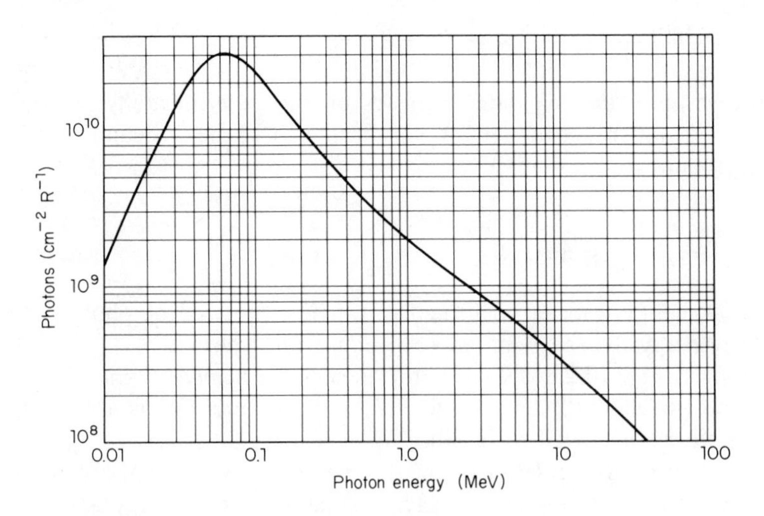

Figure 10-3. Photon number fluence required to produce 1 R.

and operated to ensure ion collection from a precisely defined air volume under the conditions specified in the definition. At the low-energy end of the spectrum (40 keV), photon absorption coefficients are becoming large and absorption of the beam is appreciable as it passes through the collimators and traverses the measuring volume. Corrections can be applied but they become undesirably large as photon energies decrease.

As photon energies increase into the MeV range, the ratio photon mean free path/electron path length decreases. Values of this ratio for air are about the same as those shown for water in Fig. 8-10. Above 2–3 MeV, an air chamber made large enough to contain the electron ranges will be very unwieldly and will introduce undesirably large attenuations into the photon beam. At the higher energies the proper measurement of exposure becomes difficult or impossible and the concept is abandoned in favor of that of energy locally absorbed.

10.04 Absorbed Dose

Quantities and units not restricted to the absorption of photons in air are obviously required. When a substance other than air is put into the photon beam shown in Fig. 10-1, it will be exposed to the same energy fluence that passed through the air volume. In general the mass energy transfer coefficient of the new material will not be equal to that of air. Each of the three main absorption coefficients are functions of atomic number and of photon energy. If by chance the new mass energy transfer coefficient equaled that of air at one particular energy, the equality would not be maintained at other energies unless the two atomic compositions were identical. The energy deposited in a mass Δm of the new material will be

$$\Delta\epsilon = \psi\left(\frac{\mu_\kappa}{\rho}\right)_m \Delta m \tag{10-4}$$

For a heterogeneous beam the total energy absorbed will be given by an integral of this expression over the energy range of the photons. A comparable integration is of course required in the case of air absorption.

A quantity known as *absorbed dose* or, more simply, *dose* can be defined in terms of the energy imparted by ionizing radiation to a mass element Δm.

$$D = \frac{\Delta E_D}{\Delta m} \tag{10-5}$$

where the Δ's have their previous significance. The unit of dose is the rad.

$$1 \text{ rad} = 0.01 \text{ J kg}^{-1} = 100 \text{ ergs g}^{-1} \tag{10-6}$$

The rad can be applied to any substance, including air, and is not restricted to a particular kind of radiation.

Only the energy locally deposited in Δm is to be included in the calculation of dose. Energy expended in bremsstrahlung production and the energy of Compton photons is excluded. Kerma on the other hand is defined as

$$K = \frac{\Delta E_\kappa}{\Delta m} \tag{10-7}$$

where ΔE_κ is the sum of the *initial* kinetic energies of all the charged particles liberated in mass element Δm. Some of these particles may subsequently produce bremsstrahlung which will be absorbed outside Δm. At high energies there may be an appreciable difference between ΔE_D and ΔE_K. At energies of a few MeV and below, the two will be practically equal. Note, however, the difference in units between D and K; although they are dimensionally identical, one is expressed in rads and the other in ergs g^{-1} or in J kg^{-1}. To fulfill the definition strictly, dose must be determined from a measurement of absorbed energy; but temperature increases are so small in the usual dose range that the precision calorimetry required is scarcely suited to routine use.

Illustrative Example

Calculate the temperature rise expected in a calorimeter receiving a dose of 1000 rads, assuming a specific heat of 1.00 cal g^{-1} (essentially water).

$$10^3 \text{ rads} = 10^5 \text{ ergs g}^{-1} = 10^{-2} \text{ J g}^{-1} = \frac{10^{-2}}{4.18} = 2.39 \times 10^{-3} \text{ cal g}^{-1}$$

$$\text{temperature rise} = \frac{2.39 \times 10^{-3}}{1.00} = 0.00239°\text{C}$$

Absorbed dose is routinely determined from a measurement of exposure (ionization) and a knowledge of the two mass energy transfer coefficients, Eq. (10-3) and (10-4).

Figure 10-4 shows the absorbed dose in rads per roentgen exposure for some materials of biological interest. The high values for bone arise from the large photoelectric absorption of Ca and P at low energies.

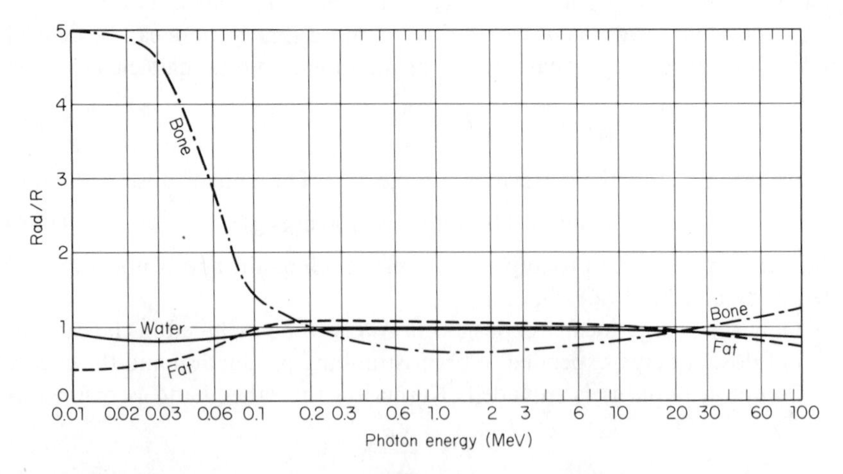

Figure 10-4. Exposure–dose relations for some materials of biological interest.

10.05 Air-Equivalent Ion Chambers

Standard air chambers are essentially laboratory devices scarcely suited to routine exposure measurements. These are usually made with *air-equivalent* chambers. The confining walls of a chamber must be made of a rigid solid to sharply define the enclosed gas volume. The inner wall surface must be an electrical conductor and a second conducting electrode must be present to complete the electrical circuit. Air-equivalence is approached by making the chamber out of materials whose atomic composition matches that of air as closely as possible.

Figure 10-5A shows in cross section an air-equivalent *thimble chamber*, named from the characteristic shape of the enclosed air volume. Figure 10-5B is a greatly exaggerated cross section of the wall of a chamber designed to be used with photons of maximum energy $h\nu$. Photons interacting with air before entering the chamber wall will produce an electron density and

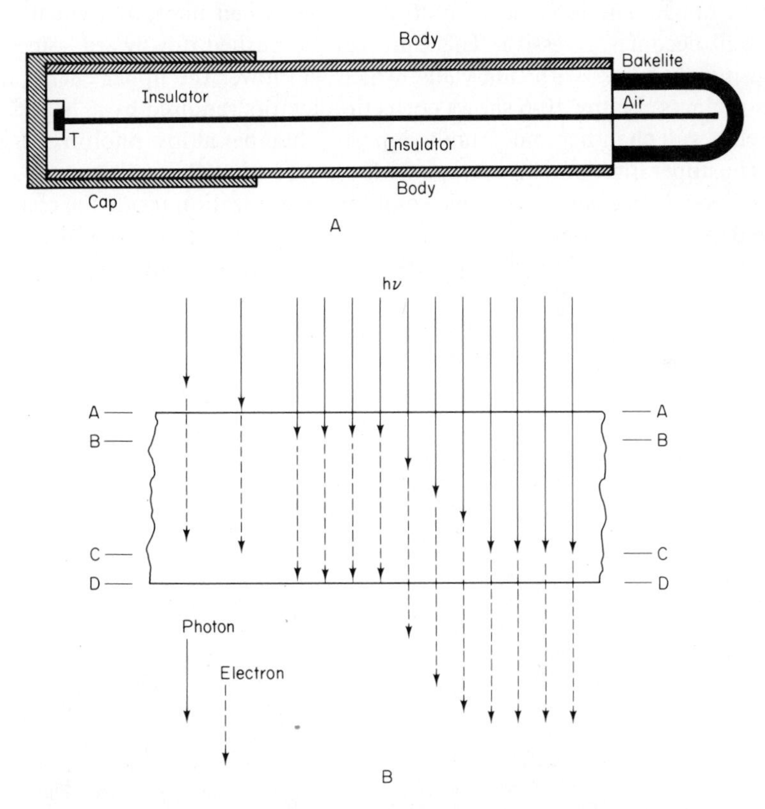

Figure 10-5. (A) Cross section of an air-equivalent thimble chamber. (B) Photon and electron paths in the chamber wall.

an electron spectrum that is characteristic of air only. However, the wall thickness is made sufficient to ensure that the most energetic electron, of energy *hv*, cannot penetrate to produce ionization in the chamber air. Below depth *C*, Fig. 10-5B, every free electron has been released in the wall.

Wall absorption takes place in a solid of density about 1.0 g cm^{-3} instead of in air of 0.001293 g cm^{-3} so there will be an increasing density of free electrons down to depth *C*. The energy spectrum of the electrons released in the wall will also differ from that produced by air absorption even though the wall material is made as air-equivalent as possible. The electron density and the energy spectrum below *C* are entirely characteristic of the wall material and are constant, except for a negligibly small decrease due to photon attenuation. There is *electronic equilibrium* between the photons and the released electrons. Ionization in the chamber gas is produced only by wall electrons which bear a constant relation to the photon fluence.

An air-equivalent chamber designed as described can be calibrated against a standard air chamber to read exposure in roentgens over a considerable energy range. The low energy limit is reached when photon attenuation in the wall becomes excessive. This limit can be pushed downward somewhat by using chambers with thin walls, which will lower the upper energy limit of usefulness. Figure 10-6 shows correction factors required by a "standard" Bakelite wall chamber and a thin nylon wall chamber at low photon energies.

The upper limit of correct response is reached when the walls are no longer thick enough to ensure electronic equilibrium. Ionization inside the chamber then depends in a complex way on both wall and outside air ionizations and the previous calibration will no longer hold. The upper limit can be extended

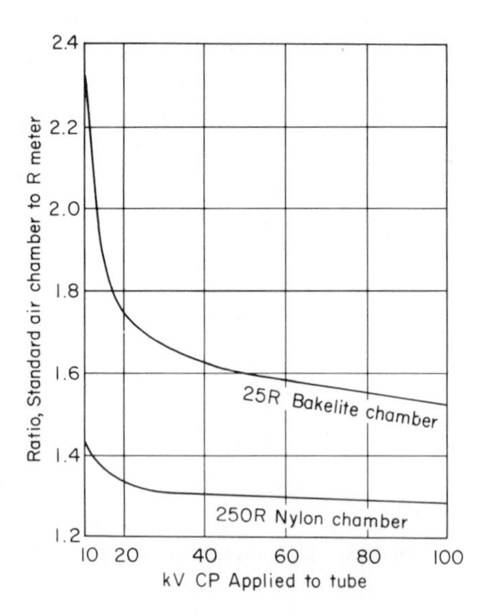

Figure 10-6. Corrections required for ion chambers exposed to the bremsstrahlung from a beryllium window X-ray tube with no added filtration. (Data from F. H. Day, *N.B.S. Journal of Research*, **41**, 295, 1948.)

by placing an auxiliary wall or cap of an air-equivalent material (usually a unit-density plastic) over the chamber. With proper caps the the range of a chamber may be raised from perhaps 400 keV to 1 MeV or even more.

10.06 Specific Gamma-Ray Constant

None of the definitions above are restricted to any particular type of radiation source and so far the discussion has been entirely general. A particularly useful quantity relates the activity of a gamma-ray source to the exposure produced by it. This quantity is known as the *specific gamma-ray constant*, Γ.

For simplicity, consider a 1-Ci source which emits only one gamma-ray energy, E MeV. Some of the decays may go directly to the ground state without gamma emission so in general the number of gammas emitted will be some fraction f of the number of decays. The rate at which photon energy flows out from the 1-Ci source is $3.7 \times 10^{10} fE$ MeV s^{-1}. The gamma-ray emissions are spatially isotropic so the energy fluence is spread uniformly over a sperical surface through the point of interest. The value of Γ is usually calculated for a distance of 100 cm from a point source, although a distance of 1 cm is sometimes used. Photon absorption along the air path from the source to the point in question is assumed to be negligible.

At 100 cm the fluence rate from the 1-Ci source is

$$\frac{d\psi}{dt} = \frac{3.7 \times 10^{10} fE}{4\pi \times 10^4} = 2.94 \times 10^5 fE \quad \text{MeV cm}^{-2}\,\text{s}^{-1} \tag{10-8}$$

and from Eq. (10-3) the rate of energy deposition at 100 cm is

$$\frac{d\epsilon}{dt} = 2.94 \times 10^5 fE \left(\frac{\mu_\kappa}{\rho}\right)_{\text{air}} \quad \text{MeV g}^{-1}\,\text{s}^{-1} \tag{10-9}$$

For convenience the units can be converted to ergs g^{-1} h^{-1} while E is retained in MeV. Then

$$\frac{d\epsilon}{dt} = 1.69 \times 10^3 fE \left(\frac{\mu_\kappa}{\rho}\right)_{\text{air}} \quad \text{ergs g}^{-1}\,\text{h}^{-1} \tag{10-10}$$

Since the absorption of 87.0 ergs g^{-1} of air is equivalent to an exposure of 1 R, we have finally

$$\Gamma = 19.4 fE \left(\frac{\mu_\kappa}{\rho}\right)_{\text{air}} \quad \text{R Ci}^{-1}\,\text{h}^{-1} \text{ at 1 m or Rhm (pronounced } \textit{rum}\text{)} \tag{10-11}$$

When more than one gamma-ray energy is involved, the specific gamma-ray constant will be the sum of terms, one for each value of E.

$$\Gamma = 19.4 \,\Sigma fE \left(\frac{\mu_\kappa}{\rho}\right)_{\text{air}} \quad \text{Rhm} \tag{10-12}$$

Thus Γ can be calculated from a knowledge of the decay scheme of the isotope and the mass energy transfer coefficients.

Illustrative Example

Calculate the specific gamma-ray constant for ^{131}I using the decay scheme shown in Fig. 10-7 and mass energy transfer coefficients from NSRDS-NBS 29.* The data from the two sources can be tabulated.

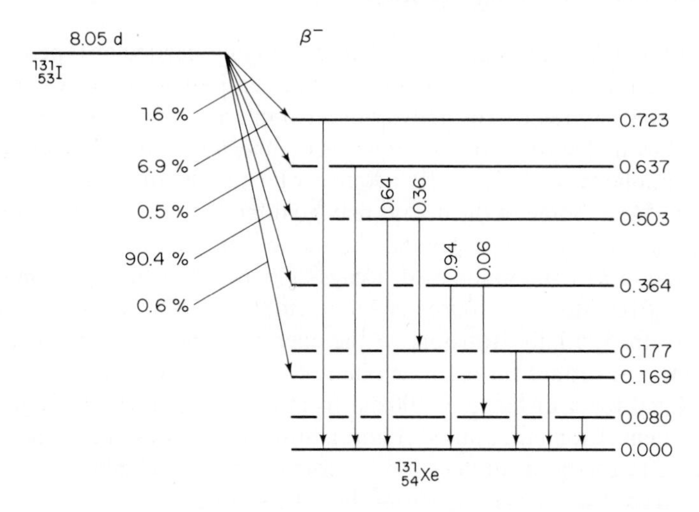

Figure 10-7. An abridged decay scheme for ^{131}I. The figure accompanying each gamma-ray transition is the fraction of the decays that start at that particular energy level and go by that particular decay channel.

E (MeV)	f	(μ_κ/ρ)	*Product*
0.723	0.016	0.292	0.00034
0.637	0.069	0.0295	0.00130
0.503	0.005 × 0.64	0.0297	0.00005
0.326	0.005 × 0.36	0.0289	0.00002
0.364	0.904 × 0.94	0.0292	0.00885
0.284	0.904 × 0.06	0.0287	0.00044
0.177	0.005 × 0.36	0.258	0.00001
0.164	0.006	0.0256	0.00002
0.800	0.904 × 0.06	0.0243	0.00010
		$\Sigma fE(\mu_\kappa/\rho)_{air}$ =	0.01113

$$\Gamma = 19.4 \times 0.01113 = 0.216 \quad \text{Rhm}$$

Note that two gamma rays with the same percent occurrence must be calculated separately and not lumped together in a single calculation with a summed energy. Only the main features of the decay scheme are shown in

**Photon Cross Section, Attenuation Coefficients, and Energy Absorption Coefficients From* 10 *keV to* 100 *GeV*. NSRDS-NBS 29 National Bureau of Standards, August 1969.

Fig. 10-7. A complete calculation would include the contributions due to X rays and any losses due to gamma-ray conversions. In most cases these refinements have only a slight effect on the value of Γ.*

10.07 Dose from Internal Emitters

The development of methods for producing radioactive isotopes multiplied many fold the number of nuclides of biological interest and importance. Research programs and the subsequent applications to clinical medicine require the administration of radioactive materials to animals and man. Benefits derived from the use of these materials must be obtained with a minimized risk of radiation injury to the patient. The proper use of radioactive isotopes require a knowledge of the radiation doses delivered by them.

Several situations require dosimetric calculations on sources internal to the body. In tracer studies with animals the activity administered must be adequate to ensure satisfactory assays on the biological samples and yet so small that radiation does not alter the ongoing physiological activity. In man the need for possible long-term effects of even small doses must be weighed against the need for diagnostic information about a suspected abnormality. Careful calculations are required for the therapeutic uses of internal emitters to ensure that an adequate but not excessive dose is delivered to the desired tissues.

In general only approximate dose calculations can be made. The radiation dose received from an internal emitter depends on the spatial and temporal distribution of the isotope from the time of its administration to its final elimination. If this distribution were known in detail, the need for a tracer study would disappear. The decay scheme of almost every nuclide of interest is now known rather precisely; this knowledge is not always matched by biological information. Fortunately high precision is seldom needed and in almost every case adequate dosimetric calculations can be made.

Each nuclide entering into a living organism is metabolized and eliminated in its own characteristic fashion. Excretion acts to lower tissue concentrations quite independently from the reductions due to radioactive decay. Excretion is usually a complex function of concentration and time and some approximation is required to make the calculations mathematically manageable.

One generally satisfactory approximation assumes that excretion proceeds at a rate that is proportional to the concentration of the active substance. This not unreasonable assumption leads to an exponential law of biological elimination similar to that governing physical decay, with a biological half-life T_b and a biological decay constant λ_b. With both exponential

*Dillman, L. T., *J. Nuc. Med.*, Supp 2, **10**, March, 1969.

factors operating, a tissue concentration will decrease according to

$$C = C_0 e^{-\lambda_p t} \times e^{-\lambda_b t} = C_0 e^{-\lambda_e t} \tag{10-13}$$

where the *effective decay constant* $\lambda_e = \lambda_p + \lambda_b$. The corresponding half-lives are related by

$$T_e = \frac{T_p \times T_b}{T_p + T_b} \tag{10-14}$$

The physical half-life T_p is the immutable characteristic of the nuclide; T_b, which must be determined from experimental data, can be expected to vary with the physical and chemical form, the route of administration, and the animal species.

10.08 Tissue Dose from Beta Particles

Let a tissue contain at zero time a concentration of C_0 microcuries per gram of an isotope emitting beta particles with an *average* energy of $\bar{E}_\beta$ MeV. The calculation of tissue dose received from beta particles is based on the assumption that the energy imparted to a tissue is equal to the beta-particle energy released in it. Most beta particles will have a range of only a few millimeters in tissue and so the assumption is quite good for most body organs and tissue masses. It may be unsatisfactory for a small volume located in a region where there are strong concentration gradients of the active material.

At zero time the rate of energy deposition in the tissue is $3.7 \times 10^4 C_0 \bar{E}_\beta$ MeV g^{-1} s^{-1}. With an effective decay constant λ_e the energy deposition at any later time t will be

$$dE = 3.7 \times 10^4 C_0 \bar{E}_\beta e^{-\lambda_e t}\, dt \quad \text{MeV g}^{-1} \tag{10-15}$$

The total energy received from time 0 to t is obtained by integrating Eq. (10-15) between these time limits. This gives

$$\begin{aligned} E &= 3.7 \times 10^4 \frac{C_0 \bar{E}_\beta}{\lambda_e}(1 - e^{-\lambda_e t}) \quad \text{MeV g}^{-1} \\ &= 5.92 \times 10^{-2} \frac{C_0 \bar{E}_\beta}{\lambda_e}(1 - e^{-\lambda_e t}) \quad \text{ergs g}^{-1} \end{aligned} \tag{10-16}$$

From the definition of the rad, the tissue dose is

$$D_\beta = 5.92 \times 10^{-4} \frac{C_0 \bar{E}_\beta}{\lambda_e}(1 - e^{-\lambda_e t}) \quad \text{rads} \tag{10-17}$$

It is usually more convenient to express the decay by the half-life and to use the hour as the unit of time. With these changes,

$$D_\beta = 3.08 C_0 \bar{E}_\beta T_e (1 - e^{-0.693t/T_e}) \quad \text{rads} \tag{10-18}$$

These expressions are valid only to the extent that the assumptions made in deriving them are satisfied. Accurate values of $\bar{E}_\beta$ are available from a number of sources but C_0 and T_e must be determined from biological data which may not apply exactly to the situation in hand.

Illustrative Example

Calculate the maximum dose due to beta particles from the complete decay and the dose delivered in the first 24 h to the thyroid gland of a patient given 15 μCi of ^{131}I. The thyroid gland is estimated to weigh 20 g and to have a prompt uptake of 40% of the administered iodine.

$$C_0 = \frac{15 \times 0.40}{20} = 0.3\ \mu\text{Ci g}^{-1}$$

$$T_p = 8.05\ \text{d} \qquad T_e\ (\text{estimated from previous data}) = 8.0\ \text{d}$$

$$\bar{E}_\beta\ (\text{from tables}) = 0.183\ \text{MeV}$$

$$D_{\beta\infty} = 3.08 \times 0.3 \times 0.183 \times 8.0 \times 24 = 32.4\ \text{rads}$$

When $t = 24\ \text{h} = 1\text{d}$,

$$(1 - e^{-0.693 \times 1/8}) = (1 - 0.917) = 0.083$$

$$D_{\beta 1} = 32.4 \times 0.083 = 2.68\ \text{rads}$$

10.09 Tissue Dose from Internal Gamma-Ray Emitters

The simplifying assumption that energy released equals energy imparted cannot be made for the calculation of gamma-ray dosage. A gamma-ray emitter in any part of the body will deliver some dose to all parts of the body. Geometrical considerations reduce the energy fluence from a distant emitter by the inverse-square law and an additional attenuation comes from photon absorption in intervening tissues. The latter factor is a complicated function of the photon energy and the composition of the tissues.

Gamma-ray dosimetry can be handled by the method of *absorbed dose fraction* described by Brownell, Ellett, and Reddy.* Absorbed dose fraction ϕ is defined as

$$\phi = \frac{\text{photon energy absorbed in target}}{\text{photon energy emitted by source}} \tag{10-19}$$

Values of ϕ for a "standard man" have been calculated by Snyder, et. al† using ellipsoids and other mathematically manageable shapes to approximate the human body.

*Brownell, G. L., W. H. Ellett, and A. R. Reddy, *J. Nuc. Med.*, Supp. 1, February, 1968.

†Snyder, W. S., M. R. Ford, G. G. Warner, and H. L. Fisher, Jr., *J. Nuc. Med.*, Supp. 3, **10**, August, 1969.

Photon energy emitted is calculated from a factor Δ which gives for each gamma ray the total energy in units of gram rads per microcurie hour of decay. Assume an activity of 1 μCi emitting a gamma ray of energy E MeV with a fractional occurrence f, and imagine the activity to continue for 1 hr. In 1 hr the total energy emitted will be $(3.7 \times 10^4) \times 3600fE$ MeV or $(3.7 \times 10^4) \times 3600 \times (1.6 \times 10^{-6})fE = 213fE$ ergs $\mu\text{Ci}^{-1}\ \text{h}^{-1}$. Since 1 rad = 100 ergs g^{-1},

$$\Delta = 2.13fE \quad \text{gram rads } \mu\text{Ci}^{-1}\ \text{h}^{-1} \tag{10-20}$$

A Δ value can be calculated for each gamma-ray energy; these values are then summed and multiplied by the pertinent number of microcurie hours to obtain the total photon energy emitted. The number of microcurie hours of a decaying isotope is calculated from the integral

$$\mu\text{Ci h g}^{-1} = \int_0^t C_0 e^{-\lambda_e t}\, dt$$

where the time must be taken in hours and λ_e in hours^{-1}. The result of the integration is

$$\mu\text{Ci h g}^{-1} = \frac{C_0}{\lambda_e}(1 - e^{-\lambda_e t}) = 1.44 C_0 T_e (1 - e^{-\lambda_e t}) \tag{10-21}$$

Illustrative Example

Calculate gamma-ray dose to the thyroid gland arising from the isotope within it, using the data of the previous illustrative example. The absorbed dose fraction interpolated from Snyder, et al., p. 43, for this geometrical configuration is 0.0312. From the previous calculation of Γ for ^{131}I it appears that only three gamma rays will make an appreciable contribution to the dose.

$$\Delta_1 = 2.13 \times 0.069 \times 0.637 = 0.093$$
$$\Delta_2 = 2.13 \times 0.904 \times 0.94 \times 0.364 = 0.657$$
$$\Delta_3 = 2.13 \times 0.904 \times 0.06 \times 0.284 = 0.032$$
$$\text{Total} = 0.782 \text{ g rad } \mu\text{Ci}^{-1}\ \text{h}^{-1}$$

For the complete decay of the isotope, $t = \infty$ and

$$\mu\text{Ci h g}^{-1} = 1.44 \times 0.3 \times 8.0 \times 24 = 83.0$$
$$\text{energy emitted} = 0.782 \times 83.0 = 65 \text{ g rads g}^{-1}$$
$$D_{\gamma\infty} = 65 \times 0.0312 = 2.0 \text{ rads}$$
$$D_{\gamma 1} = 2.0 \times 0.083 = 0.16 \text{ rads}$$

The total radiation dose received by the thyroid will be the sum of the beta and the gamma contributions, 34.4 and 2.84 rads, respectively. Gamma-ray contributions from isotope located outside the thyroid have been neglected. Much of this material will be at such a distance from the point of interest that it will not add significantly to the totals. For example, with ^{131}I the absorbed dose fraction, bladder to thyroid, is only about 10^{-5}.

10.10 Dose Equivalence

Exposure and dose are physical concepts only, without reference to the production of any effects beyond ionization and energy degradation. In principle at least a direct physical measurement will permit an expression of radiation quantity in either roentgens or rads. When biological systems are irradiated, a simple measure of ionization or even energy absorption will seldom suffice to characterize the radiation effect.

The biological effectiveness of a given quantity of radiation will depend on the linear energy transfer and other factors such as dose rate may be involved. When biological systems as large as the human body are irradiated with external beams, there will also be an unavoidable variation of dose with depth. In water the maximum value of HVL for energy transfer from photons is about 55 cm, and in the region of 1 MeV where most irradiations are carried out the value is only about one-half of that. When irradiations are carried out close to a source, geometrical requirements will unavoidably reduce the energy fluence with the depth of penetration. These spatial effects can be reduced by using multiple entrance ports but even so the dose distribution pattern may have significant variability.

A term *relative biological effectiveness* or RBE has been used as a factor to relate the effectiveness of any radiation source to that of a standard source. Custom has usually accepted the standard source as an X-ray beam generated at about 200 kVP and hardened by a moderate Cu–Al filter. Absorption coefficients do not change rapidly with energy in this region and small variations in the source will not introduce serious errors. RBE is formally defined as

$$\text{RBE} = \frac{\text{dose of standard radiation}}{\text{dose of radiation in question}} \tag{10-22}$$

for equal biological effects. A term *RBE dose* has also been used for the quantity (RBE $\times$ absorbed dose of the radiation in question).

No particular biological effect is given in the definition, so the effect used must be specified with each value of RBE. Note that relative doses for equal biological effects must be used in determining RBE. It must not be calculated from relative effects at equal doses.

The ICRU recommended in 1964 that the term RBE be reserved for use in radiation biology only. In the field of radiation health protection they suggested a new term, *dose equivalent*, or DE. Dose equivalent is obtained by multiplying the absorbed dose D by a number of modifying factors.

$$\text{DE} = \text{D} \times \text{QF} \times \text{DF} \times \ldots \text{ rem} \tag{10-23}$$

QF is a quality factor which takes account of differences in linear energy transfer and DF is a *distribution factor* relating spatial variations in dose. Provision is made for introducing other factors if they seem indicated. The

ICRP has recommended a set of QF values to be applied to radiations with different values of LET, Table 10-1.*

TABLE 10-1
ICRP Recommended values of QF

L_∞ (water), keV μm^{-1}	QF
3.5	1
7	2
23	5
53	10
175	20

DF values depend on the penetrating ability of the primary radiation and on the size of the test object.

Dose equivalents are expressed in *rem*. From the definition it is obvious that a dose in rem cannot be determined by physical measurements only, but must include factors of biological effectiveness that must be experimentally determined on living systems. These factors are not well established for all types of radiation, all energies, and all biological effects. Regulations governing the use of radiation sources are usually put in terms of rem units along with the currently accepted values for converting from physically measured doses.

10.11 Neutron Flux Density Measurements

Determinations of neutron flux densities and the tissue doses resulting from them are far more complicated than are the corresponding measurements and calculations for photons. Only a brief outline of the methods can be given here. Special sources must be consulted for details†.

Neutrons with kinetic energies of 1 keV and above can be measured from the proton recoils which they produce in an ion chamber filled with a hydrogenous counting gas such as methane, CH_4. Collisions transfer kinetic energy effectively from neutrons to protons, Sec. 9.06, and the ionization subsequently produced by the recoiling proton will be proportional to the

***Recommendations of the International Commission on Radiological Protection*, 1965. ICRP Publication 9, Pergamon Press, Oxford.

†*Neutron Fluence, Neutron Spectra, and Kerma.* ICRU Report 13, 1969, International Commission on Radiation Units and Measurements, 7910 Woodmont Ave., Washington, D.C. 20014.

Protection Against Neutron Radiation. NCRP Report No. 38, 1971. National Council on Radiation Protection and Measurements, NCRP Publications, P.O. Box 30175, Washington, D.C. 20014.

neutron energy. A measurement of the ion current can then be interpreted in terms of the neutron flux density. Chambers designed for neutron measurements are commonly made with spherical symmetry in order to obtain an isotropic response to the scattered components of the beam.

Thermal neutrons are usually measured through a capture mechanism such as $^{10}B(n, \alpha)^7Li$ in a chamber filled with BF_3 gas. Electronic circuits can be adjusted to respond almost exclusively to the large inonization pulses produced by the ejected alpha particle and the recoiling Li nucleus, effectively rejecting the smaller pulses resulting from gamma-ray absorption. It can be shown that a detector which obeys a $1/v$ absorption law will have a response proportional to the neutron flux density, Eq. (9-8).

Some knowledge of the energy distribution in a neutron beam can be gained by using a $1/v$ detector with and without a moderator such as a shell of paraffin surrounding the counter. Without the moderator the detector response will be proportional to the low-energy components of the beam. The moderator will thermalize all fast neutrons without seriously attenuating the flux by absorption, and any increased response will be an indication of the magnitude of the fast component.

The *cadmium ratio* is another indicator of the presence of fast neutrons in a thermal flux. A measurement is made with and without a shield of Cd placed around a $1/v$ detector. Without the shield the $1/v$ response will emphasize the low energy contributions from the beam. The Cd shield absorbs all the neutrons with energies below the *cadmium cutoff*, Fig. 9-1, and the detector will then respond preferentially to those neutron energies just above it. Cutoff energy depends somewhat on the Cd thickness but it is in the neighborhood of 1 eV.

10.12 Radiation Doses Due To Neutrons

The conversion of a neutron flux density measurement to a tissue dose is far from simple. Seldom, if ever, is the energy spectrum of a neutron fluence known with precision. Even if it were, the calculation of absorbed dose requires a detailed knowledge of a number of absorption coefficients, each of which is a function of energy. Since some of the secondary radiations are quite penetrating, some of the contributions to the total absorbed dose will depend on the size of the absorbing body. Finally, each type of absorbed energy must be modified by the proper quality factor in order to obtain the dose equivalent.

Detailed calculations have been carried out by W.S. Snyder and his collaborators using a *Monte Carlo* technique, in which a large number of individual "case histories" of neutrons are traced to final absorption of all radiations or escape from the body. Details of each energy transfer are governed by the appropriate interaction probability. These calculations have been completed for a number of geometrical configurations and a wide range of neutron energies.* Figure 10-8, depicting the results of calculations for an infinite slab of tissue, is fairly representative of the values obtained for finite-sized phantoms.

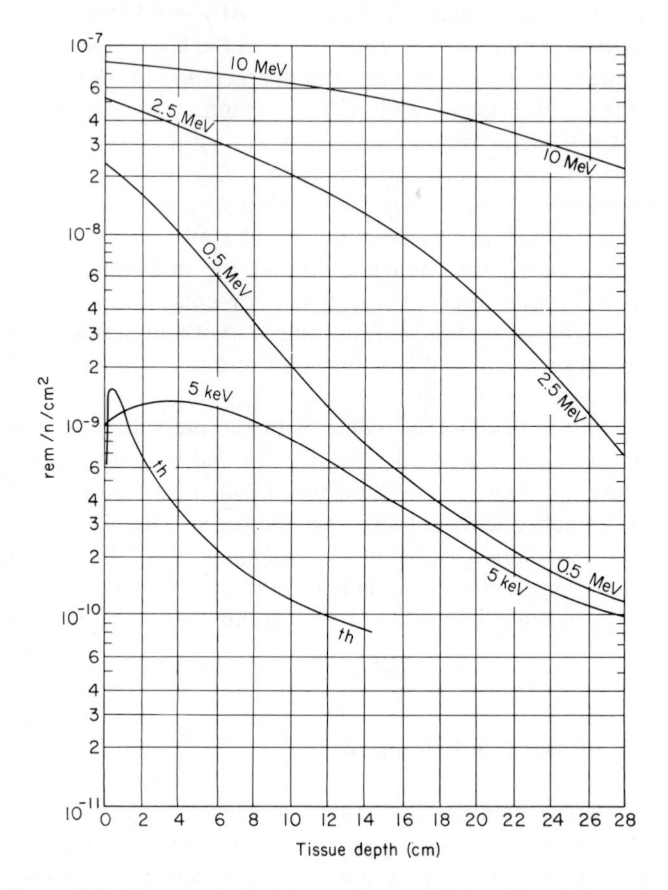

Figure 10-8. Dose equivalent in rem in an infinite slab of tissue from neutron absorption. (Adapted from calculations of W. S. Snyder and J. Neufeld, *Brit. J. Radiol.*, **28**, 342, 1955)

REFERENCES

Attix, F. H. and W. C. Roesch, ed, *Radiation Dosimetry*, Vol. I, II, and III. Academic Press, New York, N.Y., 1966.

*Protection Against Neutron Radiation, Appendix B.

BACQ, Z. M., AND P. ALEXANDER, *Fundamentals of Radiobiology*, 2nd ed. Pergamon Press, New York, N.Y., 1961.

GLASSER, O., E. H. QUIMBY, L. S. TAYLOR, J. L. WEATHERWAX, AND R. H. MORGAN, *Physical Foundations of Radiology*, 3d ed. Paul B. Hoeber, Inc., New York, N.Y., 1961.

HINE, G. J., AND G. L. BROWNELL, ED., *Radiation Dosimetry*. Academic Press, New York, N.Y., 1956.

ICRU Report 14. *Radiation Dosimetry: X Rays and Gamma Rays with Maximum Photon Energies Between* 0.6 *and* 50 *MeV*. Sept., 1969.

Report 16. *Linear Energy Transfer*. June, 1970.

Report 17. *Radiation Dosimetry: X Rays Generated at Potentials of* 5 *to* 150 *kV*. June, 1970.

Report 19. *Radiation Quantities and Units*. July, 1971. International Commission on Radiation Units and Measurements, 7910 Woodmont Ave. Washington, D.C. 20014.

JOHNS, H. E., *The Physics of Radiology*, 2nd ed. Charles C. Thomas, Pub., Springfield, Ill., 1961.

11

Natural Radioactivity

11.01 Discovery and Exploitation

Uranium had been mined for use as a coloring agent in the glass industry long before Becquerel announced the discovery of radioactivity in uranium minerals. Mines in Joachimsthal, Czechoslovakia, had been particularly active and immediately provided large quantities of ore for study. Chemical separation procedures quickly led to the identification and purification of a number of previously unknown elements. Many of the new elements were serially radioactive; the daughter of a radioactive parent would be radioactive, and its daughter in turn, perhaps for several generations. Gradually separate isotopes were identified and all the series relationships were established.

Three radioactive series occur in nature. Each is headed by a long-lived nuclide, which is a necessary requirement if the still-active isotopes were laid down at the time of formation of the earth's crust. A starting element whose half-life is short compared to the age of the earth will have decayed long since to undetectable amounts. Each of the three series has one gaseous member, an isotope of radon, which is the heaviest member of the inert gas family helium, argon, neon, . . . , radon. Each of the three series ends with a stable isotope of lead.

The three series found in nature are headed by $^{238}_{92}U$, $^{235}_{92}U$, and $^{232}_{90}Th$. They are known as the $(4n + 2)$, $(4n + 3)$, and the $(4n)$ series from the fact that the mass number of each member of the series will be given by the respective designation when n is given integral values. The *missing* or $(4n + 1)$ series was observed only after its precursor $^{241}_{94}Pu$ was produced artificially. The longest half-life in the series is 2×10^6 y $^{237}_{93}Np$, which is short compared to the age of the earth. Unique features of the $(4n + 1)$ series are the absence of a gaseous member and its termination at $^{209}_{83}Bi$ instead of at an isotope of lead. Series members $^{221}_{87}Fr$ and $^{217}_{85}At$ have no stable isotopes and were unknown until they were artificially produced.

It seems necessary to assume that the naturally occurring radioactive nuclides are to be found only in a relatively thin layer of the earth's crust. Most of the very considerable energy released in the decays will be converted to heat in material close to the emitters. The measured heat flow from the interior of the earth to the crust can be accounted for if the surface concentrations of radioactive nuclides are maintained for a depth of only about 20 miles. More radioactive material than this would lead to an excessive production of heat and a gradually warming earth.

Since their discovery, members of the uranium series have been the most important of the three. Uranium ores are widespread, with a mean concentration in the earth's crust of about 4×10^{-6} grams per gram of rock. Only a few deposits are rich enough to warrant exploitation at the present time. In seawater the uranium concentration varies from 0.3×10^{-6} to 2.3×10^{-6} g l^{-1}, varying with the salinity.

Although clinical and investigative medicine and industry are using ever-increasing amounts of artificially produced radioactive products, the importance of the natural radioactive elements has not decreased. Indeed, the production of the artificial nuclides depends for the most part on nuclear reactors operating on uranium as a fuel. With the increased demands on uranium for the production of power and nuclear weapons, world requirements have increased and workable deposits are being sought in all parts of the earth.

11.02 The Uranium Series

Only the important uranium series will be discussed here. The decay characteristics, shown in Table 11-1 and Fig. 11-1 are incomplete, only the most common radiations being listed. Tables of nuclear constants must be consulted for the complete decay modes. The total decay to the final stable product ^{206}Pb involves the emission of eight alpha particles and six negative betas. There are two possible decay modes at ^{214}Bi, which is said to exhibit *branching*. Beta emission accounts for 99.96% of the decays; the less probable mode, shown dotted, is only 0.04%. For this nuclide the *branching ratio* is 99.96/0.04 = 2500.

If native undisturbed uranium ore is in a relatively impermeable matrix, all members of the decay chain will be retained in it. In addition to the daughter products a substantial amount of ^{4}He will be present as a result of the alpha-particle emissions. Since all the decay constants are known, chemical

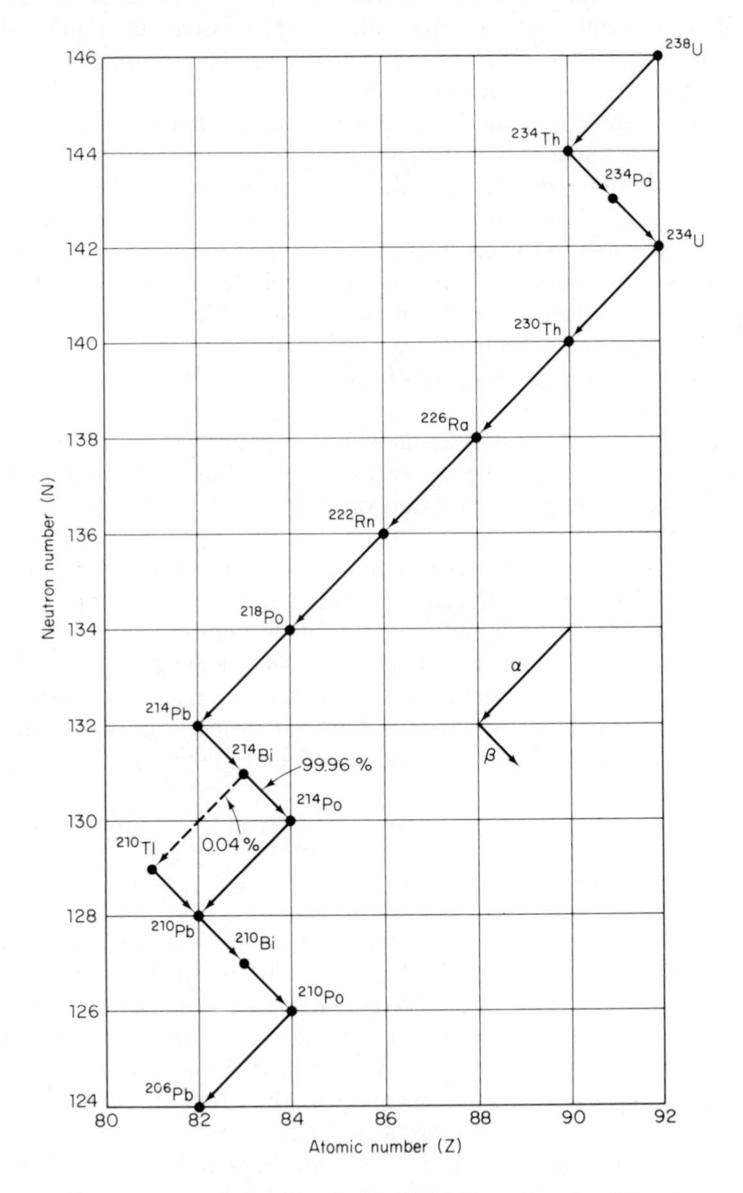

Figure 11-1. Decay relationships in the ^{238}U ($4n + 1$) series. Note the branching at ^{214}Bi and the reunion at ^{210}Pb.

TABLE 11-1
THE URANIUM SERIES*

Element	*Symbol*	*Half-life*	*Energy, MeV* α	β	γ	*Percent* γ *occurrence*
92 uranium	^{238}U	4.5×10^9 y	4.20	—	0.048	23
90 thorium	^{234}Th	24.1 d	—	0.19	0.09	4
91 protoactinium	^{234m}Pa	1.17 m	—	2.29	1.0	0.2
92 uranium	^{234}U	2.5×10^5 y	4.77	—	0.05	0.6
90 thorium	^{230}Th	8.0×10^4 y	4.68	—	0.068	24
88 radium	^{226}Ra	1602 y	4.78	—	0.186	4
86 radon	^{222}Rn	3.82 d	5.49	—	0.5	0.07
84 polonium	^{218}Po	3.05 m	6.00	—	—	—
82 lead	^{214}Pb	26.8 m	—	0.65	0.24	4
83 bismuth	^{214}Bi	19.7 m	5.51	1.5	0.61	47
84 polonium	^{214}Po	160 μs	7.69	—	0.8	0.014
82 lead	^{210}Pb	21 y	—	0.016	0.046	4
83 bismuth	^{210}Bi	5.0 d	—	1.16	—	—
84 polonium	^{210}Po	138 d	5.30	—	0.80	0.001
82 lead	^{206}Pb		Stable			

*Only the main features of the decays are given.

assays for the $^{238}U/^{206}Pb$ ratio can be used to estimate the length of time the parent uranium has been locked in the matrix. If any nonradiogenic lead is present, the dating assay will lead to incorrect results. An assay for the $^{238}U/^{4}He$ ratio can also be used as a measure for the time of sequestration but obviously will give correct results only if there has been a negligible escape of the gaseous elements ^{222}Rn or ^{4}He. Analyses of suitable samples by either method lead to values for the age of the earth that are in good agreement with estimates obtained in quite different ways.

Radium-226 was quickly recognized as an element suitable for use as a standard of radioactive decay. It was relatively available, had a long half-life, and had readily measurable radiations. The unit of activity, the *curie*, was then defined as the disintegration rate of one gram of pure ^{226}Ra. Madam Curie prepared several highly purified samples for distribution to various standardizing laboratories.

Careful measurements showed that one gram of ^{226}Ra had a decay rate of about 3.7×10^{10} disintegrations per second, or 2.22×10^{12} dpm. Although this definition was practically useful, the exact value of the curie fluctuated somewhat with improved methods of sample preparation and activity assay.

In order to avoid an uncertain activity unit the International Commission on Radiation Units and Measurements (ICRU) has redefined the unit of activity, the curie, Ci, as exactly

$$1 \text{ Ci} = 3.7 \times 10^{10} \text{ s}^{-1}$$

This definition removes the curie unit from the fluctuations inherent in experimental values and leads to a variable activity for 1 g of radium-226. Variations are not large and for most purposes the activity of 1 g of radium may be taken as 1 Ci.

In interpreting the results of radioactive assays, care must be taken to distinguish between the number of radioactive emissions and the number of nuclear disintegrations upon which the unit is based. In some cases these numbers are equal; in other cases they are not. One must know from the decay scheme the *percent occurrence* of the radiation that is being used in the assay. For example, if the 1.29-MeV gamma ray is being used in the assay of ^{41}Ar, Fig. 5-4, one must know that this emission follows only 99.2% of the nuclear disintegrations.

11.03 Series Decay Relationships

Equation (5-2) for the rate of radioactive decay applies only to the simple case of a single nuclide decaying to a stable daughter product. When radioactive series are involved, a nuclide may be produced by the decay of a heavier parent and at the same time may be decaying to produce a still lighter daughter. The head of a series, since it is not presently being formed, will decay according to the simple negative exponential of Eq. (5-2) but all subsequent populations will depend on all those that preceded it in the decay chain. A series of expressions known as the Bateman equations has been developed to describe the relationships in serial decays.

Assume that at $t = 0$ there are N_{10} atoms of the original parent nuclide and that there are no daughter products present. We adopt the following notation:

Nuclide	*Decay constant*	*Atoms at time t*
Parent nuclide	λ_1	N_1
First daughter	λ_2	N_2
Second daughter	λ_3	N_3
Third daughter	λ_4	N_4

The Bateman equations for the series of four nuclides are

$$N_1 = N_{10}e^{-\lambda_1 t} \tag{11-1}$$

$$N_2 = N_{10}(a_1 e^{-\lambda_1 t} + a_2 e^{-\lambda_2 t}) \tag{11-2}$$

$$N_3 = N_{10}(a_3 e^{-\lambda_1 t} + a_4 e^{-\lambda_2 t} + a_5 e^{-\lambda_3 t}) \tag{11-3}$$

$$N_4 = N_{10}(a_6 e^{-\lambda_1 t} + a_7 e^{-\lambda_2 t} + a_8 e^{-\lambda_3 t} + a_9 e^{-\lambda_4 t}) \tag{11-4}$$

where

$$a_1 = \frac{\lambda_1}{\lambda_2 - \lambda_1} \qquad a_2 = -\frac{\lambda_1}{\lambda_2 - \lambda_1} \tag{11-5}$$

$$\left.\begin{aligned} a_3 &= \frac{\lambda_1 \lambda_2}{(\lambda_2 - \lambda_1)(\lambda_3 - \lambda_1)} \qquad a_4 = \frac{\lambda_1 \lambda_2}{(\lambda_1 - \lambda_2)(\lambda_3 - \lambda_2)} \\ a_5 &= \frac{\lambda_1 \lambda_2}{(\lambda_1 - \lambda_3)(\lambda_2 - \lambda_3)} \end{aligned}\right\} \tag{11-6}$$

$$\left.\begin{aligned} a_6 &= \frac{\lambda_1 \lambda_2 \lambda_3}{(\lambda_2 - \lambda_1)(\lambda_3 - \lambda_1)(\lambda_4 - \lambda_1)} \qquad a_7 = \frac{\lambda_1 \lambda_2 \lambda_3}{(\lambda_1 - \lambda_2)(\lambda_3 - \lambda_2)(\lambda_4 - \lambda_2)} \\ a_8 &= \frac{\lambda_1 \gamma_2 \lambda_3}{(\lambda_1 - \lambda_3)(\lambda_2 - \lambda_3)(\lambda_4 - \lambda_3)} \qquad a_9 = \frac{\lambda_1 \lambda_2 \lambda_3}{(\lambda_1 - \lambda_4)(\lambda_2 - \lambda_4)(\lambda_3 - \lambda_4)} \end{aligned}\right\} \tag{11-7}$$

When the required initial conditions are satisfied, these equations are perfectly general and can be used by inserting the appropriate values of the time and the decay constants. Note that each of the equations is in terms of N and not activity A. The latter must be obtained by multiplying each value of N by the corresponding value of λ as in Eq. (5-5).

In a few special cases, simpler approximate relations can be used in calculations involving the first daughter, N_2.

1. Very long-lived parent $\lambda_1 \ll \lambda_2$. In this case λ_1 is so small compared to λ_2 that we may set $e^{-\lambda_1 t} = 1.00$ for almost all values of t. This is equivalent to assuming a negligible decrease in the amount of the parent during the time span under consideration. With this approximation Eq. (11-2) becomes

$$N_2 = N_{10}\frac{\lambda_1}{\lambda_2}(1 - e^{-\lambda_2 t}) \tag{11-8}$$

As the time increases from zero, the term in the parenthesis increases toward its limiting value of 1.00. N_2 increases correspondingly toward its limiting equilibrium value $(N_2)_{eq}$.

$$(N_2)_{eq} = N_{10}\frac{\lambda_1}{\lambda_2} \tag{11-9}$$

Eventually a constant N_{10}/N_2 ratio is attained and then maintained. From Eq. (11-9) the corresponding activities will then be equal, a necessary condition for parent–daughter equilibrium.

A case in point is the buildup of 3.82-d radon from an initially pure sample of ^{226}Ra, Fig. 11-2. One-half of the final activity is attained in 3.82 days, 75% in 7.64 days, and so on. According to Eq. (11-8), equilibrium is only attained after an infinite time but 99% of the equilibrium value is reached in

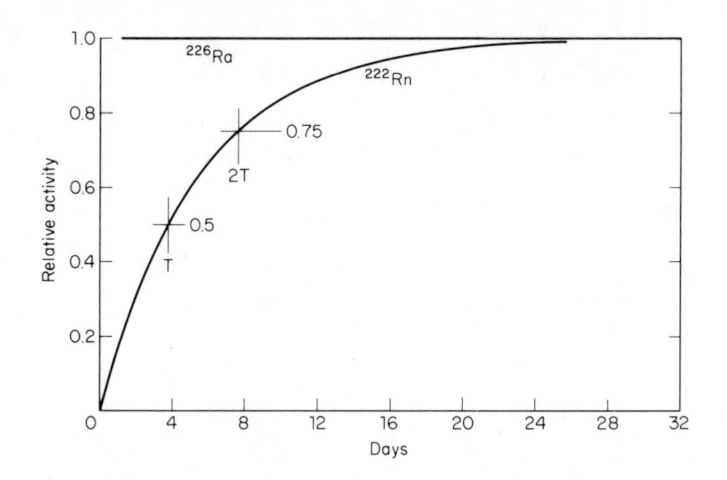

Figure 11-2. Relative activities in the case of a parent with a very long half-life. The buildup of ^{222}Rn from ^{226}Ra.

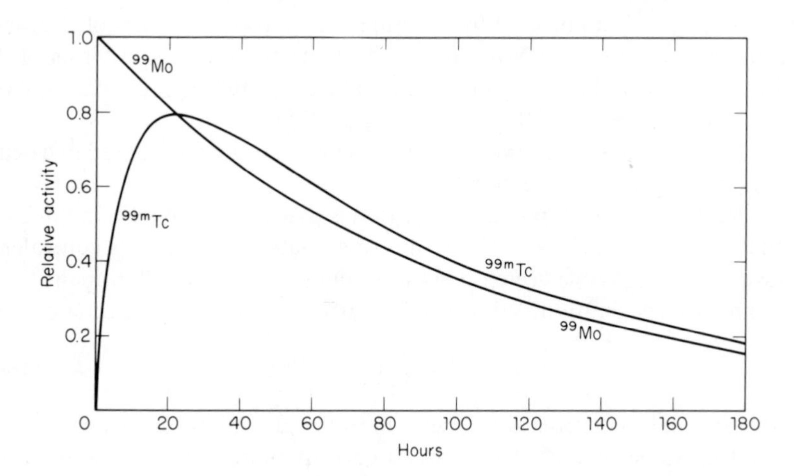

Figure 11-3. Relative activities when the half-life of the parent is long compared to that of the daughter.

less than seven half-lives (23.5 days in the case of ^{222}Rn). During this time the activity of the 1600-y parent has changed only a negligible amount.

2. Comparatively long-lived parent $\lambda_1 < \lambda_2$. After some time has elapsed, the second term in the parenthesis of Eq. (11-2) will become negligible compared to the first term. Then

$$N_2 = N_{10}\frac{\lambda_1}{\lambda_2 - \lambda_1}e^{-\lambda_1 t} \tag{11-10}$$

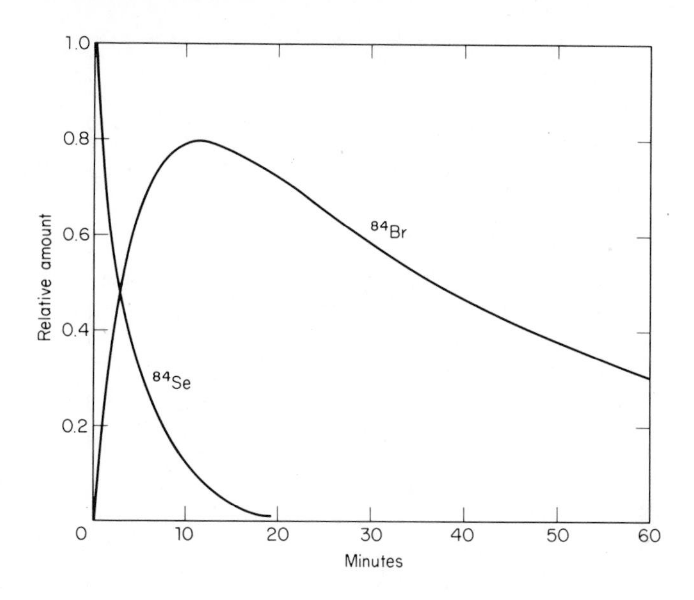

Figure 11-4. Relative *amounts* of parent and daughter when the parent has a comparatively short half-life.

or

$$N_2 = N_1 \frac{\lambda_1}{\lambda_2 - \lambda_1} \tag{11-11}$$

Here an equilibrium ratio is established between N_1 and N_2 and from that point on both parent and daughter decay at the rate characteristic of the parent. The production of the medically useful isotope ^{99m}Tc from ^{99}Mo serves to illustrate this behavior, Fig. 11-3.

3. Comparatively short-lived parent $\lambda_1 > \lambda_2$. In this case the parent decays rapidly, the first term in the parenthesis of Eq. (11-2) becomes negligible compared to the second; after this the daughter continues to decay with its own characteristic half-life. An example is shown in Fig. 11-4 where the relative amounts of 3.3-m ^{84}Se and 31.8-m ^{84}Br are plotted. The individual exponential decays can be seen more clearly in the semilog plot of Fig. 11-5. The parental decay is linear in this plot and when its contribution to the daughter nuclide becomes imperceptible, the latter decay also appears as a straight line.

11.04 Radon and Its Daughters

A four-member series starting with ^{222}Rn is responsible for much of the potential radiation hazard associated with the mining of uranium ore. Radon

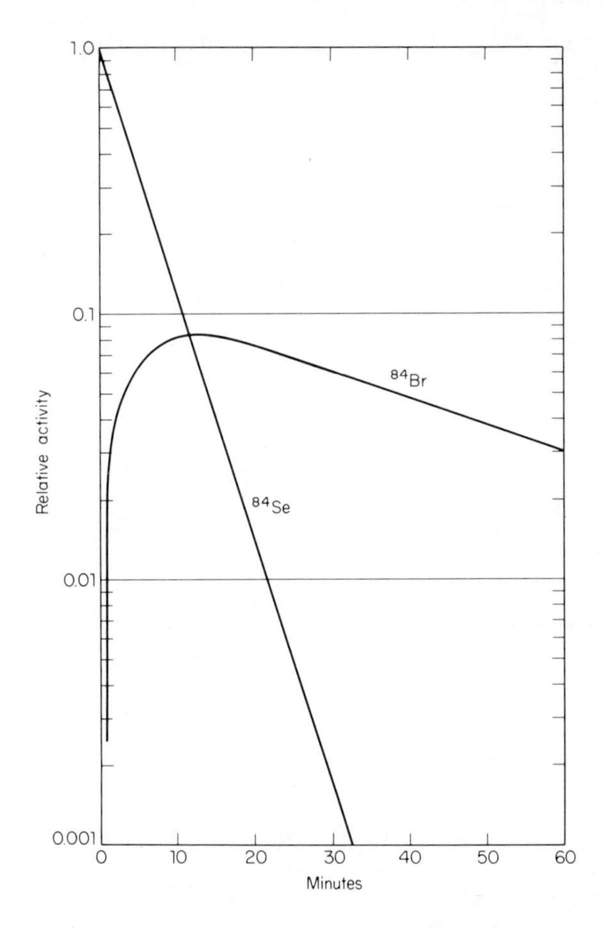

Figure 11-5. A semilog plot of the activities in the ^{84}Se–^{84}Br decays emphasizes the exponential nature of the individual transformations.

gas diffuses out from the ore deposits into the mine atmosphere where it may be inhaled by the occupants of the mine. Some of the inhaled radon will decay in various parts of the respiratory system, producing there its radioactive daughters; some radon will pass the lung–blood barrier and enter the bloodstream.

The primary radiation hazard arising from the inhalation of radon comes from the alpha-particle emission of the radon itself and its daughters. An examination of Table 11-1 shows that ^{218}Po, ^{214}Pb, ^{214}Bi, and ^{214}Po are the daughter nuclides primarily involved. Alpha emission from ^{214}Bi occurs in only 0.04% of the decays and can be neglected, as can the beta emissions from it and ^{214}Pb. The buildup and decay of these nuclides are important, however, in determining the amount of the strong alpha emitter ^{214}Po. The

^{214}Po daughter, ^{210}Pb, has a long half-life and its concentration builds up so slowly that the subsequent alpha emissions from ^{210}Po need not be considered except for very long-term exposures.

The series members through ^{214}Po come rapidly into equilibrium with an initially daughter-free sample of ^{222}Rn, Fig. 11-6, and then all decay with the characteristic half-life of ^{222}Rn. It is useful to calculate the energy released in

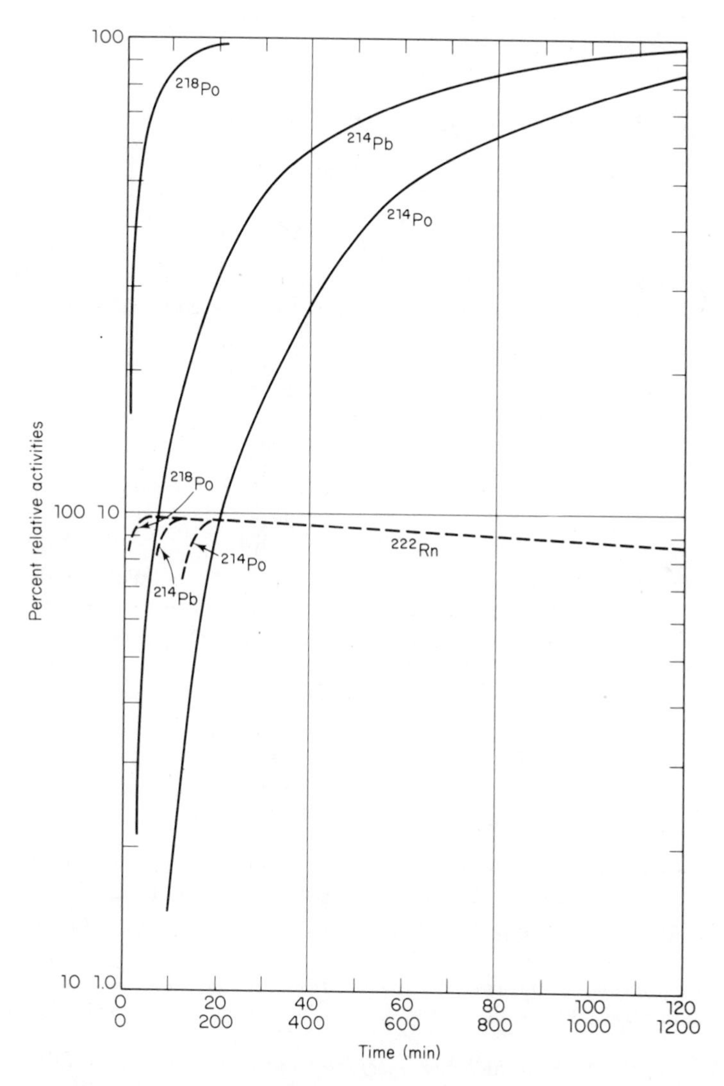

Figure 11-6. Buildup and decay of the daughters from an initially pure sample of ^{222}Rn.

the total decay of the pertinent radon daughters that are in equilibrium with the radon.

Illustrative Example

Calculate the potential energy that will be released by alpha particles in the total decay of the radon daughters that are originally in radioactive equilibrium with 1 pCi = 10^{-12} Ci of ^{222}Rn. The decay parameters are

$$\lambda(^{218}\text{Po}) = \frac{0.693}{3.05} = 0.227\ \text{m}^{-1}$$

$$\lambda(^{214}\text{Pb}) = \frac{0.693}{26.8} = 0.0259\ \text{m}^{-1}$$

$$\lambda(^{214}\text{Bi}) = \frac{0.693}{19.7} = 0.0352\ \text{m}^{-1}$$

$$\lambda(^{214}\text{Po}) = \frac{0.693}{2.7 \times 10^{-6}} = 2.56 \times 10^{5}\ \text{m}^{-1}$$

In radioactive equilibrium each species will decay at the same rate, which for 1 pCi is 2.22 dpm. The original numbers of undecayed nuclei will be

$$N_0(^{218}\text{Po}) = \frac{\Delta N/\Delta t}{\lambda} = \frac{2.22}{0.227} = 9.78$$

$$N_0(^{214}\text{Pb}) = \quad '' \quad = \frac{2.22}{0.0259} = 85.7$$

$$N_0(^{214}\text{Bi}) = \quad '' \quad = \frac{2.22}{0.0352} = 63.0$$

$$N_0(^{214}\text{Po}) = \quad '' \quad = \frac{2.22}{2.56 \times 10^{5}} = 10^{-5}$$

All ^{218}Po nuclei will undergo two subsequent alpha decays with the release of 6.00 + 7.69 = 13.69 MeV, while each nucleus of ^{214}Pb and ^{214}Bi will eventually become ^{214}Po, which releases 7.69 MeV of alpha energy. The final balance sheet shows

$$\begin{aligned}
^{218}\text{Po} &= 9.78 \times 13.69 = 134\ \text{MeV} \\
^{214}\text{Pb} &= 85.7 \times 7.69 = 658 \\
^{214}\text{Bi} &= 63.0 \times 7.69 = 485 \\
^{214}\text{Po} &= 10^{-5} \times 7.69 = 000 \\
\text{Total alpha energy} &= 1277\ \text{MeV}
\end{aligned}$$

The value calculated in the illustrative example, rounded off to 1.3×10^{3} MeV pCi^{-1}, forms the basis for the "working level" concept developed to assess the potential radiation hazard in uranium mining operations.*

**Control of Radon and Daughters in Uranium Mines and Calculations on Biologic Effects.* Public Health Service Publication No. 494, U.S. Department of Health, Education, and Welfare.

Guidance for the Control of Radiation Hazards in Uranium Mining. Report No. 8 (revised) of the Staff of the Federal Radiation Council, 1967, U.S. Government Printing Office, Washington, D.C. 20402.

11.05 Medical Uses of Radium

For many years after its discovery and isolation, radium-226 was used almost exclusively in the field of medicine. Some 10% of the extracted radium went into the production of self-luminous paints and for sources used in industrial radiography; the remainder was applied to the treatment of cancers and a variety of other disorders.

The ^{226}Ra used in radiation therapy was sealed into small *needles* usually made of gold or platinum. Needle walls were customarily 0.5 mm of platinum or its absorption equivalent, which is sufficient to absorb all the alpha particles and most of the betas. Radiation effects were due almost entirely to the gamma rays from the radium daughters, principally ^{214}Pb and ^{214}Bi. By the time the needles were put in use the radium was in equilibrium with its daughters, at least down to 21-y ^{210}Pb. The long-term buildup of ^{210}Pb and ^{210}Po did not alter the radiation characteristics of the source since neither nuclide has an appreciable output of penetrating radiation.

Most needles prepared for therapy contained from 1 mg (1 mCi) to 25 mg of elemental ^{226}Ra usually as the bromide or the chloride. One or more needles were surgically implanted in the malignant tissue that was to be irradiated, according to a treatment plan developed by the radiologist. Radiation exposure was customarily expressed in millicurie-hours, or mCi h, which makes for simple calculations since the strength or *curiage* of a ^{226}Ra source remains constant over any treatment time.

Radium as such was not always used in therapy. Radium was expensive and dangerous to handle; because of its long half-life the needles must be removed when treatment is complete. A leaking needle will release ^{222}Rn together with its daughters to the environment and will continue to be a source of contamination until the needle is repaired.

Much use was made of ^{222}Rn rather than ^{226}Ra. Perhaps 0.1–1.0 g of a soluble radium salt such as $RaBr_2$ was put in solution in a *radon plant*. Radon gas evolved by the Ra decay was pumped off from the solution into tiny capillary tubes or *seeds*. These seeds could be made up into sizes specified by the radiologist for the particular treatment in hand. By the time the seeds were implanted, the ^{222}Rn was in equilibrium with all daughters down to ^{210}Pb, Fig. 11-6. The radon source then had radiation characteristics identical with those of a radium source but with an activity that decayed with a 3.8-d half-life. In some cases a second surgical procedure was avoided by allowing the seeds to remain in position for complete decay.

The millicurie-hour value of a short half-life source such as ^{222}Rn must be calculated from the basic activity relation, Eq. (5-5). An integration of this equation over a time t hours gives

$$\text{mCi h} = \frac{A_0}{\lambda}(1 - e^{-\lambda t}) \qquad (11\text{-}12)$$

where A_0 is the initial source strength in mCi and λ is the decay constant in h^{-1}. In the case of ^{222}Rn, $\lambda = 7.55 \times 10^{-3}\ h^{-1}$ which lead to an infinite-time exposure of 134.2 mCi h mCi^{-1}. Equation (11-12) with appropriate values of λ applies equally well to the short half-life isotopes that have now largely replaced radium and radon in radiation therapy.

Radiation has been recommended and used in the treatment of a variety of diseased states and many radium preparations have been advertised and sold to the public. Many of these uses were of dubious value or were actually harmful. With the present more complete understanding of the biological effects of radiation the chance of misuse has greatly diminished.

One practice that has been discontinued involved the use of thorium-containing preparations as *contrast media*. These preparations were administered to patients in order to enhance the contrast between various tissues for radiography or X-ray fluoroscopy. Some long-lived members of the thorium decay series remained in the body for long periods of time and the continued irradiation produced, in some cases, radiogenic cancer. Thorium-containing compounds have been entirely replaced as contrast media by nonradioactive preparations.

11.06 Industrial Uses of Radium

Soon after radium was discovered, its ability to induce fluorescent light in a number of substances was recognized. Mixtures of various radium preparations with zinc sulfide came into use for the manufacture of luminous instrument dials. Usage rapidly outran any understanding of the hazards inherent in radiation. The resulting tragedy was compounded by the fact that protracted irradiation does not, in general, produce immediately apparent signs of injury. By the time an overexposure is recognized, irreparable damage may have been done.

The most famous example of overexposure to radium involved a group of women who were painting luminous intrument dials in a New Jersey factory in the early 1920's. Lacking an understanding of the hazards involved, the workers used techniques that led to the ingestion of relatively large quantities of radioactive materials. Most of the preparations used were crude products containing a number of active nuclides in addition to ^{226}Ra. This fact has complicated retrospective attempts to calculate the radiation doses received by the victims.

After several years exposure, cases of cancer appeared among the dial painters and the connection with the use of radioactive sources finally was recognized. Intensive studies were carried out on the group and data from these studies have been of great help in establishing permissible working levels. Artificially activated materials now available provide a variety of nuclides suitable for use in self-luminous dials and methods of safe handling and application are now understood.

Industry has made increasing use of large sources of penetrating radiation for the nondestructive inspection of many objects. A large sealed source of ^{226}Ra might be placed on one side of a welded joint, with a photographic plate on the other side. The resulting radiograph would show a flaw or a thin section as a dark area where the gamma radiation was less absorbed than in the heavy metal. In this application also, nuclides such as ^{60}Co have generally replaced the large radium sources previously used. The use of ^{60}Co is not without hazard but it has no gaseous phase and no family of radioactive daughters which may become dispersed in case of an accident.

11.07 Specific Activity of Radiation Sources

We have seen, Sec. 3.07, that a small source area is desirable in order to obtain a uniform radiation intensity over as large a target area as possible. The same consideration applies to the use of radioactive sources for patient treatment, radiobiological research, or industrial radiography. A source that occupies only a small volume produces a uniform radiation field and sharp radiographic images and also minimizes the intensity loss due to radiation absorption in the source itself.

High specific activity materials must be used to obtain high radiation output in a small-volume source. In many cases artificially produced isotopes have an intrinsic specific activity far greater than those available with naturally radioactive isotopes. In Sec. 5.03 the intrinsic specific activity of ^{60}Co was calculated to be just over 1 Ci mg^{-1}. A 1-Ci source of ^{226}Ra would weigh 1 g and would occupy about 1000 times the volume of a ^{60}Co source of equal activity. It is not always possible to produce carrier-free isotopes and so the limit of specific activity is seldom attained. However, many sources are available with size/activity ratios much more favorable than those obtainable with natural materials.

11.08 Nonseries Radioactive Nuclides

All but a few members of the naturally occurring radioactive series are heavy elements, $Z = 81$ or greater. There are, in addition, a few lighter, unrelated radioactive nuclei, located apparently at random throughout the periodic table. Table 11-2 is a partial listing of some of the pertinent properties of nuclides whose activities have been well established.

In addition to alpha-particle and negatron emission, electron capture decays have been identified. Electron capture has not been observed in series members. With two exceptions the half-lives of the nonseries members are

TABLE 11-2
SOME NATURALLY OCCURRING NONSERIES RADIOACTIVE NUCLIDES

Symbol	Nucleus Z	N	A	Percent Abundance	Half-Life Years	Emissions	Transition Energy, MeV
H	1	2	3		12.3	β^-	0.0186
C	6	8	14		5740	β^-	0.156
K	19	21	40	0.0119	1.3×10^{9}	β^-, EC, γ	1.5
V	23	27	50	0.25	6×10^{15}	β^-, EC, γ	2.2
Rb	37	50	87	27.8	5×10^{10}	β^-	0.27
In	49	66	115	95.8	6×10^{14}	β^-	0.49
Te	52	78	130	34.5	8×10^{20}	$\beta\beta$	?
La	57	81	138	0.09	1×10^{11}	β^-, EC, γ	1.8
Ce	58	84	142	11.1	10^{16}	α	?
Nd	60	84	144	23.9	2×10^{15}	α	1.9
Sm	62	85	147	15.1	1×10^{11}	α	2.3
Gd	64	88	152	0.2	1×10^{14}	α	2.2
Lu	71	105	176	2.6	3×10^{10}	β^-, α	1.0
Re	75	112	187	63	4×10^{10}	β^-	0.003
Pt	78	112	190	0.01	6×10^{11}	α	?
Pt	78	114	192	0.78	10^{15}	α	?
Bi	83	126	209	100	2×10^{18}	α?	?

surprisingly long in view of the transistion energies that are available. It is probable that the transitions are strongly hindered by parity and angular momentum requirements.

With the exception of ^{3}H and ^{14}C the half-lives of all the nonseries decays are compatible with their nascence some billions of years ago (4–20 Gy) at the time a "big bang" of creation has been postulated. Tritium and carbon-14 must be in continuous production in order to be present in observable concentrations. Each of these nuclides is being formed by various components of the extraterrestrial cosmic radiation. Examples of these reactions are

$$^{14}_{7}\text{N} + {}^{1}_{0}n \longrightarrow {}^{14}_{6}\text{C} + {}^{1}_{1}p \qquad (11\text{-}13)$$

$$^{2}_{1}\text{H} + {}^{1}_{0}n \longrightarrow {}^{3}_{1}\text{H} + \gamma \qquad (11\text{-}14)$$

There have been some recent additions to the list of nonseries radioactive nuclides and still more may be discovered in the future. The detection of the very low levels of activity involved is difficult and improved techniques may reveal some activities presently unknown.

Three of the nuclides listed in Table 11-2, ^{3}H, ^{14}C, and ^{40}K, are constituents of all living systems and as such are of particular biological interest. No living tissue can escape the continuous low dose-rate irradiation from these internal emitters.

Illustrative Example

Calculate the radiation dose delivered in 1 y by ^{14}C to a "standard" 70-kg man containing 16 kg of carbon. Assume a specific activity of 7.2 pCi g^{-1} of carbon, an average beta energy of 0.0493 MeV, and a uniform tissue distribution.

$$7.2 \times (3.7 \times 10^{-2}) \times (3.15 \times 10^{7}) \times (1.6 \times 10^{4}) = (1.34 \times 10^{11}) \text{ dpy}$$

$$(1.34 \times 10^{11}) \times 0.0493 \times (1.6 \times 10^{-6}) = 1.06 \times 10^{4} \text{ ergs y}^{-1} \text{ absorbed}$$

$$\frac{1.06 \times 10^{4}}{7 \times 10^{4}} = 0.15 \text{ ergs g}^{-1} \text{ y}^{-1} \text{ absorbed}$$

$$\frac{0.15}{100} = 0.0015 \text{ rads or } 1.5 \text{ millirads y}^{-1}$$

A similar calculation for ^{40}K leads to a yearly dose of about 15 millirads. More exact values can be obtained by considering the exact elemental content of individual tissues instead of assuming a uniform distribution. The dose due to tritium is also readily calculated and is found to be only a small fraction of that due to ^{14}C.

11.09 Radioactive Dating

The presence of naturally radioactive nuclides and the apparently immutable values of the decay constants permit the calculation of some very important time intervals. One of the earliest applications of radioactive dating techniques was to the determination of the age of minerals from U/He or U/Pb ratios. The amount of He production from the ^{238}U decay chain can be calculated as a function of time from the known constants of the alpha-particle emitters. If no helium gas escaped from the system, a comparison of the ^{238}U and the ^{4}He concentrations serves to determine the age at which the mineral specimen was laid down.

From the data in Table 11-1 it is evident that only four half-lives need to be considered in age determinations from the $^{238}U/^{206}Pb$ ratio. All other decays are so rapid that for age determinations instant equilibrium can be assumed and the number of these nuclides still undecayed can be neglected.

Illustrative Example

Calculate the amount of ^{206}Pb formed from 1 g of ^{238}U in 1 million years. From the data in Table 11-1 we have

$$\lambda_1(238) = 1.54 \times 10^{-10} \text{ y}^{-1} \qquad \lambda_3(230) = 8.65 \times 10^{-6} \text{ y}^{-1}$$

$$\lambda_2(234) = 2.77 \times 10^{-6} \text{ y}^{-1} \qquad \lambda_4(226) = 4.33 \times 10^{-4} \text{ y}^{-1}$$

In 1 g of ^{238}U there are $(6.02 \times 10^{23})/238 = 2.53 \times 10^{21}$ nuclei. In 10^{6} y there will be

$$(2.53 \times 10^{21}) \times (1.54 \times 10^{-10}) \times 10^{6} = 3.9 \times 10^{17} \text{ decays of } ^{238}\text{U}$$

At the end of 10^6 y the number of ^{234}U nuclei can be calculated from Eq. (11-8).

$$N_2(234) = \frac{(2.53 \times 10^{21}) \times (1.54 \times 10^{-10})}{2.77 \times 10^{-6}}(1 - e^{-2.77})$$
$$= 1.41 \times 10^{17}(1 - 0.066)$$
$$= 1.32 \times 10^{17} \text{ nuclei}$$

When the number of ^{230}Th nuclei is calculated from Eq. (11-2), the second and third exponential terms will be found to be negligible compared to the first. Then

$$N_3(230) = 2.53 \times 10^{21}\left[\frac{(1.54 \times 10^{-10}) \times (2.77 \times 10^{-6})}{(2.77 \times 10^{-6}) \times (8.65 \times 10^{-6})}\right](e^{-1.54\times 10^{-4}})$$
$$= 4.50 \times 10^{16} \text{ nuclei}$$

Because of the disparate values of the two λ's, ^{226}Ra will be in secular equilibrium with ^{230}Th and from Eq. (11-9)

$$N_4(226) = 4.50 \times 10^{16}\left(\frac{8.65 \times 10^{-6}}{4.33 \times 10^{-4}}\right) = 9.00 \times 10^{-14} \text{ nuclei}$$

The final tally is

total number of ^{238}U decays		3.90×10^{17}
hold up at ^{234}U	1.32×10^{17}	
at ^{230}Th	0.45	
at ^{226}Ra	0.009	
total	1.78×10^{17}	1.78
number of ^{206}Pb nuclei		2.12×10^{17}

$$\text{mass of } ^{206}Pb = 2.12 \times 10^{17} \times \frac{206}{6.02 \times 10^{23}} = 6.2 \times 10^{-5} \text{ g}$$

Another dating system involves the use of ^{40}K, which has two competing modes of decay.

$$^{40}_{19}K(\beta^-) \longrightarrow ^{40}_{20}Ca \quad 89\% \tag{11-15}$$

$$^{40}_{19}K(EC, \beta^+) \longrightarrow ^{40}_{18}Ar \quad 11\% \tag{11-16}$$

Each of the daughter elements is found in nature with a well-established abundance. When radiogenic ^{40}Ar is present, the mass spectrometer will show an abnormal $^{40}Ar/^{36}Ar$ ratio which can be related to the time the ^{40}K decay has been going on in a closed system from which no gas escapes. Ages up to about 2×10^6 years have been determined by this method.

Radioactive ^{14}C produced in the upper atmosphere by cosmic radiation is distributed throughout the atmosphere and into seawaters by diffusion and turbulence. Existing primarily as $^{14}Co_2$, some of this radioactive carbon will be incorporated into all living things since there is no chemical difference between the stable and the active isotopes. During active metabolic life all living organisms will contain a small equilibrium concentration of ^{14}C. Upon death the incorporation of ^{14}C ceases and that already present decays with its characteristic half-life. The age of any old organic material can then be determined by comparing its ^{14}C concentration with that observed in living objects. On a geological time scale the half-life of ^{14}C is very short and carbon dating is limited to perhaps 50,000 years.

Dating by means of ^{14}C assays depends on the plausible but not completely proved assumption that the rate of ^{14}C production has remained constant over the period covered by the production, incorporation, and decay.

Man's increasing use of energy, particularly during the present century, has introduced perturbations into the previously existing ^{14}C concentrations. Prior to the utilization of nuclear reactors, man met most of his energy needs by burning fossil fuels, primarily coal and oil. Each of these substances was originally living organic matter and as such it contained the three carbon isotopes ^{12}C: ^{13}C: ^{14}C in equilibrium concentrations. During the long period between death and utilization, essentially all the ^{14}C disappeared by decay. When these fuels are burned, the CO_2 in the stack effluents is deficient in ^{14}C and so the existing atmospheric concentration ratios are diluted with non-radioactive carbon.

With the advent of atmospheric testing of nuclear devices, large amounts of ^{14}C were produced, primarily by reaction (11-13). The quantities thus introduced into the atmosphere were sufficient, even when diluted by large-scale diffusion, to interfere with some of the low-level assays for ^{14}C that are required in carbon dating studies.

Before the recent releases of man-made tritium, naturally occurring concentrations of about 1.5×10^{-4} disintegrations per second per gram of hydrogen were observed. As in the case of ^{14}C, this specific activity would be found in living organisms but upon death incorporation would cease, and the value would decay from the equilibrium value with the characteristic half-life. The low natural activity and the relatively short half-life of tritium have limited its use to a few special cases and to relatively short times.

11.10 Natural Radiation Background

Until man began to produce and to exploit radioactive nuclides on a large scale, the earth was nearly in equilibrium with a variety of radiations. Collectively known as *background*, these naturally occurring radiations appear to have existed with little change in intensity for a long period of time.

Uranium and thorium and all their radioactive daughters are widely distributed throughout the earth's crust. Some of the gamma rays that originate even in deep ore deposits escape absorption in the earth and reach habitable areas. Each decay series has a gaseous member which can diffuse out from its matrix to decay in the atmosphere or in rivers, lakes, and oceans. Series members may dissolve in underground water to appear in high concentrations (500 pCi ml^{-1}) in some mineral springs or as trace amounts in sources of human drinking water. Growing plants absorb, each according to its characteristic metabolic pattern, appreciable quantities of many of the naturally occurring radioactive nuclides. Humans are thus exposed to the penetrating radiations from distant external sources and to the radiations from internal emitters that have entered the body through water, food, and inhalation.

The presence in foodstuffs of members of the various radioactive series can be detected through their alpha activity. Measured alpha activities vary widely, from more than 10 pCi g^{-1} for Brazil nuts to 0.1 pCi g^{-1} in flour to perhaps 0.005 pCi g^{-1} in meat. In addition, all food products contain the beta emitters ^{3}H, ^{14}C, and ^{40}K, each in proportion to its natural abundance.

As we have seen, ^{3}H and ^{14}C are being produced continuously in the earth's atmosphere by penetrating radiation from extraterrestrial sources. Our sun is one source of this *cosmic* radiation. Other components, the *galactic* cosmic radiation, come from unidentified sources outside our solar system.

The high-energy radiations from the sun consist primarily of protons which undergo nuclear reactions in the upper atmosphere to produce a variety of secondary radiations. Charged particles ranging from mesons to completely ionized nuclei as heavy as iron have been identified in the galctic cosmic radiation. A few of the primary cosmic-ray particles have enormous energies, perhaps 10^{16} MeV, but the average energy is only a few MeV.

Some cosmic-ray components are readily absorbed in our upper atmosphere so there is a gradient of decreasing intensity toward the surface of the earth. Other components, notably the muons, interact very weakly with matter and may penetrate deep into the earth or ocean waters before interacting to produce secondaries.

No region of the earth's surface can escape cosmic radiation although the intensities vary with altitude and with geomagnetic latitude. Charged particles approaching the earth are influenced by its magntic field, which is primarily responsible also for the presence of the Van Allen belts of high radiation intensity. In these belts, which range from 500–10,000 km above the earth, the radiation intensities are incompatible with extended human occupancy. Escape "windows" exist near the two magnetic poles through which leisurely travel away from the earth could be achieved. At high velocities the radiation dose acquired during direct transit through the belts is acceptable.

Outside the protective atmosphere of the earth a space traveler has only the protection afforded by his space vehicle. Under normal solar conditions the radiation levels are low enough to permit extended missions. However, solar radiation may increase manyfold with little warning in a *solar flare* to attain levels where serious biological consequences may ensue. Radiation injury from an unpredicted solar flare appears to be one of the hazards inherent in space travel. Little protection can be expected from spacecraft shielding. Weight limitations preclude the use of shielding sufficient for certain protection from a solar flare. Partial shielding may actually increase the radiation dose to the occupants by increasing the number of low-energy secondary radiations which are more readily absorbed in tissues.

At sea level the average radiation dose due to natural background is about 100 millirads y^{-1}. This dose originates from

cosmic rays	30 millirads
local gamma emitters	50
internal emitters	20

In a few locations the dose from local deposits of gamma-ray emitters may be considerably higher than the average. In one area of Brazil and one in India, the yearly doses from local emitters may reach or exceed 1000 m rads y^{-1}. The cosmic ray contributions to the total background dose varies with altitude, reaching about 60 m rads y^{-1} at 5000 ft, 3000 at 40,000 ft, and 5000 at 60,000 ft.

11.11 Man-Made Environmental Contamination

Since the development of the self-sustaining chain reaction, human activities have injected many megacuries of radioactive nuclides into the biosphere. Many of the materials released through accidents or through the testing of nuclear devices have decayed to stable forms. A few long-lived nuclides, notably ^{90}Sr and ^{137}Cs, are still present in small but detectable amounts in presumably every human being. Radiations from these internal emitters presently deliver dose rates that are small compared to that from natural background and if there are no future releases, these contributions will steadily decrease.

The concept of *dose commitment* is useful for evaluating the long-term total dose resulting from a given event or events. Dose commitment is the integral dose received by an individual during his lifetime from the source or sources of interest. Dose commitments will vary with the tissue under consideration because of the selective deposition patterns of the individual radionuclides. For example, the beta particles from ^{90}Sr will contribute primarily to the bone marrow dose and to the cells in intimate contact with bone surfaces. Dose commitments from all nuclear testing through the year 1965 range from 50–250 m rads, depending on the tissue of interest.* These values point up the urgent world wide need to limit nuclear testing, to take every reasonable precaution against accidents in reactor operation and fuel reprocessing, and to contain properly all radioactive wastes for long periods of time.

In a counting laboratory, background is considered to include any radia-

**United Nations Scientific Committee on the Effects of Atomic Radiation.* Supp. 14(A/6314), 1966.

tion, except that deliberately introduced for the assay, that is capable of activating the counting system. Statistical considerations, Sec. 12.09 to 12.11, indicate the desirability of a low background count. It is even more essential that the background in an assay laboratory remain constant during the counting periods. A moderately high background count can usually be dealt with by increasing the assay times. Data obtained when the background is fluctuating as the result of the movement of local sources will probably be uninterpretable at any confidence level.

REFERENCES

Control of Radon and Daughters in Uranium Mines and Calculations of Biological Effects. Public Health Service Publication No. 494, Department of Health, Education and Welfare, Washington, D.C.

EISENBUD, M., *Environmental Radioactivity.* McGraw-Hill Book Co., New York, N.Y., 1963.

GEYH, M. A., "Problems in Radioactive Carbon Dating of Small Samples by Means of Acetylene, Ethane, and Benzene." *Int. J. Applied Rad. Isotopes*, **20,** 463, 1969.

LAL, D., AND H. E. SUESS, "The Radioactivity of the Atmosphere and the Hydrosphere." *Ann. Rev. Nuc. Sci.*, **18**, 407, 1968.

PROTSCH, R. AND R. BERGER, "Earliest Radiocarbon Dates for Domesticated Animals. *Science*, **179**, 235, 1973.

REYNOLDS, J. H., "Isotopic Abundance Anomalies in the Solar System." *Ann. Rev. Nuc. Sci.* **17**, 253, 1967.

RUTHERFORD, E. J. CHADWICK, AND C. D. ELLIS, *Radiations from Radioactive Substances.* The Macmillan Co., New York, N. Y., 1930.

12

Radioactive Tracers

12.01 Isotopic Similarities

The chemical and biological behavior of any atom is determined almost entirely by the structure of its orbital electrons with isotopic mass a factor of only secondary importance. Two isotopes of any element, for example, stable ^{12}C and radioactive ^{14}C, will enter into identical chemical reactions because their orbital electron structures are identical. Rate constants of some reactions may be slightly different because of the lower mobility of the more massive ^{14}C nucleus. Even in the extreme case of ^{1}H and ^{3}H, mobility differences can usually be neglected because the atoms will be incorporated into compounds of some complexity, thus making the molecular weights nearly equal. There are a few exceptions to this generalization. Heavy water, D_2O, for instance, has been found to be toxic to rodents when it makes up more than about 20% of the total water intake. Plutonium isotopes show some striking differences in organ and tissue distributions, differences which apparently must be related to local ionization conditions rather than to nuclear masses. The latter cover only a narrow range but the decay constants vary by several orders of magnitude.

When the behavior of a particular molecular configuration is to be studied by radioactive tracing, a sample is synthesized with a small fraction of one of the atomic constituents replaced by a radioactive isotope. A radioactive assay establishes the specific activity of the *tagged* or *labeled* compound and a known amount is then introduced into the system under study. A radioactive assay on a sample obtained from the system at a later time will serve to determine the fraction of the administered isotope that is present in the system.

12.02 Compartments, Pools, and Spaces

As a simple example of the method, consider the determination of blood plasma volume for which radioactive iodinated serum albumin (RISA) is commonly used. Human serum albumin which has been iodinated with a radioactive isotope such as ^{131}I will hold the isotope rather firmly so that a radioactive assay for ^{131}I will serve as an assay for the albumin and hence for the plasma.

An activity of less than 10 μCi of ^{131}I given intravenously in a small volume v ml with a specific activity of C_0 μCi ml^{-1} is sufficient to determine human plasma volume. When mixing of the injected material is known from past experience to be complete, a blood sample is drawn and a ^{131}I plasma concentration C_1 is determined by assay. If there were no loss of RISA, the plasma volume would be simply calculated.

$$(V + v)C_1 = vC_0 \quad \text{or} \quad V = \frac{v(C_0 - C_1)}{C_1} \tag{12-1}$$

In fact the vascular system is not a closed *compartment* and there will be some loss of tagged material from it into a much larger volume, the total *exchangeable albumin space*, or *pool*. In practice a series of blood samples is drawn over a period of about an hour after injection. The corresponding V values calculated by Eq. (12-1) and plotted against time can be extrapolated back to the time of injection to give the plasma volume without loss. The volume thus obtained is the rapidly exchangeable plasma pool. The total albumin space, determined from assays made over a period of several days, will be found to be perhaps twice as large.

Whenever a series of sampling and assay extends over an appreciable time, corrections must be made for loss from the system by excretion, and all assays must be corrected for radioactive decay to some common time.

12.03 Turnover Rate

When two compartments of a space are in equilibrium, the rate of loss from each must equal the rate of accretion. *Turnover rate* is defined as the quantity of a substance that passes in either direction between the compartments in unit time.

A rapidly exchangeable sodium pool can be determined by administering NaCl tagged with either ^{22}Na or ^{24}Na. By 24 hours the radioactive ions will be in equilibrium with one compartment, the rapidly exchangeable sodium pool. This compartment represents only about 80% of the total sodium *space* in the body. The turnover rate of sodium ions in the bone, brain, and perhaps other tissues is very low, if not zero, and so a simple tracing study will not serve to determine the total sodium space.

In either clinical diagnostic or investigative studies, one is usually interested in turnover rates as well as in the sizes of the biological compartments and spaces. Studies of reaction kinetics require a series of determinations of the tracer concentrations before equilibrium is attained. From the time course of these values, turnover rates are calculated from a known model of the system or an attempt is made to deduce an applicable model.

12.04 Tracer Kinetics

A considerable mathematical background has been developed for treating kinetic data but the general formulation is too formidable for treatment here. A few special cases will suffice to illustrate the principles involved.

It is assumed that a tagged compound is abruptly introduced into one body compartment, using compartment in a broad sense. It might refer to the bloodstream, to all body water, or to a single organ such as the lungs. The compound being traced is assumed to be uniformly distributed throughout each compartment that it enters but concentrations may vary widely from one compartment to another. The rate at which the tagged material leaves one compartment for another is assumed to be proportional to the concentration of the material in the first compartment. The constants of proportionality are known as *transfer coefficients*.

The following notation will be used, with the subscripts changed to apply to a series of compartments $A, B, \ldots$.

Q_a = quantity of the compound in A at time t
Q_{0a} = quantity of the compound in A at time zero
v_a = volume of A
q_a = concentration of compound in A at time t
$= Q_a/v_a$
q_{0a} = concentration of compound in A at time zero
k_{ab} = transfer coefficient from A to B
k_{a0} = irreversible transfer coefficient from a to the outside
λ_{ab} = transfer coefficient per unit volume
$= k_{ab}/v_a$

According to this notation, compartment B will receive material from A at a rate given by

$$\frac{dQ_b}{dt} = k_{ab}q_a = \frac{k_{ab}Q_a}{v_a} = \lambda_{ab}Q_a \tag{12-2}$$

12.05 Special Cases

Consider a two-compartment system, Fig. 12-1, with a constant concentration q_{0a} in compartment A, an irreversible flow into B, and an irreversible flow out of B. This case has limited application because it is not easy to maintain a constant concentration in A but it may be approximated by a system in which $v_a \gg v_b$. From the assumed conditions, $k_{ba} = 0$ and we have

$$\frac{dQ_b}{dt} = k_{ab}q_{0a} - k_{b0}q_b = \lambda_{ab}Q_{0a} - \lambda_{b0}Q_b \tag{12-3}$$

If the initial concentration $q_{0b} = 0$ at $t = 0$, this integrates to

$$Q_b = Q_{0a}\frac{\lambda_{ab}}{\lambda_{b0}}(1 - e^{-\lambda_{b0}t}) \tag{12-4}$$

Equation (12-4) has the same form as Eq. (11-8) and we see that each k/v is analogous to a decay constant. As in radioactive decay, Q_b approaches an equilibrium value along a time course similar to that depicted in Fig. 11-2.

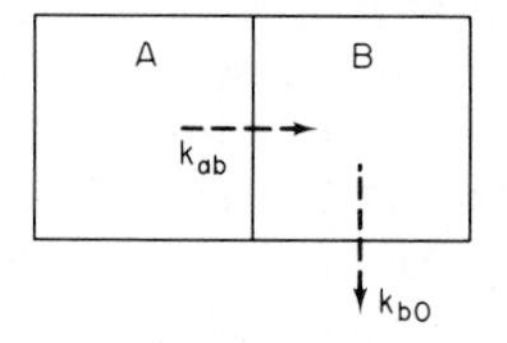

Figure 12-1. A two-compartment system with only two transfer coefficients.

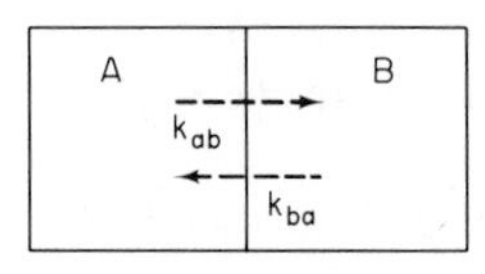

Figure 12-2. An isolated two-compartment system with no transfers to or from the outside.

As a second simple case, consider an isolated two-compartment system with no excretion ($k_{b0} = 0$), Fig. 12-2. At $t = 0$ the tagged quantities are Q_{0a} and zero, respectively. Then

$$\frac{dQ_b}{dt} = \lambda_{ab}Q_a - \lambda_{ba}Q_b \tag{12-5}$$

which integrates to

$$Q_b = Q_{0a}\frac{\lambda_{ab}}{\lambda_{ab} + \lambda_{ba}}(1 - e^{-(\lambda_{ab}+\lambda_{ba})t}) \tag{12-6}$$

Q_b approaches an equilibrium value

$$Q_b\,(\text{eq.}) = Q_{0a}\frac{\lambda_{ab}}{\lambda_{ab} + \lambda_{ba}} \tag{12-7}$$

Mathematical complexities develop rapidly as more transfer coefficients are introduced into the system model. In general there will be some movement out of compartment B, either as an irreversible excretion or as a transport to a third compartment. The introduction of transfer coefficients to account for such movements leads to equations with multiple exponential terms and with each coefficient a complicated function of the system parameters.

12.06 Data Analysis

The mathematical treatment of a simple idealized model leads to a prediction of compartmental concentrations as a function of time, expressed in terms of a series of presumably known parameters, k and v. In practice, measurements give the time course of the concentrations in one or more compartments and these data are to be interpreted in terms of the parametric constants of the system. In general, solutions of the form of Eq. (12-6) are not easily inverted to provide analytical solutions for the λ values. Empirical methods, perhaps aided by analog or digital computers, are usually invoked to analyze the system performance.

We have seen that exponential factors usually appear in the equations when material transport is set proportional to the concentration in the driving compartment. A concentration–time curve, Fig. 12-3, might then be analyzed by assuming it to consist of a sum of a few exponentials and adjusting the constants in these components to obtain a best fit with the data.

Analog computers are sometimes useful in analyzing concentration–time data. A simple resistance–capacitance electrical circuit responds to a voltage change in an exponential fashion. In the analog computer several of these *RC* circuits are combined and the individual circuit constants are adjusted until the overall performance of the circuit matches that of the biological data. Biological transfer coefficients are then deduced from the values of the electrical analogs.

Digital computers may also be used in analyzing data from tracer studies. Here again the sum of a small number of exponentials with initially undetermined coefficients is assumed to account for the behavior of the biological system. A computer program is then written for the sum of the squares of a series of error terms, which is the series of differences between the predictions of the equations and the values actually observed. The computer will then determine the values of the unknown coefficients so as to minimize the sum of the error squares. These best values are then interpreted in biological terms from a system model.

Too much biological significance must not be attached to any system parameters obtained from any of the analytical procedures. Although the methods may be mathematically elegant, a considerable variation in the parameters is usually possible while keeping the predicted time course of the sum within the inevitable uncertainties in the observed data. For example, six constants are involved in the analysis of the data shown in Fig. 12-3 and a good many combinations of them will provide a satisfactory agreement with measurement.

It is not necessary to subject the results of every tracer study to an elaborate analysis. Practitioners in the field of nuclear medicine have developed a series of routine diagnostic procedures from which organ or system be-

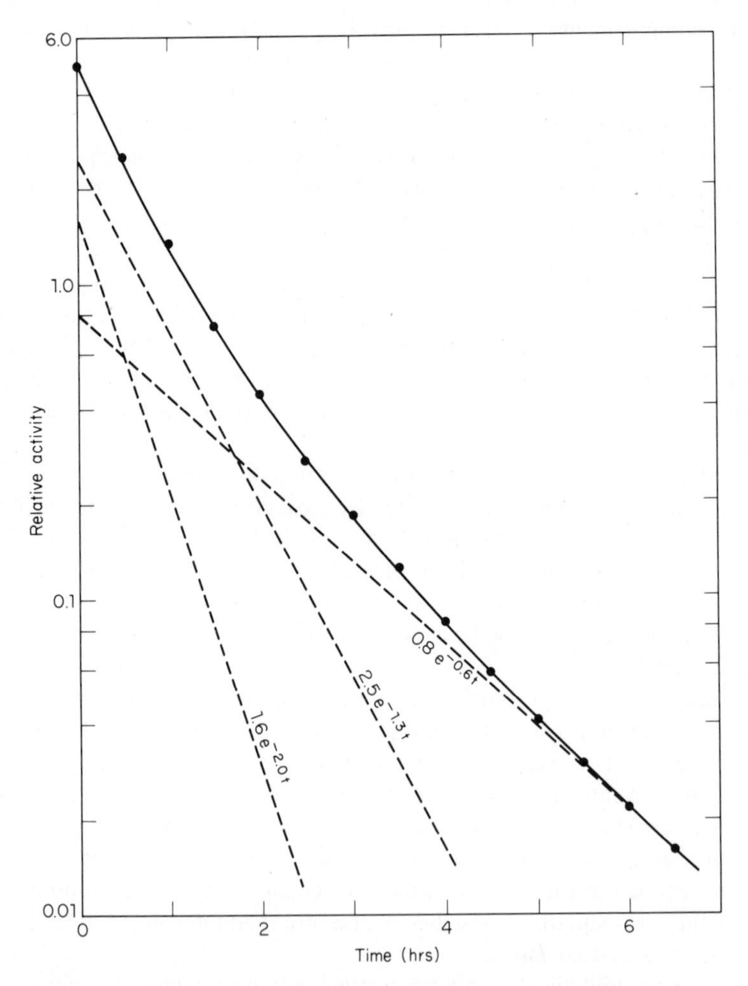

Figure 12-3. An observed concentration–time curve resolved into three exponentials: $c = 1.6e^{-2.0t} + 2.5e^{-1.3t} + 0.8e^{-0.6t}$.

havior can be simply estimated. An example is the determination of thyroid function with radioactive iodine, a procedure carried out routinely in many hospitals.

12.07 Isotope Production and Synthesis

Except in unusual circumstances a compound cannot be labeled by direct exposure to either neutrons or charged particles. An (n, γ), (n, p), or some other nuclear reaction may produce the desired isotope but when the nucleus ejects the secondary particle, recoil will almost certainly break all molecular bonds and release the radioactive atom from the original structure, Secs. 9.09 and 9.10. Radioactive isotopes usually come from the reactor or accelerator as simple inorganic compounds. ^{14}C, for example, may be in the form of some carbonate, $^*CO_3$. The tagged atom must then be synthesized into the desired compound.

Tagged compounds have been in such great demand that a wide variety are now available from commercial suppliers. Whenever possible, purchase of the desired compound is usually preferable to local synthesis since special chemical manipulations are usually required, perhaps in the presence of strong radiation fields.

The total quantity of any radioactive species is very small and if it is carrier-free, Sec. 5.03, all of it can be readily lost by adsorption on the walls of vessels. Small quantities may fail to precipitate when tabular values based on larger quantities indicate that they should. Any dilution of the radioactive isotope with a carrier, a stable isotope of the element, must be done with care to ensure that the final specific activity is high enough for the job in hand. Convenience in synthesis must usually be subordinated to yield and unusual reactions utilized to reduce the loss of active material.

12.08 Pulse Counting

In the field of radiation protection, measurements must take account of the ionizing ability of the radiations because the biological hazard will be determined largely by the total amount of tissue ionization. Survery instruments used in hazard evaluation will either be ionization chambers or will be calibrated in terms of them.

Radioactive tracer assays, on the other hand, are designed to determine the amount of a known isotope without regard to the ionizing ability of its radiations. Because of this simpler requirement, assays usually take advantage of the greater sensitivity that can be achieved with pulse counters as compared with ionization chambers. The detector used for pulse counting may be a Geiger counter tube, which can be so operated that even a single ion pair will trigger a maximal electrical response. Assays utilizing gamma-ray emissions will usually be made with scintillation detectors operated with discriminator circuits to restrict the response to ionization pulses of a limited size range.

Whatever the detector, a radioactive assay involves a determination of the rate at which a sample emits detectable radiations. To avoid undue radiation exposures of either human patients, laboratory personnel, or experimental animals, it is desirable to keep the amount of activity administered as small as is practicable. Many of the assays will, then involve the counting of relatively weak samples.

Because each individual decay is independent of all others, the emissions will occur at random times and the measured rate will be subject to the statistical fluctuations inherent in a series of random events. Particularly at low counting rates, counting schedules must be arranged to ensure that the results of an assay will have the reliability required by the job in hand.

Almost every tracer study requires only *relative* assays in which the activites are expressed as fractions of the activity administered. Some knowledge of the actual amount administered is required in order to conform to required safety regulations, Chapter 18, but in general an uncertainty of $\pm 20\%$ in this value can be tolerated. When radioactive materials are administered for therapeutic purposes, an *absolute* rather than a relative activity determination must be made in order that the desired radiation dose be delivered. Absolute activity determinations require the use of special assay methods or a calibration against a certified standard source under carefully controlled conditions.

12.09 Statistical Fluctuations

Any sample of radioactive material will decay at a characteristic average rate, and under good counting conditions a detector will intercept a constant fraction of the emitted radiations. The average counting rate $\bar{R}$ can only be measured exactly by counting over a very long period of time. Over practical counting times the observed rates will fluctuate above and below the true average rate. Statistical considerations must be invoked in order to estimate $\bar{R}$ from the measured values.

The following notation will be used:

N_t = total number of counts with sample in position (includes background)
N_b = total number of counts, background only
T_t = time sample is counted
T_b = time background is counted
R_t = gross count rate with sample in position = N_t/T_t

R_b = background count rate = N_b/T_b
R_s = count rate of sample = $R_t - R_b$
σ_t = standard deviation of the gross count rate
σ_b = standard deviation of the background count rate
σ_s = standard deviation of the sample count rate

In a strictly random system the fluctuations about the true mean are given by the *binomial* distribution function. In pulse counting we are always dealing with an integral number of events (counts) and so the unwieldy binomial representation can be replaced by the more convenient *Poisson* distribution.

$$P_R = \frac{(\bar{R}T)^{RT} e^{-\bar{R}T}}{(RT)!} \tag{12-8}$$

where P_R = probability of obtaining a count rate R in a counting time T
$\bar{R}$ = true mean counting rate

Only one parameter, $RT = N$, appears explicitely in the Poisson expression but inherent in the derivation of it is a second, the standard deviation of the mean rate, σ. σ is a measure of the "sharpness" of the count distributions around the mean. It can be shown that the standard deviation in the count of a total number of N random events is $\sqrt{N}$ and since we can assume that our time determinations are not subject to random errors, the standard deviation of a rate will be

$$(\sigma)_{\text{rate}} = \frac{\sqrt{N}}{T} = \frac{\sqrt{RT}}{T} \tag{12-9}$$

An evaluation of Eq. (12-8) for various values of N leads to the probabilities plotted in Fig. 12-4. According to this plot there is a probability of 0.317 that a single determination of R will differ from the true rate by $\pm 1\ \sigma$ or more. This probability drops to to 0.0455 for a range of $\pm 2\ \sigma$ and to 0.0027 for $\pm 3\ \sigma$. This is equivalent to saying that the chance that a single measure lies within $\pm 1\ \sigma$ of the true mean is 0.683, and so on.

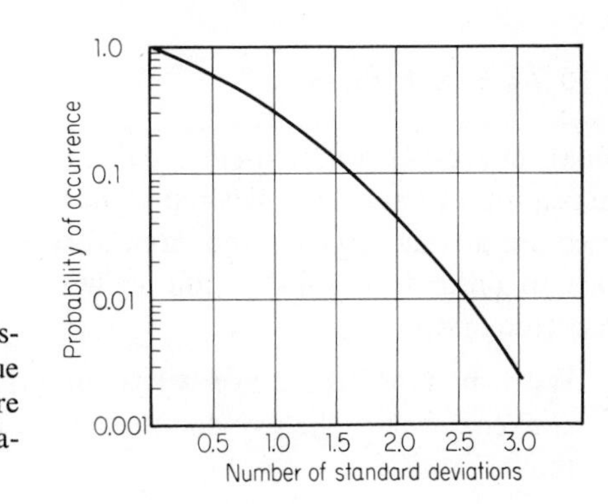

Figure 12-4. The probability in a Poisson distribution that a measured value will differ from the true value by more than a given number of standard deviations.

It is usually more useful to express the standard deviation as a fraction of the quantity being measured. This fraction, V, is known as the *fractional standard deviation* or the *coefficient of variation*. From the definition,

$$V = \frac{\sqrt{N}/T}{N/T} = \frac{1}{\sqrt{N}} \tag{12-10}$$

Any radiation detector will respond to a number of spurious, or background, counts due to radioactive material in the earth, in the surroundings and the equipment itself, and to extraterrestrial cosmic radiation. A complete assay consists of a background count N_b and a count with the sample in position, N_t. The latter count will contain contributions from the sample N_s, and from the background, and under any ordinary counting conditions $N_t = N_s + N_b$. Corresponding count rates are readily calculated and the count rate due to the sample is obtained by subtraction.

$$R_s = R_t - R_b \tag{12-11}$$

We are primarily interested in the standard deviation of the sample assay, σ_s, and since errors in the gross rate and the background rate combine in random fashion,

$$\sigma_s = \sqrt{\sigma_t^2 + \sigma_b^2} \tag{12-12}$$

The corresponding fractional standard deviation is

$$V_s = \frac{\sigma_s}{R_s} = \frac{\sigma_s}{R_t - R_b} \tag{12-13}$$

Combining these relations with Eq. (12-9), we have

$$V_s(R_t - R_b) = \sqrt{\frac{R_t}{T_t} + \frac{R_b}{T_b}} = \sqrt{\frac{N_t}{T_t^2} + \frac{N_b}{T_b^2}} \tag{12-14}$$

In practice, R_b will be known from a preliminary count and an approximate value of R_s can be estimated, either from a rough, quick count or from past experience with similar samples. A value of V_s is then selected to meet the needs of the study in hand and an assay time schedule is calculated.

12.10 Assay Schedules

Equation (12-14) can be used to calculate the counting times T_t and T_b that are required to achieve a desired value of σ_s and hence of V_s. Both of these times are at our disposal and some further requirement must be placed on them in order to obtain specific values of each. Three counting situations arise frequently.

1. With many of the automatic counters now available it may be most convenient to make two counts over equal periods of time. This requires that $T_t = T_b$.

Illustrative Example

Assume equal counting times and calculate the schedule required to obtain a fractional standard deviation of 0.02, or 2%, for a system in which $R_b = 40$ counts per minute (cpm) and R_s is expected to be about 250 cpm.

Squaring Eq. (12-14) and substituting numerical values,

$$(0.02 \times 250)^2 = \frac{290}{T_t} + \frac{40}{T_b} \quad \text{or} \quad 25 = \frac{330}{T_t}$$

$$T_t = 13.2 \text{ minutes}$$

$$\begin{aligned} &\text{total counting time} &&= 13.2 \times 2 &&= 26.4 \text{ minutes} \\ &\text{total background count} &&= 13.2 \times 40 &&= 528 \\ &\text{total gross count} &&= 13.2 \times 290 &&= 3828 \end{aligned}$$

Assume that the two counts taken for the calculated times do indeed verify that the sample count rate is nearly 250 cpm. By setting $V_s = 0.02$, we have required only a 63.8% probability that the measured value lies within 2% of the true rate, a 94.5% probability that it lies within 4% of the true value, and so on. Any choice of V_s leads only to a probability and not to a guarantee.

2. Some automatic counting equipment can be set to record the times required to obtain some predetermined number of counts. Under these counting conditions, $N_t = N_b$.

Illustrative Example

Use the same system parameters as before and calculate the counting schedule for $N_t = N_b$.

Again squaring Eq. (12-14) and substituting,

$$25.0 = \frac{290}{T_t} + \frac{40}{T_b}$$

For equal numbers of total counts the counting times will be inversely proportional to the count rates. That is, $T_b = 290T_t/40$ and so

$$25.0 = \frac{290}{T_t} + \frac{40 \times 40}{290T_t}$$

$$T_t = 11.8 \text{ minutes} \quad \text{and} \quad T_b = 85.8 \text{ minutes}$$

$$\text{total counting time} = 97.6 \text{ minutes}$$

$$N_t = N_b = 11.8 \times 290 = 3430$$

Under these counting requirements a slight reduction in T_t over that of the equal time program requires a substantial increase in T_b and in the total time of the assay.

3. It can be shown that the schedule leading to a minimum counting time $(T_t + T_b)$ is

$$T_t = \frac{R_t + \sqrt{R_t R_b}}{V_s^2 R_s^2} \quad \text{and} \quad T_b = T_t \sqrt{\frac{R_b}{R_t}} \tag{12-15}$$

Illustrative Example

Calculate the counting schedule for the same system and V_s requirement as before under minimum counting time requirements.

From Eq. (12-15),

$$T_t = \frac{290 + \sqrt{290 \times 40}}{(0.02 \times 250)^2} = 15.9 \text{ minutes}$$

$$T_b = 15.9\sqrt{\frac{40}{290}} = 5.9 \text{ minutes}$$

total counting time = 21.8 minutes

$N_t = 290 \times 15.9 = 4610$

$N_b = 40 \times 5.9 = 236$

In most assay situations, No. 3 is academic because it involves resetting automatic equipment for each sample count and background count. One background count can usually serve for several active sample counts and this relaxation acts to favor equal times rather than the minimum time as calculated.

12.11 Figure of Merit

Equation (12-14) can be written in terms of R_s, squared, and solved for T_t under the conditions of No. 1 above: $T_t = T_b$.

$$T_t = \frac{R_t + R_b}{(V_s R_s)^2} \tag{12-16}$$

For a fixed chosen value of V_s the counting time will be proportional to $(R_t + R_b)/R_s^2 = (R_s + 2R_b)/R_s^2$. The latter expression is known as the *figure of merit M* of the counting system.

When the background count is low compared to that of the sample, $M \simeq 1/R_s$, which is the trivial result that the more active the sample the shorter counting time required to attain a desired value of V_s. For samples whose activities are low compared to background,

$$M \simeq \frac{2R_b}{R_s^2} \tag{12-17}$$

Equation (12-17) demonstrates the desirability of working with a counting system which responds to a large fraction of the emitted radiations (R_s large) and which has a low background sensitivity (R_b small). Background can be reduced somewhat by placing the detector inside a heavy shield which will absorb some of the background gamma rays. The sample to be counted is placed inside the shield close to the sensitive detector volume.

When gamma rays are being counted, a substantial reduction in background can be achieved by restricting the response of the system to a limited range of photon energies. Background radiation covers a wide range of energies, while gamma rays are strictly monoenergetic. A solid scintillation detector such as sodium iodide–thallium-activated NaI(Tl) develops a *photopeak* at each gamma-ray energy, Fig. 12-5A, and the response of a lithium-activated germanium, or GeLi, detector is very narrow indeed, Fig. 12-5B. Circuit discriminators can be set to reject all pulses that are either larger or smaller than those making up the photopeak. Background activity over this narrow band of acceptance will be low, while the gamma rays from the sample are counted with good efficiency.

M is only a reflection of the counting time required to achieve a desired value of V_s with given activity conditions. M bears no relation to counter reliability, stability, or speed of response to each pulse, all of which are important characteristics of a counting system.

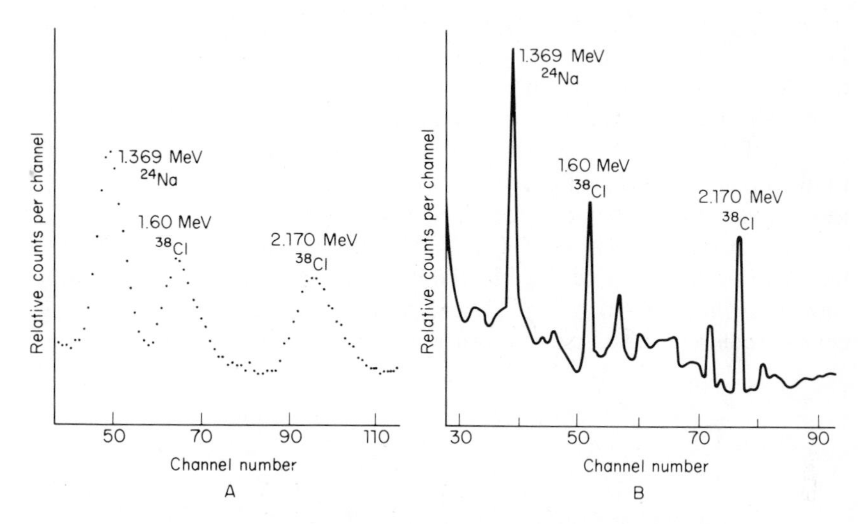

Figure 12-5. (A) A gamma-ray spectrum as recorded by a 7 × 5 in. NaI(Tl) detector and a multichannel analyzer system. (B) The same spectrum is more highly resolved with a 20 cm³ GeLi solid-state detector. (*Courtesy of D. A. Weber.*)

12.12 Carbon–Tritium Assays

Carbon-14 and hydrogen-3 are of such great importance in biology and medicine that assay techniques and instrumentation have been developed specifically for them. Each is a pure beta-emitter with maximum particle energies of only 0.156 and 0.0186 MeV, respectively. Some of the more energetic betas from ^{14}C will pass through a thin window and enter a sensitive counting volume but efficiency is low and almost all ^{3}H betas are absorbed in even the thinnest window. Although several assay methods have been used, liquid scintillation counting appears to be the most satisfactory for the kinds of samples usually obtained in biological studies.

A typical scintillating solution consists of

Primary scintillator 2,5-diphenyloxazole (PPO)	5 g
Secondary scintillator 1,4-bis-(5-phenyloxazole) benzene (POPOP)	0.1 g
Solvent toluene	1 liter

An aliquot of the solution to be assayed is added to 15–20 ml of the scintillator and the mixture is placed in a glass vial positioned accurately with respect to a pair of photomultuplier (PM) tubes. Beta particles produce ionizations and excitations in all components of the mixture and much of the energy absorbed in the toluence will be transferred to the PPO. Excited PPO fluoresces with the emission of visible and UV photons which can be detected by the PM tubes. The main fluorescent energy from PPO is emitted at wavelengths somewhat shorter than those at which phototube detection is most efficient. POPOP is added as a secondary scintillator or *wave shifter* to absorb some of the PPO scintillations and then emit its own characteristic fluorescence at somewhat longer wavelengths.

Each of the two isotopes ^{14}C and ^{3}H emits a continuous negatron spectrum and so the pulses detected by the PM tubes will have a size distribution similar to that shown in Fig. 12-6,A. Many solutions of biological origin contain colored substances or other molecular forms which will absorb

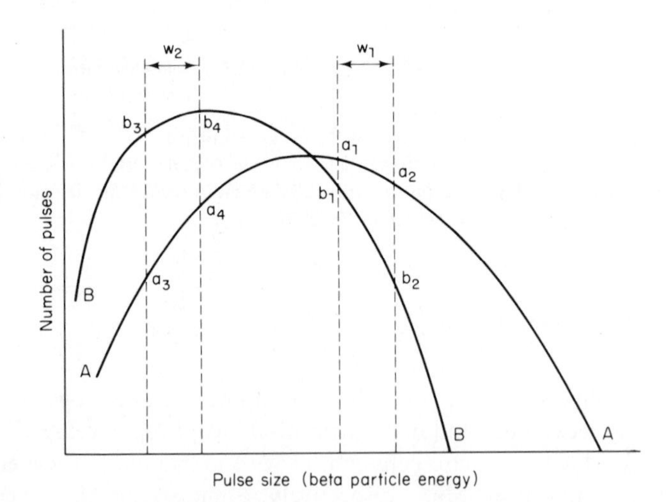

Figure 12-6. (A) Unquenched and (B) quenched spectral distribution of pulses due to beta particles in a scintillating solution. Counting is restricted to the energy windows W_1 and W_2.

some of the PPO and POPOP scintillations and will then emit their own characteristic wavelengths. Most of these secondary photons have a lower energy than the originals, Sec. 16.07, and so a lower-energy or *quenched* spectrum will be seen by the phototubes, Fig. 12-6,B.

A given activity being counted over an energy range or window W_1, Fig. 12-6, will show a count rate proportional to the area under a_1a_2 in an unquenched solution but this will be decreased to b_1b_2 if there is quenching. The count rate recorded will thus depend on the amount of some quenching substance in the sample which may bear no relation to the amount of radioactivity being assayed.

A correction for solution quenching can be made by counting simultaneously from two energy windows or channels, W_1 and W_2, Fig. 12-6. From channel W_1 the greater count rate will be obtained from the unquenched solution but from channel W_2 the count rate b_3b_4 is substantially greater than the unquenched rate a_3a_4. From the channel ratios b_1b_2/b_3b_4 obtained with a series of quenched solutions of known activity, correction factors for solutions with unknown quenching can be determined.

Carbon-14 and hydrogen-3 have beta energies sufficiently different to permit simultaneous assays on a double-tagged compound. Details of this and other liquid scintillation counting techniques will be found elsewhere.*

12.13 Organ Scanning

A number of pathological conditions such as neoplasms or inflammatory processes change the permeability of the blood-tissue barriers and lead to a local concentration of an appropriate radioisotope. This effect has led to the development of a variety of clinical diagnostic procedures and to some sophisticated instrumentation for carrying them out. The patient is given a few microcuries of the desired photon emitter and at some later time the suspect organ is surveyed with a localizing detector. Liver and brain are two organs in which scanning techniques have been particularly successful.

Organ scanning requires the highest spatial resolution possible and many types of collimators, Fig. 12-7, have been designed to obtain it. The highly collimated detector is arranged to scan over the desired portion of the body. A pen synchronized with the scanning detector makes an ink dot on a recording paper for each group of 100 counts or so. A concentration of radioactive material at some point in the organ will be represented by a concentration of ink dots, Fig. 12-8.

There are many variants of the basic scanning principle. In one widely used version, the *Anger camera*, a lead collimator is placed in front of a

*Parmentier, J. H. and F. E. L. TenHaaf, "Developments in Scintillation Counting Since 1963." *Int. J. Applied Rad. and Isotopes*, **20**, 305, 1969.

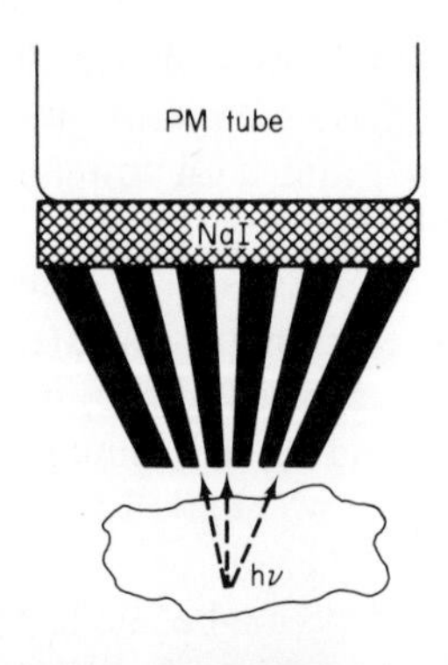

Figure 12-7. A heavy collimator of high-Z material localizes the response of the NaI(Tl) crystal detector.

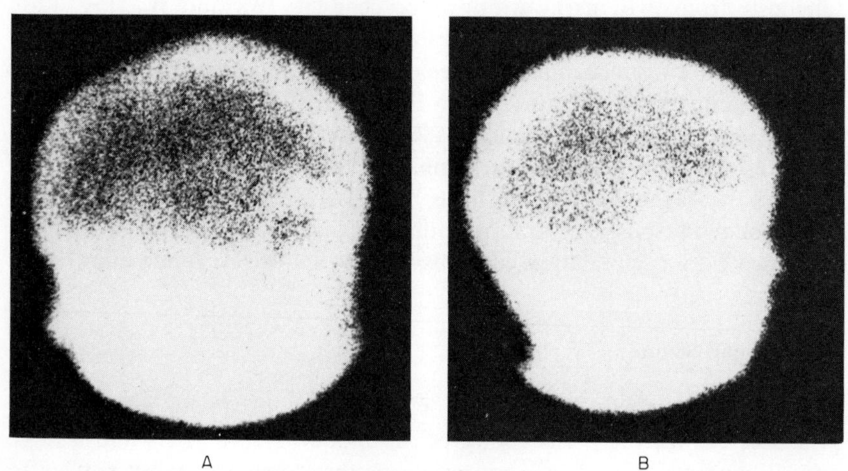

Figure 12-8. Lateral brain images produced on a cathode-ray tube by a gamma-ray camera unit following (A) 10 mCi and (B) 3 mCi of ^{99m}Tc pertechnate. Each distribution was considered normal. (*Courtesy of D. A. Weber.*)

solid scintillator, such as NaI(Tl), in the form of a thin slab. As many as 19 photomultiplier tubes have been used to view the scintillator from the rear, Fig. 12-9. A photon entering one of the channels of the collimator will produce a scintillation in a highly localized area of the NaI. This will be "seen" by all the 19 PM tubes, each with a characteristic intensity determined by its position with reference to the point of origin of the light pulse. A computer circuit interprets the relative signal strengths in terms of the geometrical location of the light pulse origin and directs a mechanism to record the pulse at the corresponding point on the recorder. The result is an effective scan over the area covered by the collimator without any relative motion between detector and patient.

A new field, known as nuclear medicine, has developed as a result of the

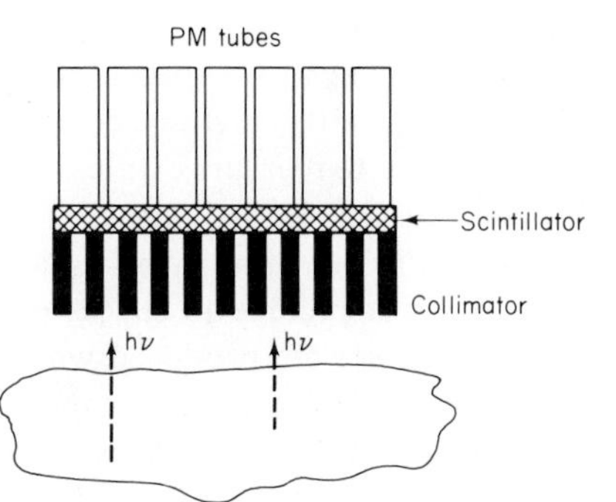

Figure 12-9. In the Anger camera arrangement a high-Z collimator is combined with a crystal scintillator viewed by a number of photomultiplier tubes.

diagnostic possibilities afforded by the wide variety of photon emitters now available. When isotopes of a desired element are not available with the proper characteristics, it is sometimes possible to substitute another element having an identical structure in the outer electronic orbits but with a different inner core. Tracing then depends on chemical similarities rather than on identities. Thus selenium is sometimes substituted for sulfur and either cesium or rubidium for potassium.

12.14 Autoradiography

Photographic emulsions are used in four different ways for the detection of radiation. Radiography is a well-established technique in medicine and dentistry and is also used extensively in industry for the inspection of metal sections with strong sources of gamma radiation. In a highly developed country such as the United States the medical profession is the largest single consumer of photographic materials.

Personnel monitoring is customarily carried out with dental-sized photographic films placed in *film badges* that are equipped with filters designed to make the response to photons more nearly tissue-equivalent. The presence in the emulsions of the high-Z elements Ag and Br cause the response to deviate substantially from that of tissues at energies below 90 keV or so. Thermoluminescent dosimeters are replacing photographic film badges for some applications but the latter provide the only permanent record of radiation exposure that can be stored for reference many years later.

Nuclear-track emulsions are used to detect and identify the tracks of individual ionizing particles. Alpha particles, protons produced most usually as a recoil from a fast neutron collision, and electrons have quite different values of LET and consequently can be readily identified by the track of ionization left in suitable emulsions. Emulsions have been invaluable in studying the properties of some of the newly discovered particles such as mesons and hyperons.

Nuclear-track emulsions are usually 50–100 μm thick instead of the more usual 8–10 μm; therefore special processing methods are needed for the thick materials. With carefully controlled procedures, nuclear-track emulsions can be used to measure exposure to fast neutrons. Quantitation requires the careful and tedious counting of proton tracks at high magnification. Here again technique approaches an art so that work in an established laboratory is the best prelude to successful application.

Autoradiographs are produced when a photographic emulsion is brought into contact with or in close juxtaposition to a tissue slice bearing a beta-emitting radioactive nuclide. After a suitable exposure time, processing will reveal blackened areas corresponding to the areas of isotope deposition. The techniques of autoradiography are legion; only general principles can be mentioned here. Special sources* must be consulted before doing serious work in the field.

Autoradiography is a powerful technique because of its high ultimate sensitivity. Electrical circuits can integrate only over a limited time but emulsions can be exposed for weeks or months, until the accumulated effect of even weak emissions has produced a measurable photographic effect.

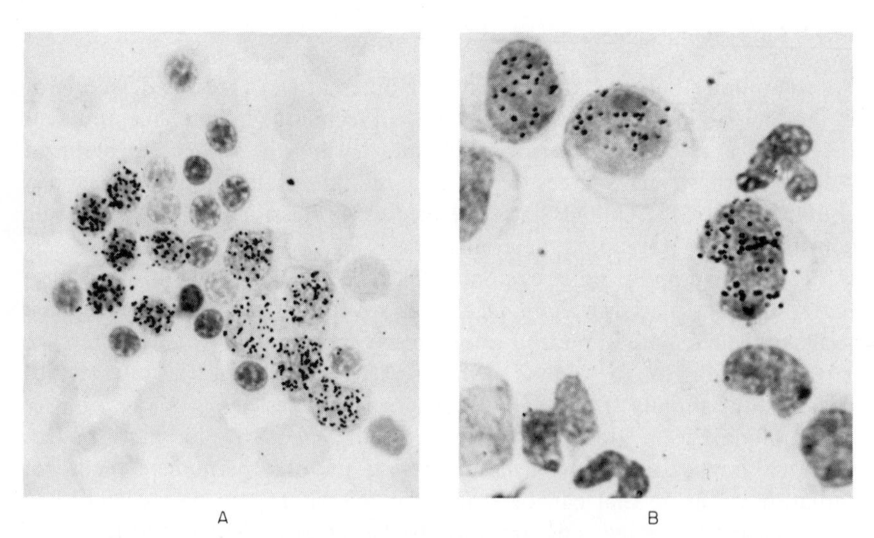

A B

Figure 12-10. Autoradiograph of erythropoietic cells incubated with ^{3}H-labeled thymidine. (A) Normal human bone marrow cells. (B) Peripheral blood cells of a patient with chronic myelocytic leukemia. (*Courtesy of V. P. Bond, T. M. Fliedner, and E. P. Cronkite.*)

*Gude, W. D., *Autoradiographic Techniques*. Prentice-Hall, Inc., Englewood Cliffs, N.J., 1968.

Emulsions are available covering a wide range of sensitivities with the common materials requiring perhaps 10^5 beta particles per square millimeter to produce a measurable optical density.

Thin tissue sections must be used if good spatial resolution is required, in order to locate accurately the source of the beta emissions. Beta particles originating within a thick specimen will suffer multiple scattering before striking the photographic emulsion and so will broaden the image. Tritium betas have such a short range that they can only emerge from a thin outer layer, and high-resolution autoradiographs can be produced from thick sections. Figure 12-10 is a good example of the high spatial resolution that can be obtained with tritium autoradiographs.

REFERENCES

ANGER, H. O. "Whole Body Scanner, Mark II." *J. Nuc. Med.*, **7**, 331, 1964.

BLAHD, W. H., ED., *Nuclear Medicine*, 2nd ed. McGraw-Hill Book Co., New York, N.Y., 1971.

Medical Radioisotope Scanning, Vol. I. Proceedings of a symposium on medical radioisotope scanning held by the International Atomic Energy Agency in Athens, April 20–24, 1964. From IAEA, Vienna, 1964.

Uses of Isotopically Labeled Drugs in Experimental Pharmacology. Proceedings of an international conference sponsored by the International Atomic Energy Agency, Chicago, Ill., 1964. From IAEA, Vienna, 1964.

WAGNER, HENRY M., ED., *Principles of Nuclear Medicine*. W. B. Saunders Co., Philadelphia, Pa., 1968.

13

Radiation Effects in Homogeneous Independent Systems

13.01 Macromolecules and Cells

Radiation absorbed in a living system initiates a complex series of reactions which sooner or later may become manifest as an alteration in the normal functioning of the system. Some effects, particularly at high doses, may appear promptly after the radiation has been received. In humans, nausea and vomiting follow within a few hours after an exposure of all or a large fraction of the body to doses of several hundred roentgens. A radiogenic cancer, on the other hand, may not become evident for 20 years or even more after a radiation exposure.

Radiation effects in an organism as complex as a mammal are the end result of the interplay of a vast number of interrelated organs and systems, each of which may have been injured by the radiation. The direct unraveling of the injury mechanisms in such a complex organism is indeed a formidable task.

Radiation effects have been intensively studied in relatively simple systems such as macromolecules, bacterial and yeast suspensions, and, more recently, cultures of mammalian cells. With these test objects the effects of statistical fluctuations in a nonuniform population can be reduced by working with large numbers of a single genotype. The use of synchronized cell populations reduces the variability due to the changes in radiation sensitivity that occur during the mitotic cycle. Studies can be continued for several generations postradiation with comparative ease.

Idealized models can be developed from studies on simple, homogeneous populations in which the radiation effect in any one individual is independent of events in other members. One of the models assumes a sensitive volume or "target" within each cell or molecule. An ionizing event or "hit" within this volume initiates changes which lead to an observable effect, be it a genetic change, inactivation, or cell death. The target concept has been useful in explaining some radiation effects but it is an incomplete picture that must be applied with caution to the more complex organisms such as mammals.

13.02 Target Concept

As far back as 1924 Crowther proposed a cellular model assumed to contain a radiation-sensitive volume or *target.* His data on the inhibition of cell division by radiation could be explained by assuming a single sensitive volume about the size of a centromere within each cell, with inhibition following even a single ionization within the sensitive volume. In principle the model could be readily expanded to include cells with multiple targets which might require more than one ionization for inactivation.

Lea extended the simple model to take into account the effects of radiations with different values of linear energy transfer and applied the target theory to a wide variety of biological systems. His book* is a classic in the field.

Later experimental evidence indicates clearly that all biological effects cannot be accounted for by assuming direct action on a sensitive target. Indirect action, in which the primary radiation effect takes place outside the affected system is well-known. This is not surprising. Most living organisms contain a large proportion of water, and from pure chance alone many and perhaps most of the ionizations will take place in water molecules. Some of the effects observed will be due to the actions of the radiolytic products of water rather than to an ionizing event in a sensitive volume. It is not easy to distinguish between the two types of mechanisms which may have quite similar dose-response relations.

13.03 Hit Probability

Ionizing radiation is an essentially nonspecific agent which transfers energy to the molecules of an absorbing medium without regard to the chemical status of the individual atoms. Nonspecificity will not obtain when the quantum energies available are nearly equal to the energies of the electronic excitations or to the energies of the interatomic vibrations and rotations. At the low energies that characterize the ultraviolet, visible, and infrared portions

*Lea, D. E., *Actions of Radiations on Living Cells*, 2nd ed. Cambridge University Press, Cambridge, 1955.

of the spectrum, resonance absorptions will preferentially emphasize some interactions at the expense of some others. Only the higher energies are considered in the present discussion.

Even in a test object as small as a bacterium or a single cell there will be many ionizations that do not result in any eventually detectable signs of a radiation effect. With any reasonable radiation dose, however, the number of ionizations is very large and there will be a small but finite probability that a detectable event (usually an injury) will be produced in one of the test objects. If the irradiated sample consists of a large population of a single genotype, each individual in the population will be equally at risk, with a mean probability of injury p. This fact permits the use of Poisson statistics in determining the number of injuries to be expected from a given radiation dose.

For this situation the Poisson distribution may be written as

$$P_x = \frac{p^x e^{-p}}{x!} \tag{13-1}$$

where P_x is the probability that exactly x events will be produced in an individual member of the population. $x!$, or x *factorial*, is the continued product $1 \cdot 2 \cdot 3 \cdots \cdot x$. The mean probability of an event, p, will be linearly proportional to the number of ionizations and hence to the dose D. Explicitly

$$p = kD \tag{13-2}$$

where k is a constant whose value depends on the cell type and the phase in the mitotic cycle at which the radiation is received and on the nature of the radiation. Some of the terms given by Eq. (13-1) are tabulated below in units of dose. Other terms may be written as required.

$$\begin{array}{c|cccccc} x & 0 & 1 & 2 & 3 & \cdots \\ P_x & e^{-kD} & e^{-kD}(kD) & e^{-kD}\dfrac{(kD)^2}{2} & e^{-kD}\dfrac{(kD)^3}{6} & \cdots \end{array} \tag{13-3}$$

The sum of all terms taken over all possible values of x, including zero, must equal certainty, or a probability of unity. That is,

$$e^{-kD}\left[1 + kD + \frac{(kD)^2}{2} + \frac{(kD)^3}{6} + \cdots\right] = 1 \tag{13-4}$$

13.04 Multihit Theory

Consider specifically a system with an initial population of N_0 identical individuals, in which we shall measure the inactivation (inhibition of proliferation) at a radiation dose D. We assume that the inactivation of a cell requires m events or hits, where m will be an integer. Then any cell receiving m hits or more will fail to proliferate. The total probability of inactivation will be, from Eq. (13-3),

$$P_m = e^{-kD} \sum_{x=m}^{\infty} \frac{(kD)^x}{x!} \tag{13-5}$$

It is usually more convenient to deal with the surviving fraction N/N_0 rather than with the number of cells killed. N/N_0 will be the sum of all the probabilities for *less* than m hits. This sum will be just the terms that were excluded from Eq. (13-5).

$$\frac{N}{N_0} = e^{-kD} \sum_{x=0}^{m-1} \frac{(kD)^x}{x!} \tag{13-6}$$

When a single hit is sufficient for inactivation, the surviving fraction becomes a simple exponential function of the dose since only one term will be needed in Eq. (13-6). Then

$$\frac{N}{N^0} = e^{-kD} \tag{13-7}$$

This is characterized by a straight line on a log (N/N_0)–D plot, Fig. 13-1. At a particular dose D_0 such that $kD_0 = p = 1$, the surviving fraction will be equal to $e^{-1} = 0.368$, which is commonly rounded off in most discussions to 0.37. D_0 is the only parameter needed to describe the single-hit inactivation. Experimentally, many cell types follow the predictions of single-hit theory with D_0 values ranging from 30–300 rads. Figure 13-1 is plotted for $D_0 = 200$ rads, whence $k = 0.05$ rad^{-1}.

When more than one hit is required for inactivation, corresponding terms will appear in Eq. (13-6) and the survival fraction will no longer be a simple exponential function of dose. On a semilog plot the survival curve will have a shoulder at low doses and will then tend toward linearity as the doses increases. Figure 13-1 shows a plot for the case of a two-hit requirement.

Multihit survival curves cannot be characterized by a single parameter. The terms in common use are diagramed in Fig. 13-1.

D_0 is now the additional radiation dose required to reduce a population to $e^{-1} = 0.37$ of its value along the essentially exponential portion of the survival curve. In Fig. 13-1, $D_0 = 210$ rads for the two-hit requirement.

D_{37} is the dose required to reduce the initial population to 0.37 of its value. The effect of the shoulder leads to a value of $D_{37} = 420$ rads in Fig. 13-1. n is the *extrapolation number* obtained by linearly extrapolating back to zero dose the high-dose portion of the survival curve. Since this curve continues to bend downward with increasing doses, the exact value of n will depend somewhat on the dose range covered by the data. The extrapolation number will always be greater than 1.0. Figure 13-1 shows a value of $n = 3.8$.

D_Q is a quasi-threshold dose, evaluated at the point where the linear extrapolation line intersects the 1.00 surviving fraction line. Since the slope of

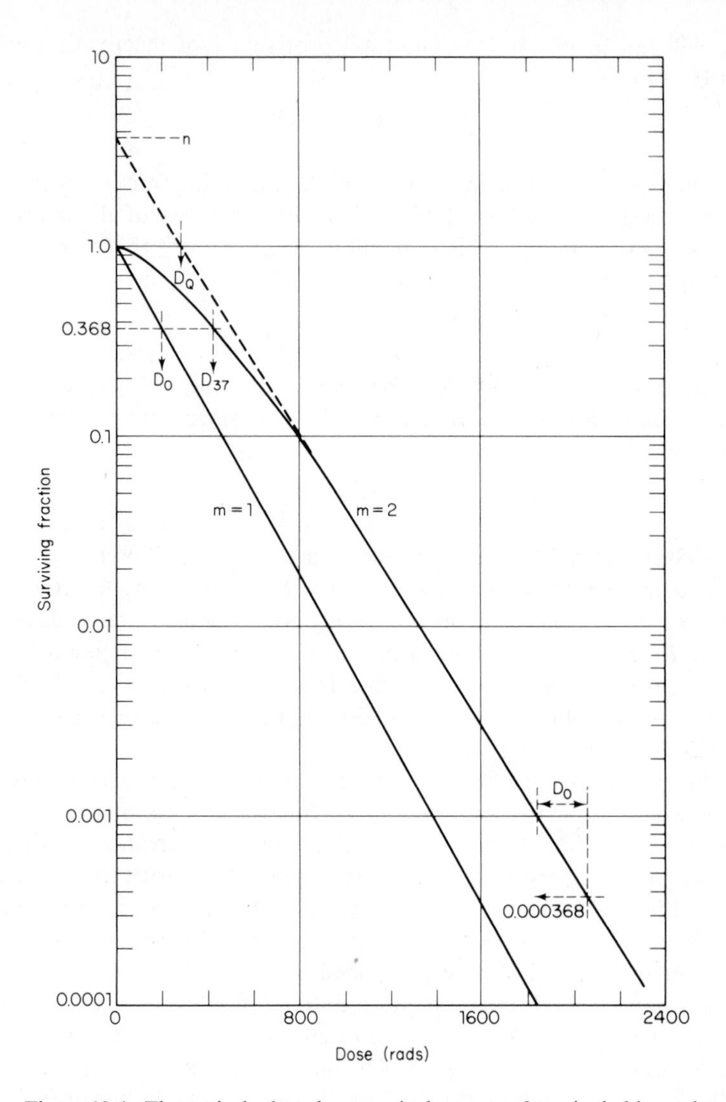

Figure 13-1. Theoretical dose–log survival curves for single-hit and double-hit requirements for inactivation. One parameter, D_0, suffices to describe the single-hit relationship; four parameters, D_0, D_{37}, D_Q, and n are needed in the two-hit case.

the extrapolation line is determined almost entirely by the last, or $(m - 1)$, term in Eq. (13-6), D_Q is a measure of the effect of the shoulder that results from the terms of lower order.

All the parameters listed arise from a theoretical treatment of a simplified cell model. In practice, survival curves are drawn as best fits to plots of experimental data. The values of the parameters derived from these plots are useful for comparing various biological systems and experimental conditions but they must not be interpreted rigidly in terms of cellular characteristics. For example, the value of the extrapolation number increases with the value of m assumed in the theory but the two are not necessarily equal, as is evident from the plot in Fig. 13-1.

13.05 Multitargets

A plausible cellular model may require that more than one target be damaged in order to produce a particular effect. For example, a genetic abnormality might be produced by a single premitotic hit on a chromosome but inhibition of proliferation might require a hit in each member of a chromosome pair.

For simplicity, consider that there are n critical targets of equal sensitivity and that a single hit in each of them is a necessary and sufficient requirement for inactivation. From Eq. (13-6) the probability of survival of one of the critical targets is e^{-kD} and hence its chance of being hit is $(1 - e^{-kD})$. The chance that each of n independent targets of equal sensitivities will be hit is $(1 - e^{-kD})^n$ and the probability of escape from the n-target requirement is $1 - (1 - e^{-kD})^n$. Thus,

$$\frac{N}{N_0} = 1 - (1 - e^{-kD})^n \tag{13-8}$$

The power term in Eq. (13-8) can be expanded in a series of terms according to the binomial theorem.

$$\frac{N}{N_0} = 1 - \left[1 - ne^{-kD} + \frac{n(n-1)e^{-kD}}{2} - \frac{n(n-1)(n-2)e^{-kD}}{2 \times 3} + \cdots\right] \tag{13-9}$$

When kD is large, the terms beyond ne^{-kD} become negligible. Then

$$\frac{N}{N_0} = ne^{-kD} \tag{13-10}$$

and

$$\log \frac{N}{N_0} = \log n - kD \tag{13-11}$$

According to Eq. (13-11), a log (N/N_0)–D plot should be linear at high values of D and an extrapolation of this portion of the survival curve back to $D = 0$ should give the extrapolation number n. In this case the extrapolation number should be equal to the number of critical targets.

Parameters D_0, D_{37}, and D_Q can be identified as in the multihit theory, Fig. 13-2. Multihit and multitarget theories lead to survival curves with roughly similar shapes. Experimental data may contain enough uncertainties

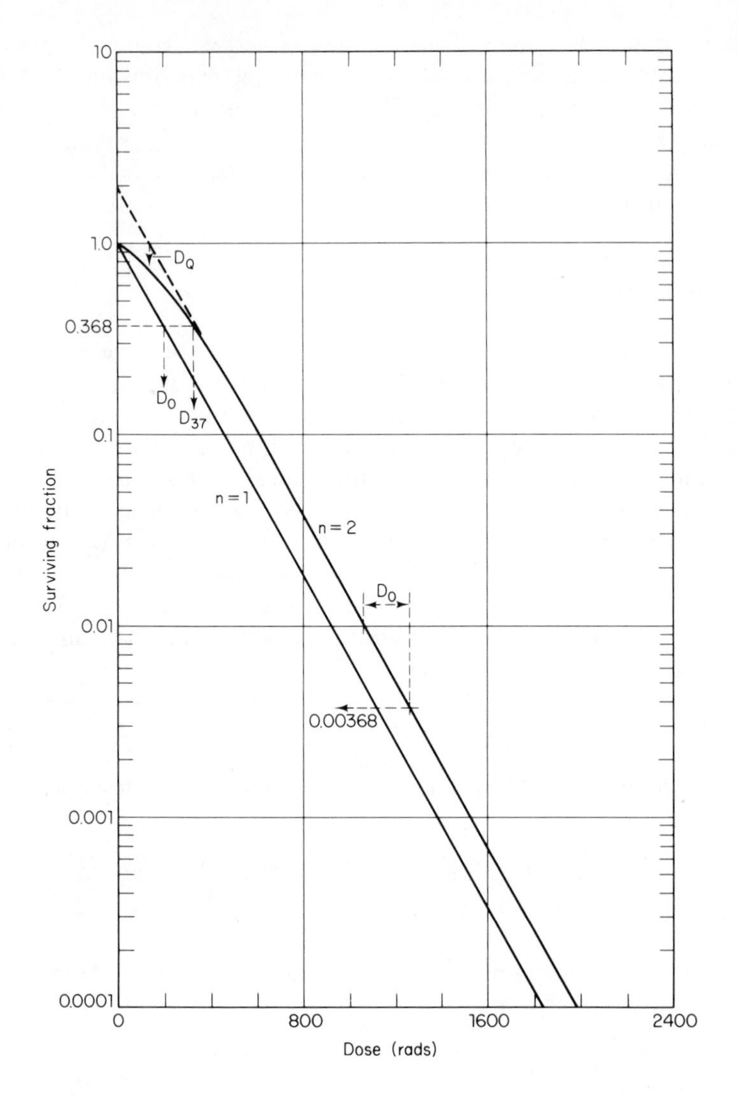

Figure 13-2. Theoretical dose–log survival curves for a single-target and a two-target inactivation requirement, with each target requiring only a single hit to produce an effective injury. Parameters are analogous to those in the multihit theory.

to prevent an absolute identification of the radiation injury mechanism in terms of either of the two theories. A combined multihit multitarget requirement can be visualized, with a particular sensitivity, or k value, associated with each effective event. Seldom, if ever does the uniformity of the biological material and the quantitation of the effect have the precision to warrant the

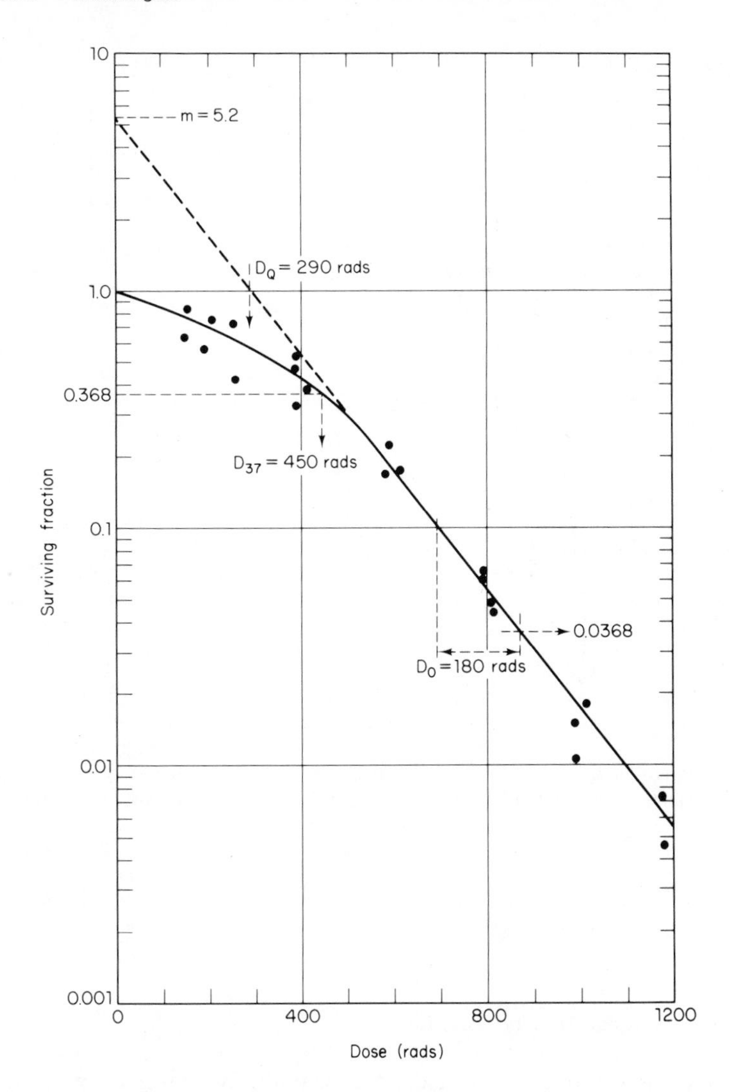

Figure 13-3. A plot of typical dose–log survival data. Gamma-ray irradiation of Chinese hamster cells in culture.

use of the more complicated mathematical treatment. Furthermore, indirect action, not incorporated into the target model, may effectively change the survival curves from those predicted. Figure 13-3 is a plot of typical data obtained with the gamma-ray irradiation of hamster cells. Parameter values are shown but no target model can be inferred from these results alone.

It is occasionally desirable to display the data on a linear dose–mortality plot. These curves will have a typical sigmoid shape with a low-dose knee

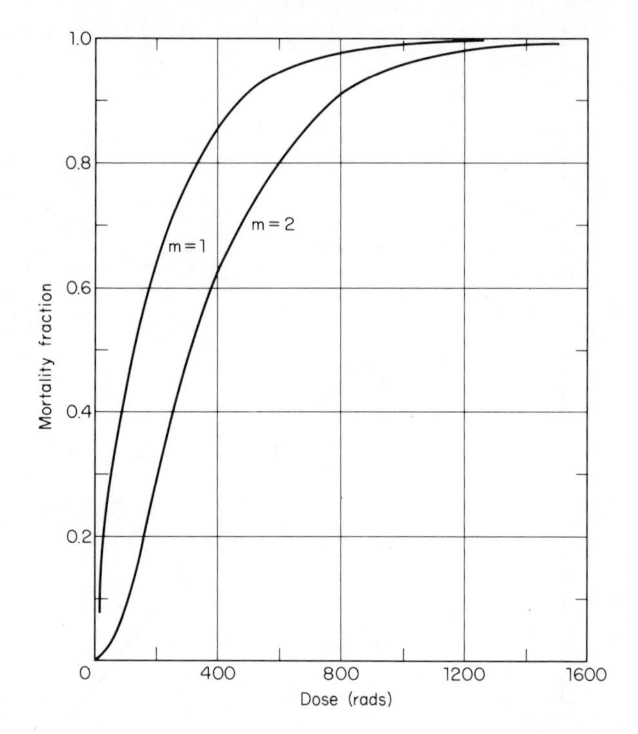

Figure 13-4. Linear dose–mortality plots of the data shown as log survival curves in Fig. 13-1. The linear plots are sigmoids with curvatures depending on the individual parameters.

and a high-dose plateau. Figure 13-4 shows linear dose–mortality plots for the log survival curves of Fig. 13-1.

13.06 High Linear Energy Transfer

We have already noted, Sec. 6.05, that a multiply charged ion such as an alpha particle traverses a very short path in biological tissues. In the example given, the 4.8-MeV alpha with a range of 35 μm in tissue will have a mean linear energy transfer of $(4.8 \times 10^3)/35 = 140$ keV μm^{-1} and a specific ionization of $(1.4 \times 10^5)/34 = 4200$ ion pairs μm^{-1} or 0.42 ion pairs Å^{-1}. Both of these figures are mean values of L_∞ averaged over the entire track of the particle. With such closely spaced ionizations the probability of at least one ionization during the traversal of a large molecule or a cell is almost unity.

For an order-of-magnitude calculation, assume a spherical molecule such as a protein or an enzyme with a molecular weight of 10^6. With an assumed density of 1.3 g cm^{-3} the molecular radius is calculated to be 67 Å. The average path length through a sphere is 1.3 times the radius and this leads to a value of $1.3 \times 67 \times 0.42 = 36$ ionizations per track. Obviously the probability of no ionization during a transit is almost zero. Under these conditions the probability of initiating a detectable event, even if multiple hits are required, will depend on the target area rather than on its volume.

Let a biological sample contain N targets per square centimeter normal to the trajectories of the incident high-L particles, and let the fluence of particles per unit area be f cm^{-2}. Then the number of inactivations will be proportional to the number of available targets and also to the number of bombarding particles. That is,

$$dN = -S_0 N\, df \tag{13-12}$$

where S_0 is a constant of proportionality. Equation (13-12) is a familiar form which integrates to

$$N = N_0 e^{-S_0 f} \tag{13-13}$$

Since f has the dimensions of a reciprocal area, S_0 is an area or cross section which is to be related to the target area presented to the bombarding particles. At the 37% survival point, $S_0 f_{37} = 1$.

When experimental data show a linear relationship between $\log(N/N_0)$ and f, and if the radiation is known to have a high value of linear energy transfer, a target cross section can be deduced from Eq. (13-13).

13.07 Intermediate Values of *L*

Many ionizing particles do not have sufficiently high values of L to ensure that at least one ionization will occur on the transit of a particle. In these cases, Poisson statistics may be invoked to calculate the probability of an ionization.

The term kD previously used in the Poisson treatment was introduced as the mean probability of an event, Eq. (13-2). For the present application the chance of zero ionizations during a particle transit is

$$P_0 = e^{-p} = e^{-p_1 d} \tag{13-14}$$

where p_1 is the mean number of ionizations per unit path length and d is the thickness of the target. The probability that there will be one or more ionizations during the passage of a particle P_1 is

$$P_1 = 1 - e^{-p_1 d} \tag{13-15}$$

As an example, consider a beta particle emitted by ^3H traversing the spherical molecule with radius 67 Å previously discussed. The beta particle has a mean value of L_∞ of about 5 keV μm^{-1}. The average specific ionization is then $\frac{5000}{34} = p_1 = 150$ ionizations per μm or 0.015 $Å^{-1}$. The mean number of ionizations per particle transit is $p_1 d = 1.3 \times 67 \times 0.015 = 1.4$. The probability that there will be one or more ionizations per transit is $P_1 = 1 - e^{-1.4} = 0.75$ instead of the value of 1.00 that is applicable to an alpha-particle passage.

The maximum value of the target cross section S_0 will be measured when the ionization probability per passage is 1.00. When this probability is less than unity as calculated above, the experimentally determined cross section S will be smaller according to

$$S = S_0(1 - e^{-p_1 d}) \tag{13-16}$$

If p_1 is known from the nature of the radiation, the S values can be corrected to obtain S_0.

A series of cross section measurements using radiations with various known values of p_1 can be used to estimate the target shape, volume, and molecular weight. A plot of log S against p_1 can be extrapolated toward high values of p_1 to obtain a value of the maximum cross section S_0. Target thickness d is obtained from the slope of the plot and from these data the target shape, volume, and molecular weight can be estimated.

Data obtained by Deering* on the inactivation of invertase and ribonuclease are typical. Values obtained from a series of deuteron irradiations were

Invertase	$S_0 = 6.9 \times 10^{-13}$ cm^2	$d = 3.7 \times 10^{-7}$ cm
Ribonuclease	$S_0 = 1.4 \times 10^{-13}$	$d = 2.6 \times 10^{-7}$

If a circular cross section is assumed, the S_0 value for invertase corresponds to a radius of 4.7×10^{-7} cm. If the molecule were spherical, it would have a mean thickness of 1.3 times the radius, or 6.1×10^{-7} cm in the present case. This value is incompatible with the measured d value of 3.7×10^{-7} so the sensitive volume is probably not spherical.

Next one might consider the possibility of a cylindrical sensitive volume with the deuteron beam coming in perpendicularly to the axis of the cylinder. The average path length through a circular target is $d = \pi r/2$, which leads from the invertase data to a radius of 2.35×10^{-7} cm. The data are then satisfied by a cylindrical sensitive volume of radius 2.35×10^{-7} cm and a length of 1.5×10^{-6} cm. These values lead to a sensitive volume of 2.5×10^{-19} cm^3 and a mass of 3.3×10^{-19} g. The molecular weight of this target is $(3.3 \times 10^{-19}) \times (6.02 \times 10^{23}) = 2 \times 10^5$, a value that is not inconsistent with other determinations. Similar analyses of the data for the ribonuclease molecule lead to a spherical target structure with a molecular weight of 27,000.

Target shapes and sizes deduced in this way are suggestive but are not

*Deering, R. A., "Inactivation Cross Section of Dried Invertase and Ribonuclease as a function of LET." *Rad. Research*, **5**, 238, 1956.

conclusive. It would not be surprising to find different cross sections for different end effects since each of the latter may originate at a specific portion of the total structure. For example, the radiation effects on chymotripsin can be determined by an assay for its ability either to clot milk or to digest casein. The former leads to a target with a molecular weight of 28,000; the latter assay, to a weight of 48,000. Quite different parts of the molecule appear to be involved in the two actions.

13.08 Ionizations Random in Volume

Mean values of linear energy transfer following the absorption of medium- or high-energy photons will be less than 1 keV μm^{-1}. Ionizations then take place at relatively widely spaced intervals along the tortuous paths of the electrons. The concept of a measurable cross section can no longer be maintained. Target volume rather than target area now becomes the pertinent parameter.

An argument parallel to that used in Sec. 13.06 leads to an analogous relation

$$N = N_0 e^{-Vi} \tag{13-17}$$

where V is the sensitive volume of the target and i is the number fluence of the ionizing events per unit volume. When experimental data show a linear dependence of $\log (N/N_0)$ on i, a mean target volume can be calculated. At the 37% survival point, $Vi_{37} = 1$.

In the multihit situation shown in Fig. 13-3 the sensitive volume can be calculated from the slope of the linear portion of the survival curve. From the plot, $D_0 = 180$ rads and we have $i_{37} = (1.8 \times 10^4)/[1.6 \times 10^{-12}) \times 34 \times 1.3] = 2.5 \times 10^{14}$ cm^{-3}. The sensitive volume $V = 1/i_{37} = 4 \times 10^{-15}$ cm^3.

When dose–survival data are available at both high and low values of L, both sensitive volume and cross section can be calculated and target shape estimated.*

13.09 Linear Energy Transfer and RBE

A biological effect requiring only one hit on a single target will be most effectively initiated by low-L radiation. At high L-values some ionizations will be wasted in an "overkill" of cells already doomed by a single ionization. When more than one hit (either multihit or multitarget) is required, the maximum efficiency should obtain at some higher value of L, determined by the degree of multiplicity involved.

*Setlow, R. B. and E. C. Polland, *Molecular Biophysics*. Addison-Wesley Pub Co., Reading, Mass., 1962.

Values of RBE, Sec. 10.10, are calculated by reference to the effectiveness of medium-energy photon irradiation, which will have a mean L value of about 1 keV μm^{-1}. For a multihit process the RBE should be 1.00 at low values of L, rise to a maximum, and then decrease as overkill becomes important at high values of L. This behavior is shown in Fig. 13-5 for the inhibition of proliferation of Chinese hamster cells.*

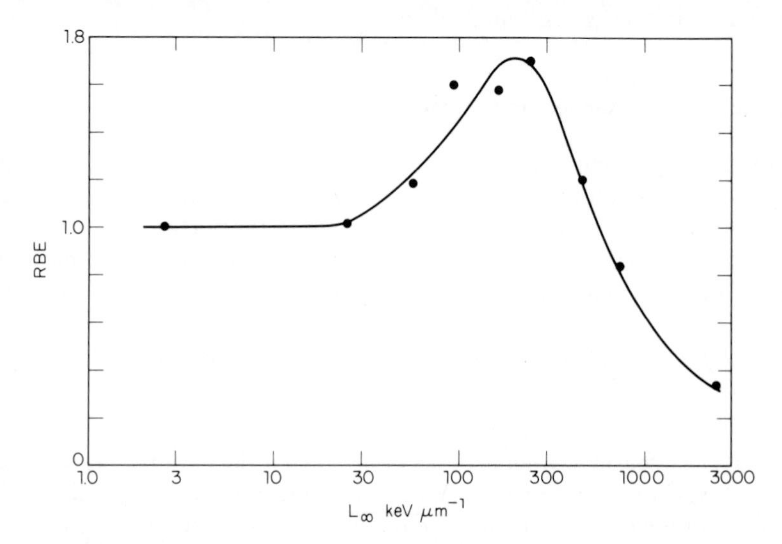

Figure 13-5. Variation of RBE for the inhibition of proliferation of Chinese hamster cells, plotted as a function of the mean L values of the radiation. (Calculated and plotted from data in Skarsgard, L. D., B. A. Kihlman, L. Parker, C. M. Pujara, and S. Richardson, "Survival, Chromosome Abnormalities, and Recovery in Heavy-Ion and X-Irradiated Mammalian Cells." *Rad. Research*, Supp. 7, 208, 1967.)

REFERENCES

Casarett, A. P., *Radiation Biology*. Prentice-Hall, Inc., Englewood Cliffs, N.J., 1968.

Elkind, M. M., and G. F. Whitmore, *The Radiobiology of Cultured Mammalian Cells*. Gordon and Breach, Inc., New York, N.Y., 1967.

Katz, R., B. Ackerman, M. Homayoonfar, and S. C. Sharma, "Inactivation of Cells by Heavy Ion Bombardment." *Rad. Research*, **47**, 402, 1971.

*Skarsgard, L. D., B. A. Kihlman, L. Parker, C. M. Pujara, and S. Richardson, "Survival, Chromosome Abnormalities, and Recovery in Heavy-Ion- and X-Irradiated Mammalian Cells." *Rad. Research*, Suppl. 7, 208, 1967.

POWERS, E. L., J. T. LYMAN, AND C. A. TOBIAS, "Some Effects of Accelerated Charged Particles on Bacterial Spores." *Int. J. Rad. Biol.*, **14**, 313, 1968.

SCHAMBRA, P. E., G. E. STAPLETON, AND N. F. BARR, ED., "Space Radiation Biology." *Rad. Research, Suppl.* 7, Academic Press, New York, N.Y., 1967.

WHITMORE, G. F. AND J. E. TILL, "Quantitation of Cellular Radiobiological Responses." *Ann. Rev. Nuc. Sci.*, **14**, 347, 1964.

14

Chemical Effects of Radiation

14.01 Ionization, Excitation, and Free Radicals

We have been using the terms *excitation* and *ionization* without inquiring into any effects that these higher energy states might have upon the molecular structures in which they occur. It is now time to examine these consequences in detail.

Ionization is not a process uniquely initiated by radiation. In all living tissues there will always be some ions and excited states simply because the tissues will be at a temperature well above absolute zero. We may use water, contained in all tissues, as an example. Because of fluctuations in energy due to thermal agitations a few water molecules will acquire energies greater than that of the H—OH bond. These bonds will then be broken to form the ion pair H^+ and OH^-. Production and recombination according to $H_2O \rightleftharpoons H^+ + OH^-$ will be continuous and an equilibrium ion concentration will be established. A similar equilibrium between excitations and de-excitations will also prevail.

Radiation impinging upon some neutral molecule may produce a positive ion and a free electron.

$$M_1 \longrightarrow M_1^+ + e^- \qquad (14\text{-}1)$$

For many years it was assumed that the electron would promptly recombine with the positive ion with the emission of low-energy photons or would combine with a neutral molecule to form a negative ion.

$$M_2 + e^- \longrightarrow M_2^- \qquad (14\text{-}2)$$

Reactions of the type of Eq. (14-2) do indeed take place. It is now known that in addition the free or *hydrated electron* may enter directly into chemical reactions without first forming a negative ion. Such a reaction might be

$$M_2 + e_{aq} \longrightarrow M_3^- + M_4 \tag{14-3}$$

Ion recombination will usually leave the neutral molecule in an excited state, denoted by *. Alternatively, excited states may be formed directly by the absorption of energy insufficient to produce ionization. An excited neutral structure may return to the ground state by the emission of photon radiation, by the degradation of the energy into heat through increased vibrations and rotations, or by *dissociation*. Dissociation may produce a pair of *free radicals*.

$$M_1^* \longrightarrow M_5^0 + M_6^0 \tag{14-4}$$

The symbol ^{0} denotes a free radical, which is an electrically neutral structure containing one electron which is not spin-paired with another.

An excited water molecule, for example, may dissociate into

$$H_2O^* \longrightarrow H^0 + OH^0 \tag{14-5}$$

H^0 will have only a single electron which is necessarily unpaired. OH^0 will contain 9 electrons and one of these must also be unpaired. Spin-pairing leads to structures with reduced energy content and increased stability, and free radicals can be expected to be chemically reactive because of their higher energy content. Free radicals can only be removed from a system by combination with other free radicals. For example, we might have

$$H^0 + M_5^0 \longrightarrow M_7 \qquad OH^0 + M_6^0 \longrightarrow M_8 \tag{14-6}$$

Radiation chemistry is concerned with the products formed when radiation is absorbed by molecules. Hot-atom chemistry is the study of the reactions entered into by molecules which have an abnormally high energy content because of radiation absorption. Reactions not energetically possible under normal conditions become feasible from excited states and when free radicals are available. These studies have been facilitated by the development of the techniques of electron spin resonance, ESR, which detects and quantitates the number of molecules with unpaired electrons.

14.02 Excitation and Dissociation

The consequences of molecular excitations tend to be forgotten, primarily because almost all measurements of high-energy radiations are made in terms of ionizations. Roughly one-half of the energy derived from the absorption of ionizing radiation goes into excitations and many of these smaller energy transfers contribute to the overall changes that take place in the absorbing system. Excitations are responsible for almost all the effects caused by low-energy photons in the ultraviolet region, Chapter 16.

An atom bound into a molecular configuration and raised to an excited electronic state by radiation can return to the ground state by a variety of de-excitation channels.

1. It may return directly to the ground state with photon emission.
2. The energy may be degraded to heat through intermolecular collisions or through increased vibrational energies that are distributed throughout the absorber.
3. It may dissociate.

All of these possibilities are described in detail in Sec. 16.06. For the present a discussion of dissociation will suffice.

Figure 14-1 represents the energy–interatomic distance situation in a diatomic molecule. The general argument will apply when more than two atoms are involved. In the ground state the energy–distance relationship will be given by a curve such as A. At a great distance the system will consist of two independent atoms. Electron-sharing will develop as the two approach and at some distance s an attractive force becomes appreciable and the combining system loses energy. The system will drop into one of the ground-state vibrational levels such as a–a′, Fig. 14-1. Only a few of the lower levels are available at ordinary thermal energies. In the absence of any additional energy from the outside the interatomic distance will oscillate between a and

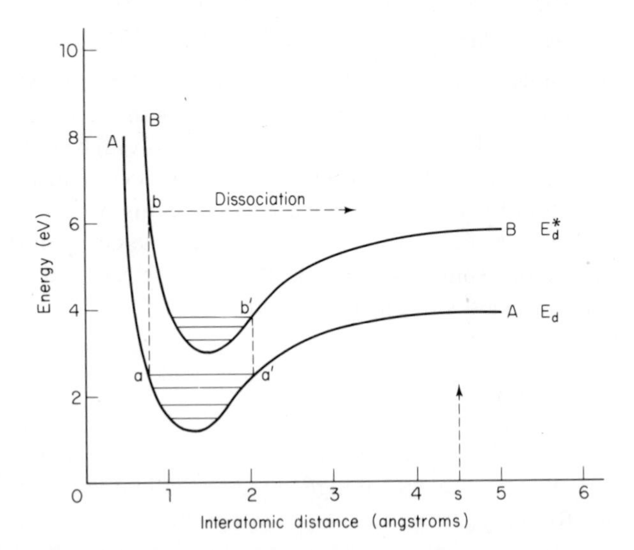

Figure 14-1. Energy–interatomic distance relationships in a diatomic molecule. Vibrational levels are shown (A) in the ground state and (B) in an excited electronic state. Each vibrational level will have a series of closely spaced rotational levels, not shown.

a′. Closer approach is prevented by the mutual repulsion of the two swarms of orbital electrons; separation is limited by the increase of the attractive force between the two shared electrons. This is the model on which the Franck–Condon theory of molecular dissociation is based.

When one of the orbital electrons is raised to an excited state, the system will be characterized by a new potential energy curve, as B in Fig. 14-1. This curve will lie above that representing the ground state and the point of minimum energy will move out to a greater distance. Infrared spectroscopy shows that the interatomic vibration frequencies are of the order of 10^{13} Hz, while the time of energy acquisition is about 10^{-15} s. All the energy of excitation will transfer to the system in about 1% of a single vibrational cycle, before the interatomic spacing can change appreciably. Thus a system at position a in the ground state will be raised to position b when the energy is absorbed.

Several modes of de-excitation are available at point b. Since the energy content is now greater than that needed for complete separation, the molecule may dissociate as shown. A typical excited state will have a half-life of the order of 10^{-8} s during which 10^5 interatomic vibrations will take place. This allows plenty of time for the system to drop down through several successive vibrational levels until it is below the dissociation energy E_d^*. At this point the only channel open is a return to the ground state by photon emission. Alternatively, de-excitation without radiation can be accomplished if the excited state at point b transforms to an isoenergetic, very high vibrational level of the ground state. Energy is then lost in a series of steps down to some thermally allowed vibrational state.

If the original excitation had taken place when the system was at some point such as a′, Fig. 14-1, there would have been less choice of de-excitation channels. Since point b′ lies will below the dissociation energy, the system must return to the ground state by radiating or by the radiationless transfer described.

We have become accustomed to describing single-bond energies in units of eV. Chemical reaction energies are more customarily expressed in kilocalories per mole of reactant. The two energy systems are readily interchangeable.

$$1 \text{ eV} = 1.6 \times 10^{-12} \text{ erg} = 1.6 \times 10^{-19} \text{ joule} = 3.82 \times 10^{-23} \text{ kilocalorie}$$

$$(3.82 \times 10^{-23}) \times (6.02 \times 10^{23}) = 23 \text{ kcal (mole)}^{-1} \qquad (14\text{-}7)$$

14.03 Chemical Yield

In a complex molecule the energy of excitation that does not lead to local dissociation according to the Franck–Condon model may be rapidly distributed throughout a considerable portion of a large molecule by bond vibrations. If the distributed energy exceeds any bond strength, dissociation may result, perhaps at some distance from the point of the original excitation. The excitation energy, when spread throughout the molecule, may be insufficient to rupture even the weakest bond. De-excitation will then be delayed until an energy fluctuation raises one bond above the dissociation energy. Energy degradation through these intermolecular collisions is more probable in liquid systems where the frequency of collisions is much greater than in gases. In solids, energy may be trapped for many years unless de-excitation is induced by heating or by some other stimulus. This trapping is put to practical use in the thermoluminescent dosimeter (TLD), usually made of either calcium fluoride or lithium fluoride.

In a pure material radiation effects can only come from *direct action* on the molecules themselves. Direct action also takes place in solution but in general this accounts for only a small fraction of the total effect. Energy transfers from charged particles are nonspecific processes and hence the number of primary interactions with each constituent of a complex system will be proportional to the amount of that constituent present. In a solution the predominant energy transfer may be to the solvent molecules, creating products such as ions and free radicals which then react with the solute. This *indirect action* is bound to be of great importance in biological systems because of the presence of water.

In simple solvent–solute systems the relative importance of the two types of actions can be determined from measurements of reaction yields as a function of solute concentration. In direct action the amount of product formed will be directly proportional to the amount or concentration of the solute. This is the basis upon which the target theory was constructed.

In indirect action, on the other hand, the amount of a reaction product will be determined primarily by the number of products originally formed from the solvent molecules. Here yield will be independent of concentration except at very low concentrations where the probability of free radical recombination is comparable to the probability of radical–solute reaction.

Illustrative Example

Calculate the average ion concentration along the path of an energetic electron in a dilute aqueous solution.

From Sec. 6.08 the electron will have an average specific ionization of 2.8×10^4 cm^{-1}. Most of the ionizations produced by the delta rays will take place in a cylinder along the particle path within a radius of about 50 Å = 5×10^{-7} cm.

$$\text{average ion density} = \frac{2.8 \times 10^4}{\pi \times (25 \times 10^{-14})} = 3.6 \times 10^{16} \text{ cm}^{-3}$$

In water there will be

$$\frac{6.02 \times 10^{23}}{18} = 3.3 \times 10^{22} \text{ molecules cm}^{-3}$$

$$\text{ion concentration} = \frac{3.6 \times 10^{16}}{3.3 \times 10^{22}} \simeq 10^{-6} \text{ molar}$$

According to this calculation, solute concentrations of 10^{-4} molar and above will be sufficient to ensure a negligible loss of ions or free radicals by recombination. Figure 14-2 is a schematic representation of the two types of effects plotted as a function of concentration.

Chemical yield is sometimes specified as the number of molecules of a product formed per ionization. If M_i molecules of product i are formed by radiation that produces N ionizations, $(\textit{ionic yield})_i = M_i/N$. There is some evidence that about 34 eV is required in liquids, as in many gases, to form an ion pair. On this basis, N can be estimated from a knowledge of the energy absorbed in the solution and M_i is obtained by chemical assay.

It is more usual to express a chemical yield by its *G value*, which is the number of product molecules formed for each 100 eV of energy absorbed from the radiation. G values will be roughly three times the corresponding values of ionic yield. There will be a G value for each reaction product formed or each radiation-induced action, and these values will depend somewhat on the type of radiation involved. As we shall see, the relative values of G_{H_2} and

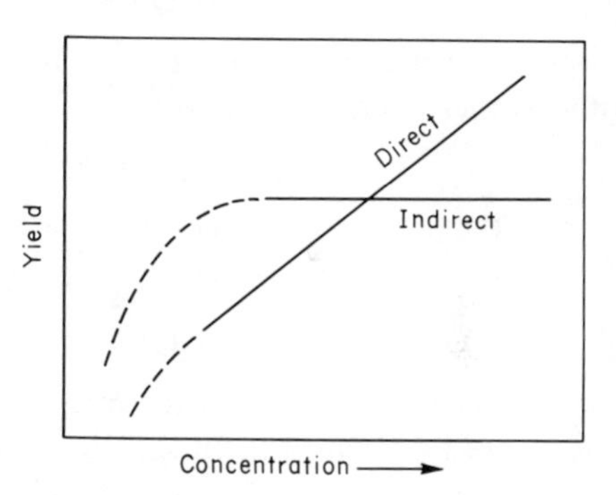

Figure 14-2. Schematic dependence of chemical yield on solute concentration at a constant radiation dose for direct and indirect actions.

TABLE 14-1
TYPICAL G-VALUES

Product or Reaction	G
Fe^{3+} from $FeSO_4$	15.6
Ce^{2+} from $Ce_2(SO_4)_3$	2.3
HCl from chloral hydrate	100
Inactivation of catalase	0.009
Inactivation of carboxypeptidase	0.55
DNA–single-strand scission	1.0
DNA–double-strand scission	0.15
DNA–H-bond breakage	60
DNA–cross-linking	0.08

$G_{H_2O_2}$, two of the radiolytic products of water, depend on the LET of the radiation. Some average G values are listed in Table 14-1.

The high value for HCl production suggests that radiation initiates a chain reaction in which one reaction leads to several others before the chain is broken. Chain-reacting systems are usually very sensitive to conditions such as temperature and the presence of impurities which may act to interrupt the chains.

Note from Table 14-1 that radiation may induce either scission or cross-linking in a large molecule such as DNA. In one structure, radiation may produce main chain breakages with a consequent reduction in molecular weight and viscosity. In another structure, radiation may lead to cross-linking, with corresponding increases in viscosity and molecular weight.

14.04 Radiolysis of Water

A vast amount of research has been done on the radiolysis of water, as befits its importance as an almost universal solvent. The initial action consists in the ionization of a water molecule with the ejection of an electron.

$$H_2O \longrightarrow H_2O^+ + e^- \tag{14-8}$$

The electron may move as much as 100 Å from the ion before it combines with another neutral water molecule.

$$H_2O + e^- \longrightarrow H_2O^- \tag{14-9}$$

Each of the ionized molecules dissociates.

$$H_2O^+ \longrightarrow H^+ + OH^0 \tag{14-10}$$

$$H_2O^- \longrightarrow H^0 + OH^- \tag{14-11}$$

Ions H^+ and OH^- play no further role. They enter the existing pool of similar ions to produce a very slight shift in the equilibrium

$$H^+ + OH^- \rightleftarrows H_2O \tag{14-12}$$

Subsequent events depend on the linear energy transfer of the radiation and on the amount of available oxygen in the solution. Consider first the case of a high-L radiation such as an alpha particle, where the primary ionizing events are only a few angstroms apart, Fig. 14-3A. The OH radicals formed by dissociation from the original positive ions will be closely spaced along the trajectory of the alpha particle. The H radicals, on the other hand, produced from the negative ion at some distance from the track, will form a sort of sheath around the actual track. Because of geometrical proximites, the favored but not exclusive reactions of the radical products will be

$$OH^0 + OH^0 \longrightarrow H_2O_2 \tag{14-13}$$

$$H^0 + H^0 \longrightarrow H_2 \tag{14-14}$$

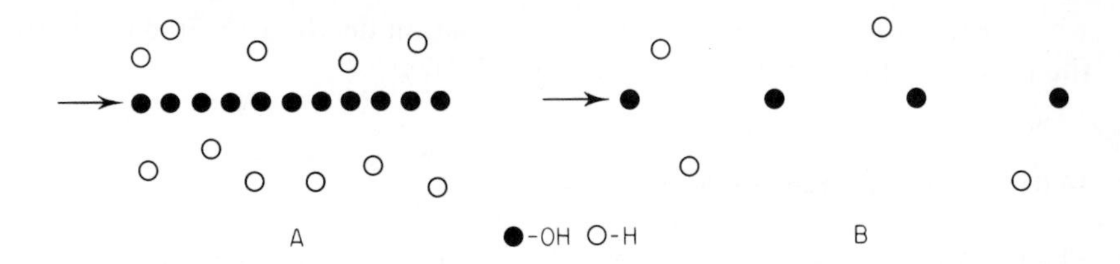

Figure 14-3. Distributions of H and OH ions along the path of (A) a high-L radiation such as an alpha particle and (B) a low-L radiation such as a beta particle.

With low-L radiation, as a beta particle, primary events will be more sparsely distributed along the path, Fig. 14-3B. The probability of recombination is now increased

$$H^0 + OH^0 \longrightarrow H_2O \tag{14-15}$$

and some free radicals will persist. These will be available for reaction with any solute that may be present. Thus, in a general way, high-L radiation tends to favor molecular products while free radicals are more apt to follow the passage of low-L radiation. Early observers noted the evolution of hydrogen from alpha-irradiated water, Eq. (14-14) and its almost complete absence where X rays were used.

Some of the free or hydrated electrons liberated in the primary ionization act directly on water molecules without the intermediate formation of negative ions.

$$H_2O + e_{aq} \longrightarrow H^0 + OH^- \tag{14-16}$$

Some of the original positive ions may form a complex with water molecules.

$$H_2O^+ + H_2O \longrightarrow H_3O^+ + OH^0 \tag{14-17}$$

followed by

$$H_3O^+ + e_{aq} \longrightarrow H^0 + H_2O \tag{14-18}$$

If the solution contains dissolved oxygen, a new oxidizing radical can be formed.

$$H^0 + O_2 \longrightarrow HO_2^0 \tag{14-19}$$

The HO_2 radicals have a relatively long life in solution (0.1 s) and may diffuse some distance from the site of formation before reacting with a solute molecule or combining:

$$HO_2^0 + HO_2^0 \longrightarrow H_2O_2 + O_2 \tag{14-20}$$

Whatever the type of radiation the radiolysis of water produces both oxidizing and reducing products. A complex series of reactions becomes possible with any solute molecule. Amost all the initial events will be completed within perhaps 10^{-4} after the radiation has been absorbed. One of the outstanding problems of radiobiology is the elucidation of the mechanisms which leads from this state to perhaps a radiation death in 15–20 days or to the production of a radiogenic cancer in 15–20 yr.

14.05 Radiation Effects in Macromolecules

The radiation products of water are of interest to the biologist primarily because of their function as intermediate products between the original ionizations and excitations and the final effect on molecules of particular biological importance. Most of the latter are large complex structures with high molecular weights. In many cases even the undisturbed chemical structure is not known exactly and radiation effects can only be inferred from changes in function, activity, or physical properties.

With the number of reactive intermediates available from the radiolysis of water a complex series of changes in macromolecular structures is to be expected. To these must be added others resulting from the direct action of radiation on the molecules themselves. Although direct action may not be the most probable effect, it is not negligible.

Measurements have been made on a number of proteins, enzymes in particular. In some cases a series of amino acid residues is observed, indicating disintegrations of the original structure into smaller fragments. Changes in physical properties may sometimes be found with a constant molecular weight, suggesting alterations in molecular configurations without rupture. Functional changes observed are probably the result of radiation effects on the side chains as well as changes in the spatial orientations of the groups because of actions on the peptide chains. There are some variations in the sensitivity to injury from radiation but the range is moderate, as indicated by the G values listed in Table 14-1.

Radiation effects on DNA are of considerable interest because of the possibility of radiation-induced alterations in the genetic codes. The G values listed in Table 14-1 cover only a few of the radiation effects that have been observed in DNA. Some of the effects, such as hydrogen-bond rupture, may have a high probability of repair. Others, such as double-strand scission, may be followed by profound changes in the structure and its function. Detailed results must be sought in the voluminous literature.

14.06 Oxidation of Ferrous Iron

The radiolysis of hundreds of chemical compounds has been studied in water and in other solvents. Only a few of the most important reactions can be mentioned here. In 1927, Fricke showed that radiation oxidizes ferrous iron to the ferric state and suggested that the reaction might be useful as a radiation dosimeter. The ferrous–ferric reaction has been intensively studied and has become one of the most useful dosimeters in the medium- to high-dose range.

As ordinarily used, the Fricke dosimeter consists of a 10^{-3} to 10^{-4} molar solution of either ferrous sulfate, $FeSO_4 \cdot 7H_2O$, or ferrous ammonium sulfate, $Fe(NH_4)_2(SO_4)_2 \cdot 6H_2O$, in water carefully purified to remove organic material. The solution is made 0.1–0.8N in H_2SO_4 to prevent the loss of reaction products by hydrolysis. Radiation converts some of the Fe^{2+} ions to Fe^{3+} and an assay for the latter ion forms the basis for the calculation of the absorbed radiation dose. The reaction is indirect through the intermediate radiation products of the solvent water. This makes the sensitivity of the system relatively independent of the solute concentration.

Values of G (Fe^{3+}) vary somewhat with the linear energy transfer of the radiation, Table 14-2. The decrease in yield at high values of L is probably due to radical recombination. More detailed G values have been complied by the ICRU.*

Assay for Fe^{3+} is done almost exclusively by absorption spectrophotometry. Ferric ions have two absorption bands in the ultraviolet, centered at 224 and 305 nm, where there is a negligible absorption by Fe^{2+}. In the spectrophotometer a monochromator selects a narrow spectral region at the absorption peak and directs this light onto a cell containing the dosimeter solution. The incident and transmitted light intensities are related by

$$I = I_0(10)^{-\epsilon c l} \tag{14-21}$$

where ϵ = a constant known as the *molar extinction coefficient* with dimensions of liters mol^{-1} cm^{-1} (ϵ is a measure of the absorbing power of the Fe^{3+} ions at the wavelength used.)

c = concentration of the absorbing ions in mols $liter^{-1}$

l = effective length of the light path through the cell containing the solution

TABLE 14-2
REPRESENTATIVE G-VALUES FOR FERROUS OXIDATION

Radiation	*G (Fe^{3+})*
10-MeV electrons	16.0
2-MeV electrons	15.5
^{60}Co gamma rays	15.6
100-kVP X rays	14.8
50-kVP X rays	14.0
^{3}H beta particles	12.9
^{210}Po alpha particles	5.1

**Radiation Dosimetry: X-Rays Generated at Potentials of 5 to 150 kV*. ICRU Report 17, June 15, 1970.

In logarithmic form, Eq. (14-21) becomes

$$\log_{10} \frac{I_0}{I} = \epsilon c l \tag{14-22}$$

where $\log_{10} (I_0/I)$ is known as the *optical density* or O.D. of the solution. The spectrophotometer will measure the optical density and from this c can be calculated since ϵ is known from other measurements. With c known, the energy absorbed can be calculated from the known G value of the reaction.

14.07 The Fricke Dosimeter

Since G (Fe^{3+}) and ϵ can be determined by completely independent methods, the ferrous sulfate system is an absolute dosimeter in the sense that it does not require a calibration against a standard instrument. Because of its many important features this system has become the standard method for determining absorbed doses for almost all studies in radiation chemistry. The dosimeter solution recommended by the ICRU is

Ferrous sulfate	1 m mol liter^{-1}	0.28 g
or		
Ferrous ammonium sulfate	1 m mol liter^{-1}	0.39 g
Sodium chloride	1 m mol liter^{-1}	0.06 g
Sulfuric acid	0.4 m mol liter^{-1}	22 ml
Water distilled from alkaline permanganate to make		1 liter

Sodium chloride is added as a scavenger for any residual organic impurities that otherwise would compete for the primary radicals. Oxygen consumed in the reaction is obtained from the air dissolved in the distilled water, which is allowed to come to equilibrium with room air. This oxygen will very slowly oxidize some of the ferrous ions to give an increasing blank reading for Fe^{3+}. A dosimeter solution should be allowed to stand for a day or so in order to ensure dissolved gas equilibrium and should be discarded when an appreciable optical density is measured in a blank solution prior to irradiation.

The yield of Fe^{3+} will be proportional to absorbed dose up to about 30 krads where oxygen depletion becomes appreciable. For most purposes the yield can be considered to be independent of the temperature at exposure but for the highest precision a positive temperature coefficient of 0.1% °C^{-1} can be assumed. G values depend somewhat on the linear energy transfer of the radiation but are independent of dose rate up to 10^8 rads s^{-1}.

At 305 nm, Fe^{3+} has a molar extinction coefficient of 2195 liters mol^{-1} cm^{-1} which sets the useful lower limit of dose at about 3000 rads when a 1-cm absorption cell is used. This lower limit can be reduced to 1500 rads by measuring the absorption at 225 nm where the extinction coefficient is 4560 liters mol^{-1} cm^{-1}. The lower useful dose limits will drop to about 300 and

150 rads, respectively, if 10-cm absorption cells are used and these are practically the sensitivity limits of the method.

At photon energies down to about 100 keV the standard Fricke solution has an absorption coefficient that is within 1% of that of soft tissue. Below this energy the sulfur atoms, $Z = 16$, increase the photoelectric absorption over that of tissue until there is a difference of 10% at 5 keV. This difference can be reduced by reducing the acid concentration but this must not drop below 0.05 mol liter^{-1}. A reduction to this level has little effect on the G value.

Illustrative Example

After irradiation, a standard Fricke dosimeter solution has an optical density of 0.138 measured at 305 nm in a 1-cm cell. Calculate the absorbed dose in rads, assuming $G\,(Fe^{3+}) = 15.6$.

$$\log \frac{I_0}{I} = 0.138 = 2195 \times c \times 1$$

$$c = 6.29 \times 10^{-5} \text{ mol liter}^{-1}$$

$$Fe^{3+} \text{ ions per gram} = \frac{6.29 \times 10^{-8}}{6.02 \times 10^{23}} = 3.78 \times 10^{16}$$

$$\text{energy absorption} = \frac{3.78 \times 10^{16}}{0.156} = 2.42 \times 10^{17} \text{ eV g}^{-1}$$

$$= (2.42 \times 10^{17}) \times (1.6 \times 10^{-12}) = 3.88 \times 10^{5} \text{ ergs g}^{-1}$$

$$\text{dose} = \frac{\text{ergs g}^{-1}}{100} = 3880 \text{ rads}$$

14.08 Reduction of Ceric Sulfate

For many applications the upper dose limit of the Fricke dosimeter is sufficient. This limit can be extended somewhat by saturating the solution with oxygen prior to irradiation but when this is necessary, it is usually preferable to shift to the use of ceric fulfate. The useful range of the ceric system extends up to some 10^8 rads.

The ceric sulfate dosimeter consists of a 10^{-5}–0.1 molar solution of $Ce_2(SO_4)_3$ in 0.1–0.8N H_2SO_4. Upon irradiation some of the Ce^{3+} ions will be reduced to the cerous state primarily by indirect action. G values for the reduction range from 2.4–2.8 depending on the L of the radiation. Assay is most conveniently carried out by absorption spectrophotometry at 320 nm where Ce^{3+} has an absorption band. The molar extinction coefficient is of the order of 5000–5800, the exact value depending on the acid concentration. The use of the ceric system is described in detail in an ICRU report.*

***Physical Aspects of Irradiation.* ICRU Report 10b. Issued as Handbook 85 of the National Bureau of Standards, March 31, 1964.

The ceric system is extremely sensitive to impurities. No stabilizer comparable to the NaCl scavenger used in the Fricke dosimeter is available here. Extraordinary care must be taken in preparing the water for the solution and only the cleanest glass vessels can be used for the irradiation.

REFERENCES

CASARETT, A. P., *Radiation Biology*, Chapter V. Prentice-Hall, Inc., Englewood Cliffs, N.J., 1968.

EBERT, M. AND A. HOWARD, ED., *Current Topics in Radiation Research*, **II**. North-Holland Publishing Co. Amsterdam, 1966.

GINOZA, W., "The Effect of Ionizing Radiation on Nucleic Acids of Bacteriophage and Bacterial Cells," in *Annual Reviews of Nuclear Science*, E. Segrè, ed. Annual Reviews, Inc., Palo Alto, Calif., 1967.

HENLEY, E. J. AND E. R. JOHNSON, *The Chemistry and Physics of High Energy Reactions*. Washington, D.C., University Press, 6411 Chillum Place, Washington, D.C., 1969.

HUTCHINSON, F., "Radiation Effects in Macromolecules of Biological Importance," in *Annual Reviews of Nuclear Science*, E. Segre, ed. Annual Reviews, Inc. Palo Alto, Calif., 1963.

SMITH, D. E., "Free Radical Reactions in Irradiated Biological Materials and Systems" in *Annual Reviews of Nuclear Science*, **12**, E. Segrè, ed. Annual Reviews, Inc. Palo Alto, Calif., 1962.

15

Radiation Effects in Mammals

15.01 Past Experiences

A surprisingly short time elapsed between Roentgen's report on his experiments with X rays and the discovery that penetrating radiations have a profound effect on living tissues. Within a month after Roentgen's first paper a patient was admitted to Cook County Hospital, Illinois, suffering from what was almost certainly a radiation injury.* Many laboratories in many countries had been studying electrical discharges in gases and were immediately able to exploit the original discovery. Medical applications, some useful and others of dubious virtue, quickly followed the technical developments. For several years, few, if any, of the workers recognized the close relationship that was appearing between radiation exposure and clinical signs of injury. Increasingly, physicists, engineers, some inventors, and physicians were acquiring injuries which eventually led to death, usually from radiogenic cancers.

Although the industrial and medical uses of both X rays and naturally occurring radioactive materials increased steadily, relatively little basic research on radiation hazards was done prior to 1940. Broadly based, fundamental studies on the biological effects of radiation first appeared as one component of the intensive effort to develop a nuclear weapon. Since that time, radiation effects have been investigated in innumerable living forms and under a wide variety of radiation conditions.

Sensitivity to radiation varies enormously from one living form to another but all that have been studied have been found susceptible to injury. It can

*Brown, P., *American Martyrs to Science Through the Roentgen Rays.* Charles C. Thomas, Pub., Springfield, Ill., 1936.

hardly be otherwise since, as we have seen, radiation can produce profound structural changes in any molecule. As more experience has been gained, radiation injuries have been found, albiet at high doses, in many nonliving substances such as some of the structural materials used in or near nuclear reactors.

The radium dial painters in New Jersey form the best documented group of early radiation injury cases.* Luminous paint production and application were introduced into the United States in 1913 and within a few years several hundred people were working in the new industry. Clinical signs of injury, unidentified for years, were first associated with radiation in 1923 and the first published report appeared in 1925. Lack of understanding of radiation hazards is not all in the far past. Recently one owner of a dial painting shop resisted remedial measures in the belief that all radiation hazard disappeared when the paint had dried.

The extensive bone necrosis observed in the dial painters was caused by alpha particles from radium and its daughter products, deposited in bone because of the chemical similarities between radium and calcium. Other specific effects included leukopenia, tumor formation, sterility, and terminal infections. These same effects, varying in relative degree, may also be seen after a gross overexposure to sources of penetrating radiation external to the body. Detailed studies of the survivors of the Hiroshima and Nagasaki bombings were promptly initiated, and are being continued under the auspices of the Atomic Bomb Casualty Commission (ABCC). A series of scientific reports has been published by the ABCC and the major findings have been summarized.†

Signs of radiation injury appeared early in the hands of many pioneer radiologists because of the common practice of checking operation of the X ray tubes by using the left hand as a test object, with a fluorescent screen held in the right hand. Dry cracked skin, brittle fingernails, and poorly healing ulcers progressed into malignancies and death, usually after many unsuccessful types of treatment.

The expression "radiation sickness" was originally used to described an acute syndrome that has plagued radiologists since the earliest days of radiation therapy. Recently there has been a tendency to apply the term to all phases of radiation injury but it is desirable to retain the original restricted definition.

In its original sense, radiation sickness refers to the nausea, vomiting, anorexia, and fatigue seen in humans after whole-body doses of 100 rads or

*Details of the injuries sustained by the dial painters can be found in *Collections of Reprints on Radium Poisoning* by Harrison S. Martland. Technical Information Service, Oak Ridge, Tenn.

†*Radiation Effects on Atomic Bomb Survivors*. ABCC Technical Report 6-73, National Academy of Sciences–National Research Council, Washington, D. C. 20418, 1973.

so. The time of onset, the severity, and the duration of the symptoms depend on the dose and the tissues receiving it. Radiation absorbed in the visceral region is particularly effective. Typically, radiation sickness may appear within a few hours after exposure and may last 24–48 hr. In cases of unknown exposure the course of the radiation sickness syndrome may be used as a rough estimate of the severity of the injury.

In many human exposures, either intentional or accidental, there are strong psychogenic components in the radiation sickness syndrome. Physiological disturbances must also be present, for comparable phenomena are seen in animal irradiations where psychogenic factors are almost certainly absent. Many animals show signs of discomfort during the administration of large radiation doses and some, notably swine and monkeys, will vomit.

A wide variety of drugs has been tried in attempts to relieve the patient discomfort associated with radiation sickness. All treatments have had indifferent success and no truly effective remedy appears to be available.

15.02 Survival Time–Dose Relations

As the acute, whole-body radiation dose is increased, the radiation sickness syndrome becomes more severe and a point will be reached at which a few individuals fail to survive the first 30 days (this is the rather arbitrary observation time used in most experiments). Continued increases in dose will produce correspondingly greater 30-day mortalities until finally 100% of the population is killed within 30 days. The doses producing from 0–100% mortality (LD_0–LD_{100}) make up the *lethal range*.

In animals the maximum number of deaths over the dose range LD_0–$2LD_{100}$ occurs at about 13–15 days after irradiation. The corresponding time for man is not known but is probably somewhat longer. This is known as the region of *infection death* because bacteremia is a prominent finding at autopsy.

As the dose is increased above $2LD_{100}$, survival time drops rather abruptly to 3–5 days and remains fairly constant until much higher doses are reached. This is the region of *gastrointestinal death*, characterized by gross damage to the intestinal lining.

At still higher doses, survival times become progressively shorter, Fig. 15-1. Signs of central nervous system (CNS) injury appear and become more intense with dose. Although there are few morphological signs of CNS injury at autopsy, this is known as the *CNS death* from the clinical signs of injury. Radiation death in animals is not instantaneous even at whole-body doses of 10^5 rads or more. In one human accident case, signs of CNS involvement were seen almost immediately after a nonuniform dose with about 10,000 rads delivered to the head. Death ensued in about 33 hr.

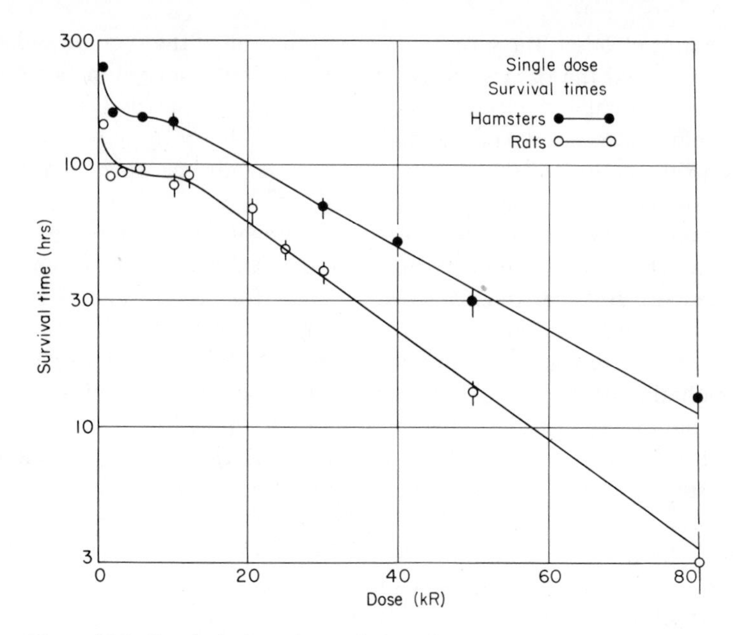

Figure 15-1. Survival time–dose relationships in hamsters and rats at supralethal doses.

In terms of absorbed energy, living tissues are remarkably sensitive to ionizing radiation. As an example, consider a whole-body dose of 10^4 rads which in man will certainly lead to death with pronounced CNS symptoms within a few hours. This dose will deposit 10^6 ergs $= 0.1$ joule or $0.1/4.18 = 0.024$ cal in each gram of tissue. If we assume a specific heat for these tissues of 1.0 cal g^{-1}, the energy absorption will produce a temperature rise of 0.024°C. In a "standard" man of 70 kg the total energy absorbed would be less than 1700 cal. Death would be delayed but would be inevitable after a dose of one-tenth that assumed here. We make the further assumption that there is either an ionization or an excitation for each 5 eV of energy absorbed. Then for a dose of 10^4 rads there will be $(10^6/5) \times (1.6 \times 10^{-12}) = 1.2 \times 10^{17}$ events per gram of tissue. One gram of tissue contains on the order of 10^{23} atoms so the fraction of the tissue atoms experiencing some direct effect of the radiation will be $(1.2 \times 10^{17})/10^{23} = 1.2 \times 10^{-6}$. The fraction of the molecules involved will be somewhat greater.

15.03 The Lethal Dose Range

Whole-body irradiation reduces the natural resistance to infection and interrupts normal immune responses. Because of these actions the course of events following whole-body doses in the LD_0–LD_{100} range is strongly influenced by infectious processes. Postradiation infections, mostly enterogenous, have been observed regularly both in experimental animals and in the Japanese exposed to nuclear weapons.

The course of an animal experiment can be grossly altered by the presence of a particularly virulent organism such as *Pseudomonas*. Antibiotics, particularly streptomycin, are of some help in combatting these infections but no known agent has a sufficiently broad spectrum to control against all invaders until natural defenses can resume normal function.

Infection is not the only cause of death in the LD_{100} range. Germ-free animals have an $LD_{50/30}$ which is only about 30% greater than that for a corresponding normal population. Survival times are somewhat lengthened in the absence of infection but again the gain is not great.

Epilation is a common finding in man after radiation doses in the lethal range. Hair loss usually starts abruptly about 2 weeks after exposure and continues for another 2 weeks. Epilation is not permanent among the survivors. Typically, regrowth starts within 3 months of the exposure. Color and texture may be different from those prior to the exposure.

Purpura is another consistent part of the radiation syndrome in the lethal dose range. Time of onset varies from 3 to 30 days, appearing earlier following larger doses. In survivors, purpura will last about 15 days. Purpura is the earliest visible sign of radiation damage to the hemopoietic system leading to a hemorrhagic tendency. A combination of purpura and epilation is almost pathognomonic for radiation injury.

Extensive hemorrhages usually follow purpura, emphasizing the severe damage to the hemopoietic system. Bleeding from the gums and mouth was most common in the Japanese but other areas were frequently involved. Bloody diarrhea was a common factor in heavily exposed Japanese. Diarrhea is regularly seen at somewhat lower doses; when accompanied by bleeding, prognosis is poor.

15.04 Variations in Radiosensitivity

The lethal dose range varies from one animal species to another and there are substantial sensitivity differences even among individual members of an inbred strain of a given species. Values of LD_{50} are sensitive to many experimental conditions, including the type of microorganisms present in the colony.

In large animals the $LD_{50/30}$ may be found to vary with the effective energy delivered by an external source of radiation. In some cases this may be due to differences in the distribution of dose with depth rather than to intrinsic differences between the two RBE's. The dose delivered by an essentially point source to any depth d in an absorbing body will vary as $e^{-\mu d}/D^2$ where D is the distance to the source of radiation. μ is a function of photon energy and so the spectral distribution as well as the dose will change with depth, even if the original source is monoenergetic. When whole-body irradiation is desired, D should be made as large as is compatible with obtaining the desired dose rate, and a penetrating radiation (μ small) should be used. Some whole-body $LD_{50/30}$ values obtained with penetrating photons at high dose rates (50 R m^{-1} or so) are given in Table 15-1. Some of the animal values are strain-dependent and so ranges must be given. Lack of data makes the values for man particularly uncertain.

TABLE 15-1
$LD_{50/30}$, VALUES FOR ACUTE WHOLE-BODY IRRADIATION

Species	*$LD_{50/30}$, rads*
Dog	350
Goat	350
Guinea pig	400
Man	300–500
Monkey	600
Mouse	500–700
Swine	550
Rat	600–800
Burro	600–800
Hamster	700
Rabbit	800

There is no correlation between values of LD_{50} and any obvious species characteristic. Size is certainly not an important factor. Basal metabolic rate, at least in the normal range, is likewise not intimately related to LD_{50}. Metabolic rates are not without influence, however, for in hibernating animals, where the rate is very low, signs of radiation injury are suppressed, to appear when more active metabolism is restored.

Variations of radiosensitivity within members of even an inbred animal strain are perhaps not surprising. Whole-body radiation represents an insult to all parts of a living organism and the result is undoubtedly a complex reaction to leukopenia, hemorrhage, bacteremia, toxic products, impaired nutrition, blocked immune responses, and shock. Variable rates of repair enter to further complicate the picture.

Age at time of irradiation is one important factor in radiation sensitivity. Young and old animals are more sensitive than those in mid-life. Details of sensitivity dependence on age appear to vary with the species.

Reasoning from the effects of X rays on rat testicles, Bergonié and Tribondeau proposed in 1906 the law to which their names are now attached.

Freely translated, the law of Bergonié and Tribondeau is: X rays act most strongly on those cells with the greatest reproductive activity, longest period of mitosis, and the least morphological and functional differentiation. The actual facts are not as simple as the law implies.

Both normal tissues and malignant growths are classified into varying degrees of radiosensitivity or radioresistance, each according to its observed response to radiation. For example, bone marrow is considered to be radiosensitive, whereas nerve tissue is radioresistant. From the radiation–chemistry viewpoint a given dose will produce the same number of free radicals per gram in either tissue, small differences of absorption excepted. Many of the same chemical structures are present in the two tissues and many of the same chemical effects are to be expected. It may well be that the amount of primary injury in the two cases is more comparable than is apparent and that differences in mitotic activity make it more obvious in one tissue than in the other.

15.05 The Gompertz Function

Mortality studies using whole-body irradiation are widely used in radiobiology in order to obtain fundamental scientific information and also to further our understanding of the effects of widespread radiation upon the human population. Various mathematical relationships have been used to express mortality data in terms of parameters of possible biological significance. The Gompertz function is one relationship that seems applicable to radiobiological data.

Assume a population consisting of N_0 individuals at zero time and N individuals at some later time (age) t. *Mortality rate* R can be defined as the fractional decrease in N in unit time where the unit of time in dt is taken to suit the problem at hand.

$$R = \frac{-dN/N}{dt} \tag{15-1}$$

In 1825, Gompertz found that a normal population could be quite accurately described if

$$\ln R = \ln R_0 + mt$$

or (15-2)

$$R = R_0 e^{mt}$$

where R_0 is the rate corresponding to N_0. The value of R from Eq. (15-2) can be put in Eq. (15-1) and the functional relationship between N and t obtained by integration. This leads to the *Gompertz function*

$$N = N_0 e^{-(R_0/m)(e^{mt}-1)} \tag{15-3}$$

Equation (15-3) solved for t gives survival time as a function of the surviving population fraction N/N_0.

$$t = \frac{1}{m} \ln\left(1 + \frac{m}{R_0} \ln \frac{N_0}{N}\right) \tag{15-4}$$

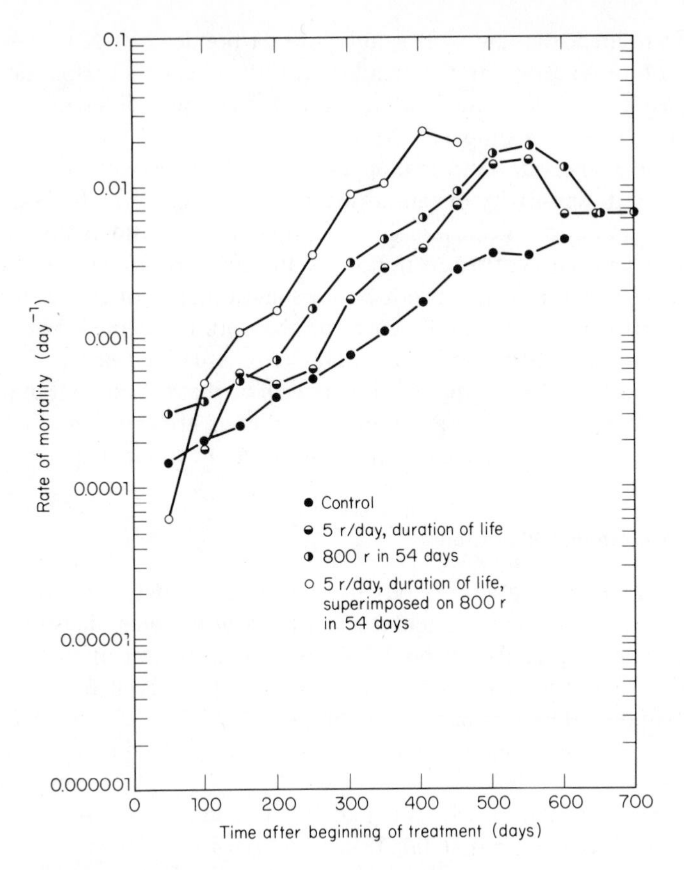

Figure 15-2. Effect of radiation on the Gompertz mortality plot for Carworth female mice. (*Courtesy of A. M. Brues and G. A. Sacher, Symposium on Radiobiology, J. J. Nickson, ed. John Wiley & Sons, Inc., New York, N.Y., 1952.*)

Procedures are available* for determining the characterizing constants m and R_0 from experimental data.

If a population obeys the Gompertz function, Eq. (15-2) requires a linear relationship between mortality rate and age. Figure 15-2 is a Gompertz plot of data from mice, for all causes of death except lymphoma. The unirradiated control population shows an essentially linear plot, as predicted. The line for mice who received 800 R early in life is displaced upward but has a slope about like that of the normal. Chronic irradiation at 5R per day gave a line with a steeper slope, and combined acute and chronic irradiation resulted in

*Davis, D. S., *Empirical Equations and Nomography*. McGraw-Hill Book Co., New York, N.Y., 1943.

both a displacement and a steeper slope. Thus irradiation effects on mortality were expressed in terms of changes in two constants, m and R_0.

15.06 Radiation Mortality Statistics

When dose–mortality studies are carried out on animal populations, the data, Table 15-2, show a dose threshold, a rather narrow dose range between $LD_{10/30}$, and $LD_{90/30}$, and a shoulder as the mortality approaches 100%. In general the percent mortality–dose curve is a sigmoid with a steep slope, Fig. 15-3, which is plotted from the data in Table 15-2. Various LD values can be taken directly from such a plot but it is usually preferable to *transform* or *rectify* the curve to obtain a linear relationship between the variables. With well-controlled conditions, dose–mortality data will have a Poisson distribution and the mathematics of normal probabilities can be applied. Direct application of the probability equations is difficult and cumbersome.

Graphical methods are regularly used to obtain the various LD values. The use of these methods will be illustrated with the data given in Table 15-2.

A plot of log dose against percentage mortality on a probability scale, Fig. 15-4, shows an essentially linear relationship as expected. A line of best fit can be determined by standard methods but it usually suffices, as in the present case, to draw the line by inspection. From the line drawn in Fig. 15-4, LD_{50} = 684 rads, as compared to 689 rads obtained from the sigmoid plot of the raw data.

The probability plot is most easily made on specially ruled coordinate paper with one axis divided according to normal probabilities of deviation from a central mean value. This type of paper was used for the plot of Fig. 15-4, which shows only a portion of the probability scale.

TABLE 15-2
ACUTE DOSE–MORTALITY DATA

Dose, rads	*$\log_{10}$ dose*	*Percentage mortality*
565	2.75205	0.0
580	2.76343	4.0
615	2.78888	7.0
650	2.81291	23
665	2.82282	28
680	2.83251	49
700	2.84510	61
715	2.85431	76
745	2.87216	85
800	2.87506	92

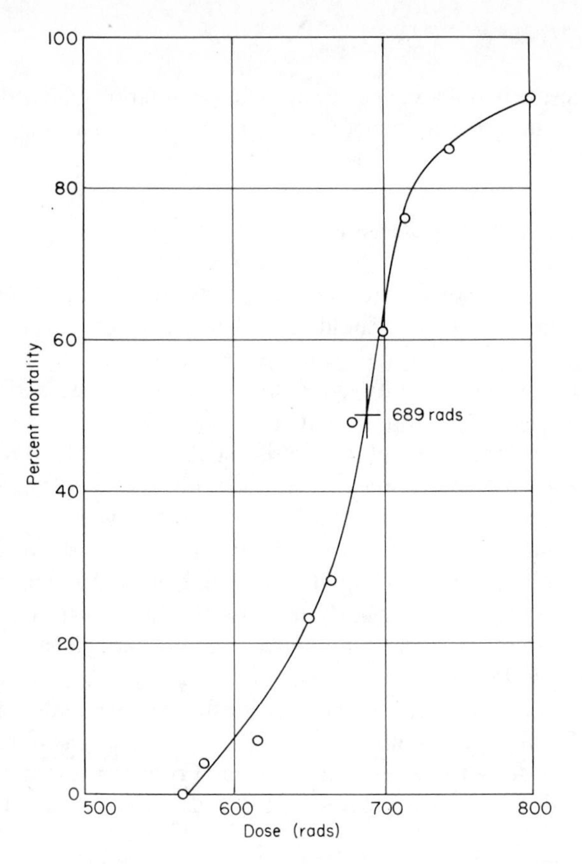

Figure 15-3. Typical sigmoid mortality curve showing an $LD_{50/30}$ of 689 rads.

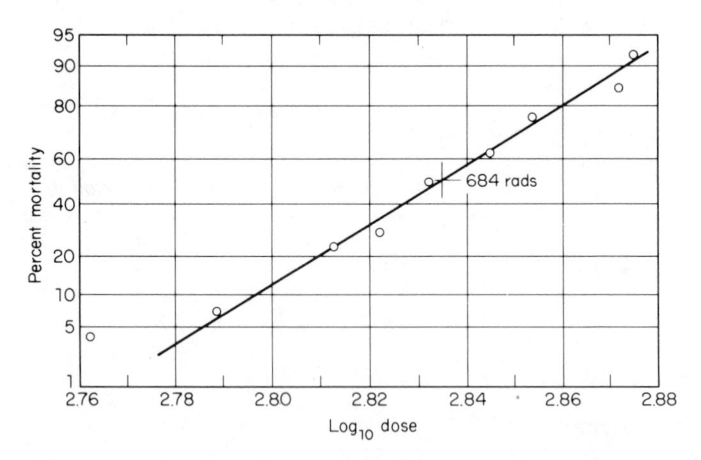

Figure 15-4. Plot of the tabulated mortality data on a probability scale. $LD_{50/30}$ = 684 rads.

15.07 The Probit Plot

Instead of using probability paper, one can convert each mortality percentage to the corresponding deviate from the curve of normal expectancy and express it in units of standard deviation. Minus signs can be avoided by adding 5.000 to each figure thus obtained. The resulting values are known as *probits*.

In practice the probits are taken from tables since the conversion calculations are complicated. Table 15-3 lists probit values* for the data given in Table 15-2.

A plot of log dose against probits will be linear, as in Fig. 15-5, and a best straight line can be drawn, usually by inspection. Then from the probit value corresponding to the desired value of LD one can read off the value of dose. In Fig. 15-5, $LD_{50} = 686$ rads since probit $50\% = 5.00$. Methods are available for determining the line of best fit and the constants for this line† but these methods lie beyond the scope of the present text.

TABLE 15-3
PROBIT VALUES

Dose, rads	*Percent*	*Probit*
565	0	—
580	4	3.249
615	7	3.524
650	23	4.261
665	28	4.417
680	49	4.975
700	61	5.279
715	76	5.706
745	85	6.036
800	92	6.405

15.08 Approximate Methods

Several other types of functions have been found useful in rectifying the sigmoid curves usually found with mortality data. For many sets of data some of these methods give as good results as those based on formal statistical considerations and have the advantage that the descriptive constants are more

*Probit values taken from Fisher, R. A., and F. Yates, *Statistical Tables for Biological, Agricultural, and Medical Research*. Table IX, Oliver and Boyd, London, 1949.

†Kempthorne, O., ed., *Statistics and Mathematics in Biology*. Iowa State College Press, Ames, Iowa, 1954.

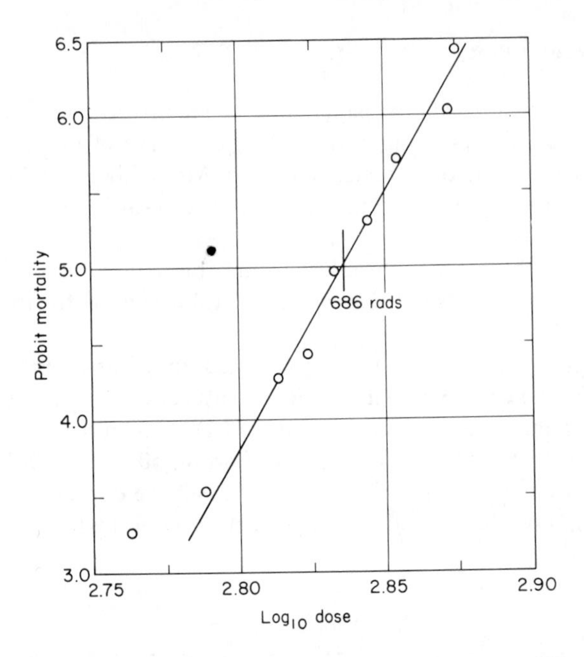

Figure 15-5. A probit plot of the mortality data. $LD_{50/30}$ = 686 rads.

easily determined. The only function considered here is the *arcsine*. This is based on the assumption that the fraction dying, f, is related to dose D by

$$f = \sin^2 [a(D - b)] \tag{15-5}$$

where a and b are constants to be determined. Equation (15-5) can be made linear by the transformation

$$y = \arcsin \sqrt{f} \tag{15-6}$$

TABLE 15-4
ARCSIN TRANSFORMATION DATA

Dose, rads	*Percent*	$\sqrt{f}$	*y, radians*
565	0.0	—	—
580	4	0.200	0.202
615	7	0.264	0.268
650	23	0.480	0.501
665	28	0.529	0.556
680	49	0.700	0.785
700	61	0.780	0.875
715	76	0.871	1.06
745	85	0.921	1.17
800	92	0.959	1.29

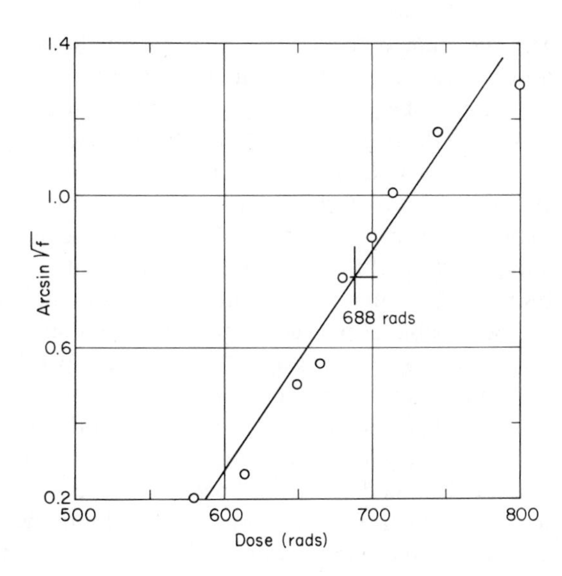

Figure 15-6. An arcsin plot of the mortality data. $LD_{50/30} = 688$ rads.

where arcsin means *the angle whose sine is*. With this transformation,

$$y = aD - ab \tag{15-7}$$

Table 15-4 shows the data of Table 15-2 transformed for plotting according to Eq. (15-7).

A linear plot of dose against y is given in Fig. 15-6. At 50% mortality $\sqrt{f} = 0.707$ and $y = 0.785$. The LD_{50} is then 688 rads.

15.09 Chronic Radiation Exposure

Except for a very rare accidental exposure, human radiation experiences will be restricted to low doses, far below the level that will produce any acute symptoms. Most of the radiation will be received at low dose rates and some by only limited portions of the body. Four sources of general population exposure can be recognized.

1. Cosmic rays originating outside our earth, gamma rays from radioactive materials in the earth's crust, and the beta particles from ^{14}C and ^{40}K internal to the human body comprise the *background radiation*. The yearly contribution from these sources amounts to about 35, 50, and 20 millirads, respectively, with some local variations.
2. Nuclear devices detonated in connection with weapons testing have injected millions of curies of radioactive fission products high into our atmosphere. Most of the activity decays before the products return to the earth's surface as *fallout*. Long-lived nuclides such as ^{90}Sr and ^{137}Cs

remain, enter the human food chain, and eventually are ingested by humans. Present dose rates from fallout are only a fraction of that from normal background but expanded testing or the large-scale use of nuclear weapons could raise the fallout contribution, perhaps to dangerous levels.

3. The expanding use of radiation in medicine and industry leads to the *occupational exposure* of more and more people, although the total number is still relatively small. Most occupational exposures involve primarily only a portion of the body and governmental regulating bodies specify allowable dose limits. These limits are set well below levels that are known to be capable of producing any sign of radiation injury.
4. *Medical exposures* arise from diagnostic and therapeutic procedures prescribed by members of the medical profession. Some diagnostic studies may deliver doses of a few rads to portions of the body. Although doses of this magnitude will produce no visible signs of radiation injury, the ratio benefit/risk should be weighed before a procedure is authorized. Therapeutic radiation may involve doses of several hundreds of rads for the purpose of destroying some unwanted tissues. Here again the benefit/risk ratio will determine whether the procedure is elected.

The effects of repair processes become evident when radiation is received at low dose rates over long periods of time. In long-term animal experiments with continuous exposure at low dose rates the usual signs of radiation injury will be absent even after doses of several thousands of rads. This is not to say that repair is complete and that no biological injury has been produced. Radiation is known to be mutagenic and carcinogenic and some life shortening is observed in animals receiving doses of the order of 1 rad or more per day.

15.10 Genetic Consequences of Irradiation

Ionizing radiation is known to produce a variety of types of genetic injury ranging from subtle point mutations to severe chromosome damage such as strand breakage, translocations, and deletions. Mutations in somatic cells may be responsible, at least in part, for the initiation of radiogenic cancers. Radiation-induced alterations in sperm or oocytes may be passed on to progeny, perhaps to remain latent for many generations before becoming manifest.

Severe genetic injuries such as chromosome translocations and deletions are incompatible with fetal development. Many conceptions involving these severely damaged structures will be eliminated even before there are any signs of pregnancy. Lesser injuries may permit full-term development and birth, with a new mutant passed on to a large number of descendants.

Consider a population in which P individuals receive a genetic injury such that a fraction x of those possessing it will not transmit it to their offspring. Assume a static population size with each family pair having two children.

The injury will be passed on to $P(1 - x)$ individuals in the first breeding after the injury was originally produced. At the nth generation, $P(1 - x)^n$ new injuries will be produced. The total number of persons bearing the characteristic, N, including the original number P, will be

$$N = P\left[1 + \sum_{n=1}^{\infty} (1 - x)^n\right] \tag{15-8}$$

The summation is readily evaluated and then we have

$$N = P\left(1 + \frac{1 - x}{x}\right) = P\left(\frac{1}{x}\right) \tag{15-9}$$

If, for example, $x = 0.1$, an original number of P affected individuals will eventually lead to a total number of $10P$.

15.11 Carcinogenesis

Many early workers demonstrated in tragic fashion that large doses of X rays are carcinogenic in man. Repeated moderate doses led to local epilation and a chronic dermatitis which was frequently followed by the appearance of squamous cell carcinoma.

Surface doses were high with the soft X rays of the early days and so it is not surprising that most of the malignancies seen then originated in the skin. As more penetrating radiations came into use, radiogenic cancers developed in deeper tissues and it now appears that any cell that is capable of division may be susceptible. Cell types have varying degrees of sensivity but probably none is immune.

Radiogenic cancer in man is indistinguishable from cancer that arises spontaneously, a fact that introduces serious legal complications. Twenty years or even more may intervene between a radiation overexposure and the appearance of a cancer, and there may be no causal relation between the two events. Difficult problems in injury compensation are bound to arise.

15.12 Life Shortening

In several carefully controlled experiments, animals receiving chronic doses of the order of 0.1 rad per day have lived somewhat longer than their non-irradiated controls. The meaning of this apparently beneficial effect of radiation is not clear. At dose rates of 1 rad or more per day, life shortening is regularly observed.

The normal life expectancy of a species, its radiation sensitivity as measured by the acute $LD_{50/30}$, and the age at which the radiation is received are among the factors acting to modify the life-shortening effect. There is no specific cause of death that can be considered pathognomonic of radiation injury. An increase in the number of malignant neoplasms and in the incidence of proliferative diseases of the blood is seen as long-term consequences of moderate overexposures. The causes of death in the rest of an irradiated animal population run the gamut of those seen in a normal population.

Various dose–injury models have been postulated and subjected to mathematical analysis with indifferent success. Direct data on man are essentially nonexistent and transfers from animal data are fraught with great uncertainties. We might speculate that 100 rads of chronic, whole-body radiation will reduce the life expectancy in man by 1% and that an equal dose delivered at a high dose rate will be more effective. Neither the figure of 100 rads nor the method by which it might be extrapolated to other doses can be vigorously defended. As in carcinogenesis some difficult problems in injury compensation appear to be inevitable.

15.13 Dose–Injury Relations in a Population

We have seen that the amount of radiant energy required to produce a lethal injury is rather small. When one considers the enormous numbers of interactions with molecular structures, however, the production of a detectable radiation injury is a relatively rare event. When low-exposure experiments are attempted, it is found that the fluctuations in the normal incidences tend to obscure abnormalities that can be attributed to radiation. Below whole-body doses of perhaps 30–50 rads a prohibitive number of animals is needed in order to obtain a response that is statistically significant; yet this dose level is many times the yearly limit currently set for the general population and is still further removed from doses which past experiences have led us to expect.

Risk estimates in the low-dose range are urgently needed by regulatory bodies. With direct experimental data in the pertinent dose range unavailable, recourse must be had to extrapolations downward from data obtained at higher doses.

Over a limited range, animal experiments may show a linear relation between dose and the probability of radiation injury, which is just the fraction of the animals at risk that show the effect in question. At high doses the incidence rates approach 100% gradually rather than in the abrupt fashion that would be required by a strictly linear relationship. At low doses, experimental data are not available and are not likely to be in the near future. There is no direct evidence for or against linearity in the low-dose domain. There may or may not be a threshold dose below which no injury is produced.

A single ionizing event may be capable of triggering a sequence of biological processes that eventually lead to an observable end point such as leukemia, a solid tumor, or death. This single unique event is only one result of the large number of ionizations and excitations produced by the absorbed radiation. The probability that a single initiating event will be produced will then be a linear function of the dose.

It is also quite possible that some end effects will require more than one precursor event to trigger the biological succession that eventually leads to the observable injury. Perhaps in these cases one event is repairable, while two are not. In these cases any number of primary events less than the critical number will be ineffective. There is nothing new in this concept. It is a restatement of the target theory presented in Chapter 13. Here the primary event is not to be thought of as an action on a specific target with a geometrically defined shape and size. Rather it is conceived to be some sort of a chemical or physical change, perhaps of genetic significance. A two-event requirement, for example, implies that the initial state of the organism and the state that irrevocably leads to overt injury are separated by one intermediate state. The consequences of these concepts will be examined in the next section.

15.14 Extrapolation Schemes

In any effective dose of radiation the number of ionizations and excitations is large and the mean probability that one of them will produce one of the critical precursor events is small but finite. These conditions permit the use of the Poisson distribution function to estimate the number of the precursor events as a function of dose. Equations (13-1) to (13-4) apply exactly to the present situation but now we are interested in the number of events rather than the number of escapes.

The *total* probability P_N for an injury that requires n precursor events is

$$P_N = e^{-kD} \sum_{x=n}^{\infty} \frac{(kD)^x}{x!} \qquad (5\text{-}10)$$

P_N is the *cumulative* Poisson distribution whose values can be found in tabulations of statistical parameters with values of kD and n as arguments.* Alternatively, P_N may be calculated by subtracting the sum of the ineffective terms from the probability of all possible values, which is unity. In this form

$$P_N = 1 - e^{-kD}\left[1 + \frac{kD}{1} + \frac{(kD)^2}{2!} + \frac{(kD)^3}{3!} \cdots + \frac{(kD)^{(n-1)}}{(n-1)!}\right] \qquad (15\text{-}11)$$

The solid curves of Fig. 15-7 show the probabilities as a function of dose when one, two, or four critical events are required to initiate some biological

*W. H. Beyer, ed., *Handbook of Tables for Probability and Statistics*. Chemical Rubber Company, 1966.

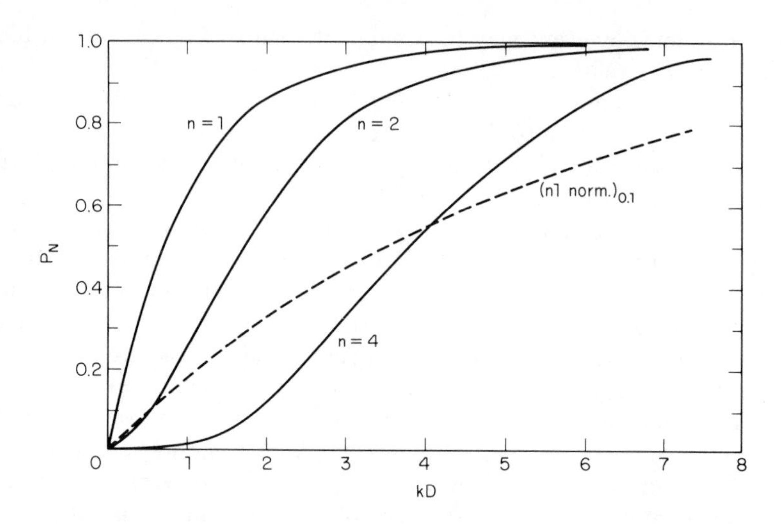

Figure 15-7. The solid curves show the computed dose–effect relations for the cases of one-, two-, and four-event requirements. The dashed curve is a one-event calculation with the parameters adjusted to equal the two-event incidence at $P_N = 0.1$.

response. Each of the relations has an approximately linear portion but each also has regions of pronounced curvature.

At some value of dose D_e, the probability of overt injury incidence $(P_N)_e$ will be known from experimental data. Below this *established point*, experimentally determined values of P_N are too uncertain for acceptance. At these lower doses, injury incidences can only be calculated from some sort of an assumed relationship. The relationship chosen will have a profound effect on the predicted values of injury incidence.

Illustrative Example

Compare the predicted values of injury incidence for biological models where $n = 1$ and $n = 2$, extrapolating downward from an established point at which $(P_N)_e = 0.1$ and $D_e = 0.53$ in arbitrary units.

From Eq. (15-11) the two functional relations are

$$P_1 = 1 - e^{-kD} \quad \text{and} \quad P_2 = 1 - e^{-k'D}(1 - k'D)$$

At the established point, $P_1 = P_2 = 0.1$ and this requirement leads to values of $k = 0.198$, $k' = 1.00$. The incidence values will then be predicted from

$$P_1 = 1 - e^{-0.198D} \quad \text{and} \quad P_2 = 1 - e^{-D}(1 - D)$$

Figure 15-8 is a semilog plot of these two functions.

According to Fig. 15-8 the single-event model predicts higher incidence

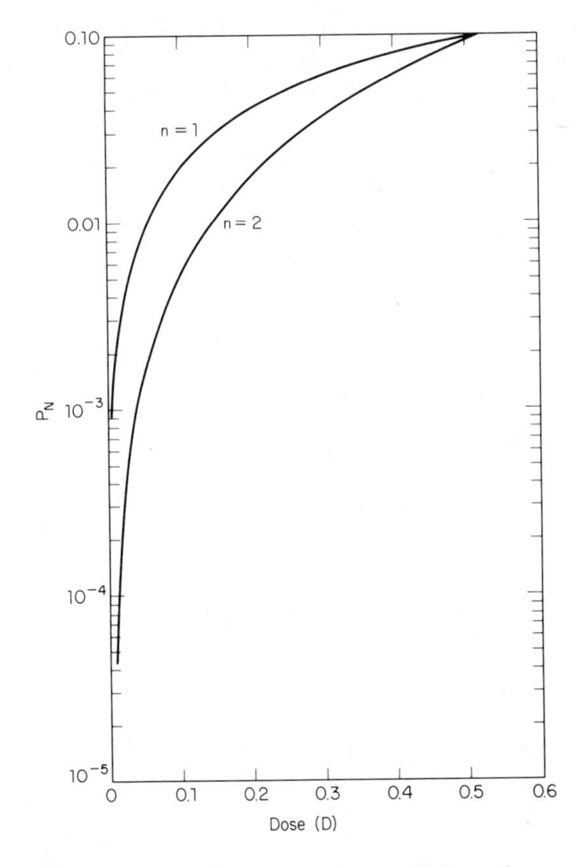

Figure 15-8. Semilog plot of the low-dose predictions of one- and two-event models normalized at $P_N = 0.1$.

values at all values of dose below D_e than are obtained from the two-event model. At very low doses the differences between the two predictions are more than an order of magnitude. The differences will be even greater if a value of n larger than 2 is assumed. Above D_e the single-event curve will lie below those representing higher n-fold requirements. The dotted curve in Fig. 15-7 shows the behavior of the $n = 1$ curve that has been normalized at the value given in the illustrative example.

15.15 Dose Threshold

The value of P_D at the established point is not known exactly and any value predicted from it to lower doses will also be subject to statistical uncertainties. Assume that a sufficiently large animal population was chosen to determine

$(P_N)_e$ with a standard deviation equal to 5% of the mean value. There is then a 95% probability that a measured value will fall within $\pm 2\sigma$ or 10% of the mean. This range is shown plotted at $P_e = 0.1$ in Fig. 15-9 where dose is expressed as a fraction of the established dose D_e. If the same number of animals is assumed at lower dosages, the magnitude of $\pm 2\sigma$ will increase as $1/\sqrt{P_N}$ as the incidence probabilities decrease.

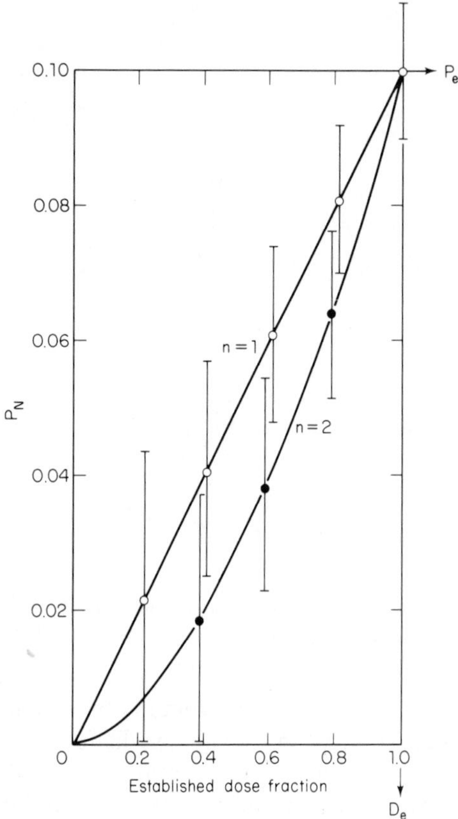

Figure 15-9. When $\pm 2\sigma$ error bars are put on the normalized curves, zero incidences are seen to be probable at low doses.

At some value of dose, the $\pm 2\sigma$ range (or any other multiple of σ that might be chosen) will include zero incidence. In the example plotted, this occurs at a value of $0.2D_e$ for a one-event model and at $0.4D_e$ when two events are assumed. At all lower doses some zero-incidence data are to be expected.

Experimental data that show some instances of zero incidence can be interpreted as evidence for a threshold dose when in fact they may be merely expressions of the variability of very small incidence values. Any inference for or against the existence of a threshold is complicated by the fact that no completely unique radiogenic injury has been identified. Each is superposed upon a natural incidence from which the contribution due to radiation must be separated by subtraction.

No unequivocal choice can be made between extrapolation methods nor has the presence or absence of a threshold been established. These choices are of more than academic interest. For protection of the population, prudence dictates the choice of a linear extrapolation downward from the limit of experimental data. This may overestimate low-dose hazards by several orders of magnitude, however, and may thus lead to excessive costs for some very desirable uses of nuclear energy. If the existence of a threshold could be proved, regulatory bodies would have some firm basis for promulgating allowable exposure limits. Lacking firm information, limits can only be set by considered judgements from past experiences.

REFERENCES

BACQ, Z. M. AND P. ALEXANDER, *Fundamentals of Radiobiology*, 2nd ed. Pergamon Press, New York, N.Y., 1961.

CASARETT, A. P., *Radiation Biology*. Prentice-Hall, Inc., Englewood Cliffs, N.J., 1968.

ERRERA, M. AND A. FORSSBERG, *Mechanisms in Radiobiology*, Vol II. Academic Press, New York, N.Y., 1960.

HEMPELMANN, L. H., H. LISCO, and J. G. HOFFMAN, "The Acute Radiation Syndrome." *Ann. Int. Med.*, **36**, 279, 1952.

National Academy of Sciences–National Research Council, Publication 452. *Pathological Effects of Atomic Radiation*, 1956.

National Academy of Sciences–National Research Council, Summary Reports. *The Biological Effects of Atomic Radiation*, 1960.

SCOTT, W. G. AND T. C. EVANS, ED., *Genetics, Radiobiology, and Radiology*. Charles Thomas, Pub. Springfield, Ill., 1959.

SHIPMAN, T. L., ET AL., "Acute Radiation Death Resulting from an Accidental Nuclear Critical Excursion." *J. Occupational Med.*, Supp. 3, 145, 1961.

16

The Ultraviolet Region

16.01 Ultraviolet—Originator of Life

Although the details are obscure, it is very probable that all life forms on earth evolved from photochemical and photobiological reactions initiated by ultraviolet photons from the sun. Certainly life as we now know it would not be possible without the reactions of photosynthesis whereby complex molecules, notably carbohydrates, are produced by the action of UV on carbon dioxide, water, and some inorganic salts.

All the actions of UV photons are not benevolent. Our continued existence on earth depends on the presence of a thin layer of ozone, O_3, high up in our atmosphere. Ozone, a strong absorber of UV photons, reduces the high UV intensity emitted by the sun to a value that is compatible with life. The balance between benefit and injury is a delicate one. On the one hand the addition of pollutants to our atmosphere may reduce the UV intensity at the earth's surface below that required for an adequate level of photosynthesis. On the other hand any marked reduction in the ozone layer which might result from reactions with quantities of oxidizable pollutants could lead to dangerously high levels of UV.

16.02 The Ultraviolet Spectrum

The lower-energy limit of the UV portion of the electromagnetic spectrum is strictly set at the limit of human vision at about 3.26 eV, corresponding to a wavelength of 380 nm. We are not concerned here with the phenomenon of vision and so we shall use the term UV to include any photons in the visible portion of the spectrum that are capable of producing electronic excitations in absorbing molecules. The region above 3.26 eV will be denoted by the term *true UV* when distinction is necessary.

At the high-energy limit the UV merges gradually into the X-ray region. We here arbitrarily choose the division at 10 eV or 124 nm. This limit effectively excludes ionizing photons and confines the UV region to that in which electronic excitations predominate.

The UV is a transition region where interactions are energy quantized and designations in eV are useful but where the techniques of optical spectroscopy are used to measure wavelengths. UV wavelengths are conveniently expressed in angstroms (10^{-10} m) or in nanometers (10^{-9} m). The E–λ relation of Eq. (1-19) may now be written

$$E(\text{eV}) = \frac{1240}{\lambda(\text{nm})} \tag{16-1}$$

Wave numbers, $\tau = 1/\lambda$ are also frequently used in this region, usually in units of cm^{-1}.

The true ultraviolet is frequently subdivided into the *near ultraviolet* (380–300 nm), the *far-ultraviolet* (300–190 nm), and the *vacuum ultraviolet* (190–1 nm), which overlaps a portion of the soft X-ray region. Air and water vapor absorb these wavelengths so strongly that all studies must be conducted in a vacuum. Data in this region are meager because of the experimental difficulties involved.

16.03 Ultraviolet Absorption

We have already made use of the fact that in a liquid the absorption of a UV beam follows an exponential law. Equation (14-21) is repeated here.

$$I = I_0(10)^{-\epsilon cl} \tag{16-2}$$

Note that in this particular case the exponential is based on (10) rather than on (e). This comes about because the term *optical density* or OD has long been defined as

$$\text{OD} = \log_{10} \frac{I_0}{I} = \epsilon cl \tag{16-3}$$

Absorption in gases and solids is usually expressed in powers of (e). The relation

$$I = I_0 e^{-\alpha x} \tag{16-4}$$

applying to an absorber of thickness x is analogous to Eq. (8-1) relating to ionizing photons.

Both ϵ and the linear absorption coefficient α are strong functions of photon energy. With very high-resolution instruments the physicist can demonstrate much fine detail in UV absorption curves, indicating the presence of many closely spaced energy states. When atoms are bound into molecules, the inner electron shells are little affected. Schematic energy-level diagrams such as those shown in Fig. 14-1 can be thought of as applying to the outer valence electrons only. Valence electrons are shared between adjacent bound atoms and the electronic energy levels in each atom are now modified by a series of vibrational energy states. Only a few of these are shown in Fig. 14-1. The strength (or energy) of these interaction bonds tends to be characteristic of the atoms directly involved but there will be some variations due to the presence of other adjacent atoms. A few representative bond-energy values are listed in Table 16-1.

As Table 16-1 indicates, vibrational energy levels lie in the infrared region at energies too low to be effectively excited by UV photons. Their presence does, however, create a vast number of possible energy states at each level of electronic excitation. This number is increased still further by the presence of molecular rotation states. Each vibrational level will have a number of energy states due to the quantized rotations of the molecule as a whole. The rotational frequencies themselves lie in the microwave, or high-radio-frequency, region of the spectrum.

The absorption of ionizing radiation can be a continuous function of energy. Although any bound electron will require an exact amount of energy for its ejection, any amount of energy in excess of this can be converted to kinetic energy because the free electron is not required to go into a precisely determined state. In the excitation region, on the other hand, an electron is transferred from one precisely quantized state to another. Excess energy cannot be converted to kinetic energy and so sharp absorption resonances are to be expected.

Many of the vibrational and rotational levels lie so close together that they are separable only with high-resolution instruments. Most biological studies are done with medium-resolution equipment using quartz prisms or diffraction gratings to provide the energy dispersion. These instruments

TABLE 16-1
INTERATOMIC BOND ENERGIES

Bond	*Energy, eV*	*Frequency, GHz*	*Wavelength, μm*	*Wavenumber, cm⁻¹*
C – C	0.124	30,000	10.0	1000
C – N	0.138	33,000	9.0	1100
C – O	0.146	35,000	8.5	1200
C = C	0.200	48,000	6.2	1600
C = N	0.203	49,000	6.1	1600
C ≡ C	0.276	67,000	4.5	2200
C – H	0.376	91,000	3.3	3000

respond to ϵ and α values averaged over many vibrational and rotational levels and so broad absorption bands rather than sharp resonances are observed. The solid curve of Fig. 16-1, representing the absorption of a solution of DNA in water, shows no trace of fine structure. Very-low-resolution absorption spectrometry has been used for many years in the *colorimeter* where broad spectral bands are selected by colored filters.

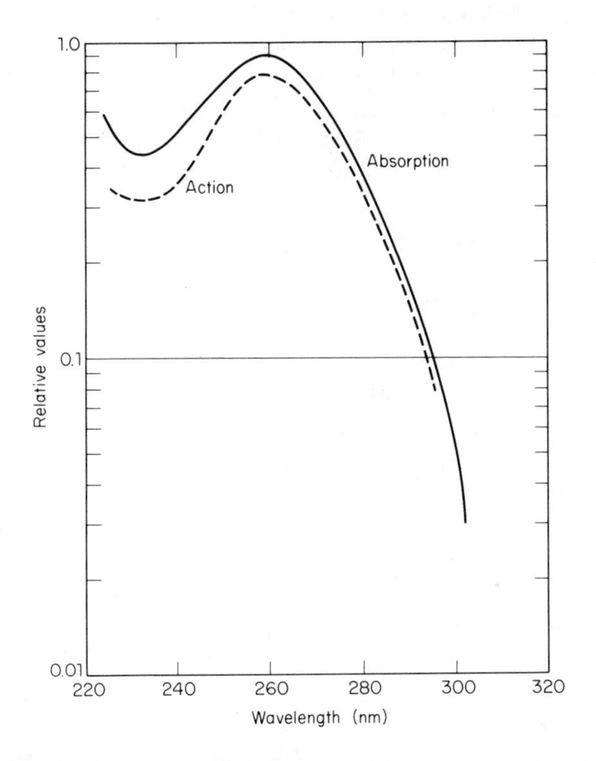

Figure 16-1. Ultraviolet absorption spectrum of a water solution of DNA, compared with the action spectrum for the formation of thymine dimers. The ordinate scales have been normalized for easy comparison.

16.04 Ultraviolet Scattering

Scattering also acts to remove UV photons from a beam. The oscillating field of a photon will *polarize* or distort and orient the electric charges in a molecule whose dimensions are small compared to the wavelength of the photon. The electric charges set in motion at the photon frequency will produce a secondary electromagnetic field at the same frequency. Energy absorbed from the original photon will be reradiated but since this secondary radiation tends to be spatially isotropic (scattered), the intensity in the direction of the original beam will be reduced.

This *Rayleigh scattering* is a nonresonant phenomenon with a cross section about proportional to λ^{-4}. Larger molecules may have several scattering centers with a combined wavelength dependence more nearly as λ^{-3}. Rayleigh scattering appears in an absorption curve as a component that increases smoothly with photon energy. A correction for scattering can be made by an extrapolated peeling-off method. A log OD–log λ plot of the observed data will show an essentially linear segment on the long-wavelength side of the absorption band, Fig. 16-2, curve A. This portion of the curve is extrapolated linearly through the absorption curve region to shorter wavelengths. Values taken from this extrapolated scattering curve are subtracted from the corresponding values on the composite, observed curve. The remainders are plotted and make a good approximation to the true absorption curve, Fig. 16-2, curve C.

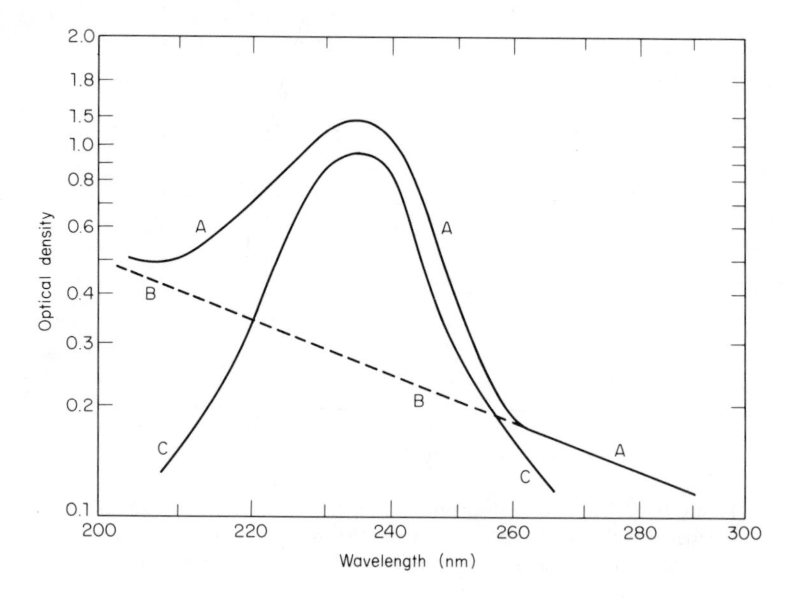

Figure 16-2. (A) An observed UV absorption curve showing a contribution due to scattering. (B) Extrapolation of the scattering component. (C) Corrected absorption curve.

16.05 Action Spectra

Studies of the specific types of reactions initiated by UV photons have led to a better understanding of the effects of ionizing radiations and have provided vast amounts of information applicable to many areas of biology. Much of this information has come from studies of *action spectra*.

An action spectrum is a delineation of the relative response of a system to different UV wavelengths. As might be expected, action spectra are closely related to the absorbing properties of the system. A fundamental rule of photochemistry states that a photon must be absorbed to be effective. Figure 16-1 shows the close correlation between the absorption of DNA and the action spectrum for the formation of thymine dimers from the monomer. Obviously thymine dimerization must play an important role in the action of radiation on DNA. Other actions are not excluded, and they are observed, but dimer formation is the predominant effect.

Quantum yield ϕ is defined as the ratio

$$\phi = \frac{\text{molecules (or cells) altered}}{\text{photons absorbed}} \tag{16-5}$$

with units of molecules per photon. A new unit of photon quantity, the *einstein*, is introduced for convenience in working with molar quantities.

$$1 \text{ einstein} = 6.02 \times 10^{23} \text{ photons} \tag{16-6}$$

With this new unit, ϕ will have the same numerical value as before but the dimensions will now be moles per einstein. The einstein is a numerical unit only; the energy content is obtained by multiplying by the quantum energy involved. From Eq. (16-6) and (16-1) it is readily shown that

$$E = \frac{(2.86 \times 10^4)}{\lambda\,(nm)} \text{ kcal (einstein)}^{-1} \tag{16-7}$$

Numerical values of ϕ cover an enormous range, from nearly 1.0 as for the production of thymine dimers to perhaps 10^{-8} for the inactivation of large yeast cells. Quantum yield is calculated in terms of the number of photons absorbed and not upon the number that are incident. Use of the latter figure gives the reaction cross section σ_r.

$$\sigma_r = \frac{\text{molecules (or cells) altered}}{\text{photons incident}} \tag{16-8}$$

If σ_a is the cross section for photon absorption,

$$\sigma_r = \sigma_a \times \phi \tag{16-9}$$

Neither of the σ's has the dimensions of an area but each expresses a probability.

Care must be taken in interpreting measured values of quantum yield because UV photons may initiate back reactions as well. For example, under appropriate conditions thymine dimers will be split back to the monomer by UV, a case of *photoreactivation* or induced repair of a radiation injury. In this case irradiation leads to a thymine-dimer equilibrium, the relative concentrations depending on the number and the wavelength of the activating photons.

16.06 Molecular Quantum Numbers

An electron that is shared between two atoms has a set of quantum numbers analogous to those specifying its energy state in an atom and it is subject to the Pauli exclusion principle. In the atomic situation the magnetic quantum number m determines the orientation of the orbital plane with respect to the direction of an external magnetic field. In the case of a shared electron the reference direction is the interatomic axis and the designation m is replaced by a new quantum number λ subject to the same restrictions that apply to m. That is, $\lambda = 0, \pm 1, \pm 2, \ldots, \pm l$, where l is the quantum number specifying the orbital angular momentum of the electron. The algebraic sum of the λ values for a molecule represents the projection of the total orbital angular momentum onto the interatomic axis. As this sum takes on successive values 0, 1, 2, 3, . . . , the states are denoted by the Greek capital letters $\Sigma, \Pi, \Delta, \Phi, \ldots$ by analogy with the Arabic $S, P, D, F, \ldots$ notation for the atom. Orbital and spin angular momenta couple, as in the atom, but J values are not given for the molecule because now the total angular momentum depends on the rotational state of the molecule as a whole.

A molecular state is customarily denoted by the angular momentum symbol preceded by a superscript denoting the level multiplicity. Thus $^1\Sigma$ refers to a singlet state, Sec. 2.08, in which the total orbital angular momentum is zero. The singlet designation shows that all electron spins are paired parallel–antiparallel. Most organic molecules have an even number of electrons and this requires the formation of singlet and triplet states rather than the doublets that are associated with an odd number of electrons.

16.07 Molecular Energy Levels

Figure 16-3 depicts some of the energy levels in an organic molecule together with some of the transitions that can follow the absorption of a UV photon. Only a few of the possible vibrational states are shown and all rotational levels have been omitted. At ordinary temperatures only a few of the lowest vibrational states have energies that are compatible with equilibrium occupancy. These are shown in the figure as light solid lines. Higher vibrational levels, shown by broken lines, will rapidly degrade by collisions with the conversion of the energy into molecular motion, or heat.

The ground state of an organic molecule is usually $^1\Sigma$, as shown. An absorbed UV photon may raise the structure to the first excited singlet state, shown as a $^1\Pi$. The electron spins remain paired but the total orbital angular momentum has increased. Several de-excitation channels are available.

In the simplest process the structure will radiate a photon with exactly the energy that was absorbed and will return to the original vibrational level in the ground state. More probably, some vibrational energy will be

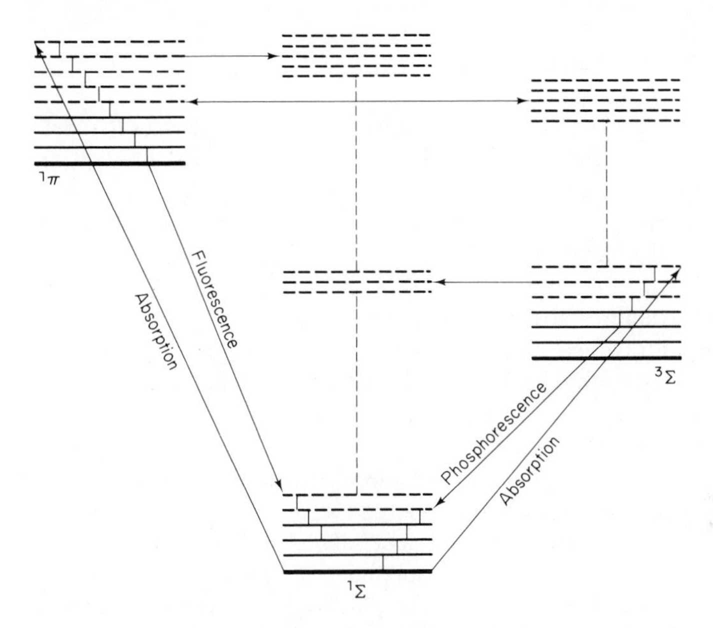

Figure 16-3. Possible intramolecular transitions following the absorption of a UV photon.

lost before photon emission can occur, and the transition may then go to a high vibrational level in the ground state. Thus these photons, which form the *fluorescent radiation*, have energies equal to or less than that of the exciting photon. Fluorescent radiation is emitted within about 10^{-8} s after the absorption.

Somewhat less energy is required to raise the molecule to the first excited triplet state shown tentatively as $^3\Sigma$ in Fig. 16-3. This transition requires only the conversion of a paired electron spin to parallel–parallel or to antiparallel–antiparallel. De-excitation may take place by photon emission as from the excited singlet but there is a degree of forbiddenness that slows the process. Photon emission by *phosphorescence* may lag behind absorption by as much as 1 s. During this time, degradation of high vibrational levels is almost certain to occur and phosphorescent radiation will have longer wavelengths than that causing the excitation.

The higher vibrational levels are very closely spaced and this permits some isoenergetic multiplicity transfers, either singlest–triplet or triplet–singlet, as shown. Vibrational degradation and photon emission can follow transfer, as before.

Radiationless de-excitation can take place from either an excited singlet or triplet state. An isoenergetic transfer may take place to one of the high vibrational levels of the ground state. This level will then degrade to a thermally acceptable level without the emission of a photon.

16.08 Chromophores

Some atomic groups have characteristic absorption energies that are so slightly altered by adjacent bonds that the presence of the absorption band is presumptive evidence of the group. These groups are known as *chromophores* or color carriers, even though most of them are found at wavelengths below the limit of human vision.

Electronic energy levels are such that almost all the saturated carbon-bonded structures absorb only at wavelengths of 200 nm and below, in the vacuum ultraviolet. Saturated compounds such as cyclohexane, C_6H_{12}, and 2, 2, 4-trimethyl pentane, $(CH_3)_3CCH_2CH(CH_3)_2$, are useful nonabsorbing solvents in the near- and far-ultraviolet regions with which the biologist is most concerned. Water is also a nonabsorbing solvent down to about 210 nm but it is a strongly polar molecule which may interact with the solute to somewhat alter the absorption curve.

Unsaturated carbon bonds generally absorb at wavelengths that are amenable to experimental measurement. *Conjugated* or alternating single and double bonds form strong chromophores. These alternating bonds may be in a ring structure such as benzene (260 nm) or in an aliphatic chain. In a double ring such as naphthalene the absorption maximum may shift to 270 nm and in a triple ring (anthracene) to 280 nm. Substituents such as —H and —CH_3 absorb only at very short wavelengths and change the other absorption bands only slightly.

When two chromophores are separated by an aliphatic group such as $=CH_2$, the two absorptions will be simply additive, indicating essentially independent electronic levels. Two chromophores bonded directly together will form a new chromophore with its own characteristic absorption.

Many structures besides conjugated bonds form chromophores at wavelengths greater than 200 nm. The disulfide group, $>C=S$ (330 nm), sulfhydryl, —SH (220 nm), and carbonyl, $>C=O$ (280 nm), are examples of important chromophores. Two great classes of biologically important molecules, proteins and nucleic acids, are strong UV absorbers with bands at 280 and 240–290 nm, respectively.

16.09 Injury and Repair of Nucleic Acids

There is an enormous amount of literature on the action of UV photons on biologically active molecules which must be consulted for details. Only a brief discussion of some results with nucleic acids is presented here.

Nucleic acids are of prime importance because they are apparently the only compounds that contain the genetic information essential for the existence and reproduction of living organisms. A small injury to genetic material may have profound effects (mutation or death), while an injury of comparable magnitude in some other type of molecule might go unnoticed. For example, a single alteration in the primary genome of a cell could have the most serious consequences because this genome might be the only one of its type available to the cell. Denaturation or even a more drastic injury to a single protein molecule will probably have no biological consequences because there will be many duplicates of the original to carry out the required function. DNA probably plays an important role in the remarkable sensitivity to radiation shown by living tissues.

Several types of DNA alteration have been observed following UV irradiation. These injuries include a local denaturation, which might appear as a point mutation in later generations; the formation of hydration products from the pyrimidine bases; the production of cross-links with proteins; and pyrimidine (thymine) base dimerization. Single-strand breaks may be repairable or may be followed by excision of a portion of a strand. Ionizing radiation can produce, in addition, double-strand breaks which may be followed by excision of a portion of one or both strands or by a strand rejoining perhaps with strand crossing. Double-strand breakage is a drastic, almost certainly fatal injury.

There is abundant evidence that some types of radiation injury are repairable and several mechanisms for this are suggested by the studies of the effects of UV on DNA.

1. Radiation may reverse the injury reaction and return the molecule to its original state. This is observed in the production of thymine dimers where UV photons also act to return the structures to the monomers.
2. If undamaged material is available, a radiation injury may be removed from the DNA strand by excision and replaced by replication of the original structure.
3. Single-strand breaks and some damage to the nucleotide bases are repairable probably by replication from undamaged molecules.
4. *Enzymatic photoreactivation* of dimers can be affected with certain enzymes when they are illuminated with the proper wavelength of UV. The energy required for the splitting appears to come from the photon.
5. A radiation injury may be ignored in replication if an uninjured copy of the molecule exists. Apparently replication prefers to use "good" material in preference to "bad" or perhaps two injured structures combine parts to form one viable unit and some unusable fragments.

Repair processes may be partly or wholly responsible for the dose-rate dependence of some radiation injuries. Repair could also lead to nonlinear dose-response relations at very low doses although such nonlinearity has not been yet demonstrated, Sec. 15.13.

16.10 Gross Effects of UV Exposure

Values of the absorption coefficients are so large that UV photons are completely absorbed in a few micrometers of human skin. Skin is the chief tissue at risk, with tanning, sunburn, and erythema the well-known results of moderate exposures to wavelengths shorter than 320 nm. Fever, chills, weakness, and signs of shock may accompany moderate overexposures. A still greater overexposure may disturb the basic heat-regulating mechanism and lead to symptoms of serious systemic trauma in addition to the local erythema at the site of absorption. The resulting hyperpyrexia can be a serious threat to life and must be dealt with promptly and skillfully.

Exposure of the unprotected eye to high intensities in the near or far ultraviolet can lead to a severe conjunctivitis. Local pain may appear promptly and become severe enough to be uncontrollable by an analgesic such as codeine. Complete symptomatic recovery takes place in 24 hr or so with little if any permanent damage.

Chronic exposure to sunlight or to strong sources of UV accelerate aging of the skin. Precancerous keratotic lesions may appear on exposed areas. Evidence is conclusive from chronically exposed persons such as farmers, sailors, or confirmed sunbathers that the incidences of basal cell epitheliomas and squamous cell carcinomas are directly related to the amount of exposure to UV.

A few individuals exhibit an unusual *photosensitivity* to sunlight. These hypersensitivity reactions arise from a variety of causes, many of them unidentified. Herpes simplex, the common "cold sores" is readily triggered by exposure to sunlight. Many drugs and chemical compounds, either ingested or applied topically, may induce photobiological reactions leading to dermatitis, urticarial lesions, and bullae. Symptoms usually appear promptly during the exposure but may persist for some time afterward.

16.11 Ultraviolet Sources

Incandescent lamps, particularly when operated at an abnormally high temperature, emit a continuous spectrum comparable to that of sunlight. The excess energy in the red and infrared regions of these sources can be removed by filtration. Incandescent sources are deficient in the deep blue and the ultraviolet and are of little use below 380 nm.

Absorption spectroscopy requires a light source that emits a continuous rather than a line spectrum. Light from the source is focused on the entrance slit of some sort of a wavelength selector or monochromator, usually a quartz prism or a diffraction grating. The selected narrow wavelength band then passes through an absorption cell and the incident and transmitted intensities are compared. With sensitive detectors and electronic amplifiers available only modest source intensities are required.

Over the visible spectrum and down to perhaps 360 nm the ordinary incandescent light bulb is a quite satisfactory source. Its output falls off badly at the shorter wavelengths where it is usually replaced by a hydrogen-discharge lamp. An arc discharge in either hydrogen or deuterium produces a continuous spectrum with usable intensities from 450–200 nm. Hydrogen lamps, with the beam taken out through a thin window to reduce absorption, are almost universally used for absorption spectroscopy in the true UV.

Higher photon intensities are needed for studying the chemical and biological effects of UV. At the longer wavelengths large sources such as carbon arcs can be used but they emit large amounts of red and infrared, and heat dissipation can be a problem. Commercially available fluorescent lamps are relatively cool sources over a limited range 600–400 nm. In these lamps an arc in mercury vapor at low pressure produces the line emission spectrum of mercury with most of the intensity appearing in the 253.7-nm line. This line excites the fluorescent coating on the inner surface of the glass envelope to produce a continuous spectrum which tends to resemble sunlight, with some mercury lines superposed.

A wide variety of mercury-arc sources are available for work at shorter wavelengths. An arc in mercury vapor at low pressure emphasizes 253.7- and 1850-nm lines and both of these will be transmitted through a quartz envelope. Only the longer wavelength will pass through Vycor glass. The so-called "germicidal" lamps provide almost monochromatic sources of 253.7-nm radiation, which is most useful in biological work since many molecular forms have absorption bands in this region.

Mercury vapor so readily absorbs the 253.7-nm line that this wavelength is almost completely absent in high-pressure arcs. These arcs, which may operate at pressures of 100 atmospheres, emphasize emissions at 365, 405, 430, and 546 nm. Each of these emissions will be a narrow band rather than a line because of Doppler effect broadening in the high-pressure source.

Another useful high-intensity source is the xenon arc, which produces a continuous spectrum useful down to about 280 nm with some superposed lines at 470 nm.

16.12 Laser Sources

Emissions from the sources just described are the result of an enormous number of independent events leading up to the production of photons that are random in time, direction, and plane of polarization. Interference cannot be demonstrated between two of these *incoherent* radiation sources because of the random nature of the emissions.

Light from a laser source is *coherent* and has many properties quite distinct from those of more common sources.

1. Laser emissions are much more nearly monochromatic than are those from incoherent sources.
2. Laser beams are emitted in extremely small solid angles with well-defined, almost plane wavefronts.
3. Enormous power densities can be attained in a laser beam.
4. Enormous electric field strengths can be attained in the electromagnetic field of a laser beam.
5. Highly plane-polarized beams are emitted by a laser.

Chromium-doped ruby, $Al_2O_3 \cdot Cr^{3+}$; various doped glasses; helium–argon mixtures; carbon dioxide; dye solutions; and some solid-state devices are prominent among the materials that have been made to lase. Basic principles of operation can be described in terms of a solid such as ruby which emits a beam at 694.3 nm.

An artificial ruby rod, precisely doped with chromium, and with highly polished, plane parallel ends, is located between two plane parallel mirrors. Mirror M_1, Fig. 16-4, is totally reflecting, while M_2 allows a portion of the light incident upon it to be transmitted.

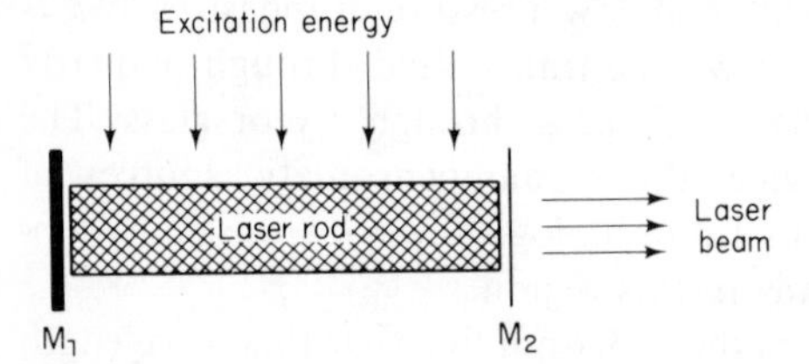

Figure 16-4. Schematic representation of a simple solid laser.

The ruby is *pumped* by flooding it with a broad spectrum from an intense external light source such as a xenon arc. Large numbers of the Al_2O_3 molecules will be raised into excited states and many of them will return to the ground state through vibrational degradation and fluorescent photon emission. A goodly number of excitations will transfer to the Cr impurities and these states are metastable, with lives in the order of milliseconds. Some of these states will decay with the emission of 694.3-nm photons but vigorous pumping can lead to a *population inversion*, in which more than one-half of the Cr atoms are in the excited state.

At some population density, amplified photon emission becomes possible. One 694.3-nm photon will trigger the emission of a second similar photon without being itself absorbed and this release process will be repeated until the population of excited states drops below the level where stimulated amplification is possible. All photons escaping from one end of the rod will be returned by the full reflecting mirror, while some of them pass through the partial reflector to form the useful beam. Laser parameters can be arranged to provide pulses ranging from about 10^{-6} to 10^{-3} s and continuous or CW operation is also possible.

Very short high-intensity pulses can be produced by *Q-switching*. *Q*, a term borrowed from the electronic circuit field, may be broadly interpreted here as "efficiency." A laser may be *Q-spoiled* or made inefficient by interrupting the light path to the totally reflecting mirror. Because of the lack of the reflected photons a high degree of population inversion can be attained without producing an amplified discharge. When the light path to the mirror is restored by *Q*-switching, the efficiency of the system is raised and a prompt massive discharge pulse results.

There are many variants of the basic laser operation which are described in specialized texts. Wavelengths available extend from over 1000 nm in the infrared to 100 nm in the vacuum ultraviolet where the quantum energy is over 10 eV. Pumping at these energies is accomplished by electron beam collisions rather than by photon absorption. In tunable dye lasers the output wavelength can be selected by an adjustment of the operating parameters.

The average power output of a laser may be measured in watts but the power in each short pulse may be in the order of many megawatts. Because of the special properties of coherent beams the output energy can be concentrated onto a very small area. Peak power densities may reach 10^6 watts cm^{-2}, which may be compared with the total power density of the sun, 0.14 watt cm^{-2} just outside our atmosphere.

The electric field intensities in a laser pulse may reach 10^7 volts cm^{-1}, a value that is comparable to the fields between interatomic electric charges. High-powered laser beams are quite capable of producing effects over and beyond those due to the thermal energy content.

Carefully controlled laser beams of modest power have been used in medicine for the repair of detached retinas by local coagulation. At higher powers, lasers have been used to reduce malignancies by thermal destruction. Uncontrolled, the high intensity from even a small laser can cause serious permanent eye injuries. Even reflected beams have energy densities that are quite capable of producing serious lesions. All but the very lowest-powered lasers must be used with extreme caution.

16.13 Maximum Permissible Exposure

High-intensity sources of ultraviolet have been in commercial use for many years and acceptable protective methods have been developed. The eye is the critical organ. Complete protection can be obtained with UV-absorbing goggles that freely transmit visible light and hence do not interfere with normal visually directed operations. All are familiar with the methods used to protect workers and the public from arc welding operations.

When the eyes are protected with goggles, situations may arise in which the skin becomes the endangered tissue. Skin exposure rates up to 0.5 μwatts cm^{-2} of 0.2537 μm radiation appear to be acceptable when extended over a normal working day of 8 hours. Somewhat higher exposures may be allowable at either longer or shorter wavelengths because the 0.2537 μm radiation is most effective in producing skin damage.

Laser beam technology and applications are so new that presently established values of maximum permissible exposure (MPE) must be considered subject to change based on future information. Several organizations, notably the Department of Defense, the American National Standards Institute (ANSI), and the Bureau of Radiological Health of the Environmental Protection Agency are involved in developing satisfactory values of MPE.

For protection purposes four classes of lasers are recognized:

1. *Exempt lasers*, which are incapable of emitting radiation levels exceeding the MPE.
2. *Low-power* lasers whose emissions may exceed the MPE for the direct beam but not for diffusely reflected exposures. These lasers shall not be capable of continuously delivering 1 watt cm^{-2} on a surface larger than 1 cm^2. Hazard control measures include education and training of operating personnel and the use of beam enclosures, eye protection, and viewing aids.
3. *High-power* lasers capable of producing dangerous exposure levels in reflected beams. These sources should be operated in limited-access areas with full beam enclosures, interlocked to prevent operation when personnel are in the exposure area.
4. *Protected lasers* require full protective housing, suitably interlocked to prevent unauthorized or unexpected laser operation. Viewing optics must be carefully designed to ensure that MPEs are not exceeded.

Goggles used for viewing laser beams must be carefully chosen to be certain that the intense radiation does not destroy the attenuating material and thus reduce the effectiveness for subsequent exposures.* Not to be forgotten are the fire hazard from the beam itself and the danger of electrocution from the high voltage capacitors commonly used for driving the pumping source.

Although these secondary hazards are very real, the eye is the critical organ at risk in laser operation. During very short exposures ($< 10^{-5}$ s), essentially no heat will be removed from the retina on which the beam is

*Elder, R. L., "Lasers and Eye Protection." *Science* **182**, 1080, 1973.

focused. For these short exposures the MPE will be based on the total energy input to the cornea, assumed to have a maximum pupillary diameter of 7 mm. For wavelengths between 0.4 and 1.4 μm an MPE of 2×10^{-7} joules seems acceptable in short pulses.

During longer pulses there will be some diffusion and removal of heat from the retina. Here the MPE can be based on power, or rate of energy input at the cornea and the value will have a weak dependence on pulse length. An MPE of $7 \times 10^{-4}\ t^{-1/4}$ joules s^{-1} (watts) is probably acceptable in the 0.4–1.4 μm region.

REFERENCES

HARRISON, G. H., R. C. LORD, AND J. R. LOOFBOUROW, *Practical Spectroscopy.* Prentice-Hall, Inc., Englewood Cliffs, N.J., 1948.

JAGGER, J., *Introduction to Research in Ultraviolet Photobiology.* Prentice-Hall, Inc., Englewood Cliffs, N.J., 1967.

KIRSCHENBAUM, D. M., ed., *Atlas of Protein Spectra in the Ultraviolet and Visible Regions.* Plenum Press, New York, N.Y., 1972.

SCHLOSSBERG, H. R. AND P. L. KELLEY, "Using Tunable Lasers." *Physics Today,* **7**, 36, 1972.

SELIGER, H. H. AND W. D. MCELROY, *Light: Physical and Biological Action.* Academic Press, New York, N.Y., 1965.

SETLOW, J., "The Molecular Basis of Biological Effects of Ultraviolet Radiation and Photoreactivation" in *Current Topics in Radiation Research,* **II**. M. Ebert and A. Howard, ed. North-Holland Publishing Co., Amsterdam, 1966.

SWANSON, C. P., ed., *An Introduction to Photobiology.* Prentice-Hall, Inc., Englewood Cliffs, N.J., 1969.

17

From Infrared to Microwaves

17.01 Rotational Energy Levels

At infrared wavelengths, quantum energies are too low to induce electronic excitations but are quite sufficient for resonance absorption by vibrational and rotational states. Every molecular structure has one or more absorption bands in the infrared. In the near infrared, quantum energies are large enough to excite interatomic bond stretching or bending, Table 16-1. Rotational energy levels lie in the far- or low-energy end of the infrared, Table 17-1.

Any molecular rotation can be resolved into components along three mutually perpendicular axes. Although the choice of axes is arbitrary, it is helpful to select them, as far as possible, along lines of structural symmetry. In general there will be a unique value of the moment of inertia I about each axis. As in classical macroscopic mechanics, I is defined as the sum

$$I = \sum mr^2 \tag{17-1}$$

taken for each atomic mass m located at a distance r from the axis. Each angular velocity ω is quantized and this quantizes each angular momentum $I\omega$ and each rotational energy $I\omega^2/2$. A few rotational energy levels, expressed in rotational frequencies and wavelengths, are listed in Table 17-1.

The zero and infinity entries for H_2 and CO_2 show that each is a linear molecule with a moment of inertia that is zero around the interatomic axis. Spatial symmetry in CH_4 and CD_4 is indicated by the three equal values in each case. The lower-frequency values for CD_4 reflect the double mass of the deuterium atom. The three distinct values for H_2O show that the molecule cannot be linear.

TABLE 17-1
ROTATIONAL ENERGY LEVELS

Structure	*Axis A*		*Axis B*		*Axis C*	
	λ, cm	*ν, GHz*	*λ, cm*	*ν, GHz*	*λ, cm*	*ν, GHz*
H_2	0.0	∞	0.0164	1825	0.0164	1825
CO_2	0.0	∞	2.64	11.7	2.64	11.7
CD_4	0.191	157.5	0.191	157.5	0.191	157.5
CD_2	0.378	79.5	0.378	79.5	0.378	79.5
H_2O	0.0375	799.2	0.0695	432.0	0.109	274.8

17.02 Infrared Absorption

Experimental work in the infrared is made difficult by the strong absorption bands exhibited by all substances in some part of the region. Glass can be used in prisms and cell windows out to a wavelength of 2.5 μm; potassium bromide is usable out to 25 μm and silver chloride cell windows are transparent to about the same limit. Few transparent solvents are available although some mineral oils are useful since they show only two strong absorption bands, due to C—H stretching and bending.

Any infrared wavelengths, from sunlight, heat lamps, or industrial sources, will be absorbed in the superficial layers of tissue. Energy transferred there to molecular vibrations and rotations will appear as localized heating. Thermal conduction and blood flow will distribute heat from the hot spot to the rest of the organism. There is no evidence for a wavelength-dependent, nonthermal biological effect due to infrared absorption. Heat production seems to be the only significant action.

The eye would be expected to be a sensitive organ because it has a limited blood supply and hence is poorly serviced for the removal of heat. Some studies have suggested that there is an increased incidence of lens opacities in the eyes of workers around massive sources of infrared such as the furnaces used in the glass and steel industries. A syndrome known as *glassblower's cataract* has been described but the evidence is not conclusive. *Eclipse blindness* is a localized retinal lesion due to wavelengths shorter than those in the infrared. Visible light entering the eye will be focused by the lens action; looking at the bright sun without protective glasses can lead to the prompt and irreversible destruction of the retina at the focal point.

17.03 The Microwave Region

The most dramatic progress in exploring and utilizing new frequency regions came from the long-wavelength end of the spectrum rather than from extensions of the infrared. Methods of communication through radio waves were developed early in the century and by the 1950's the demand for operating channels in the 10^6 to 10^8 Hz region far exceeded the supply. Television transmitters require spectral bandwidths that are available only at the higher frequencies. Radar, a development of World War II, has become an essential adjunct to safe travel by sea or air and makes heavy demands upon the high-frequency bands. Thus scientific and commercial pressures led to rapid exploitation at the high-frequency end of the radio spectrum. Frequencies were soon measured in billions of cycles per second, or gigahertz, GHz.

Radio-frequency, or RF, power is customarily generated by a three-electrode vacuum tube, or triode, at a frequency determined by the inductance L and the capacitance C of a resonant electric circuit. At frequencies of 1 GHz or so the physical dimensions of the circuit components become inconveniently small, and restrictions on the sizes of the triodes severely limit the amount of power that can be generated.

Two new types of triodes, the *magnetron* and the *klystron*, are capable of producing large amounts of power at the high frequencies. Coils and condensers are replaced by *resonant* cavities in which standing electromagnetic wave patterns serve as the frequency-determining circuit element.

With these new devices the generation of power by direct electrical means was pushed to higher and higher frequencies until they reached and overlapped the far-infrared radiations produced by molecular excitation. Experimental difficulties are still severe, however, and the frequency range from 30–3000 GHz (10–0.1 mm) is far from being fully exploited.

A frequency range lying roughly between 100 and 100,000 MHz (300–0.3 cm) is known as the *microwave* region because it was approached by a series of developments from longer wavelengths. The name would have been quite different if it had been opened up from the shorter wavelengths of the infrared.

17.04 The Correspondence Principle

Although microwaves are generated by a steadily oscillating vacuum tube, there is no doubt that their energy exchanges are governed by quantum requirements. Quantum energies in this region are small, 4×10^{-7} to 4×10^{-4} eV, and so are much less distinct than are those, say, in the X-ray region where they may reach many keV. At microwave frequencies all discussions can be made in terms of the wave properties of the radiation.

The predominance of wave properties is in accord with the *correspondence principle* enunciated by Bohr in the early days of the quantum theory. According to this principle quantum properties, while still existent, must merge into wave properties at high quantum numbers or low quantum energies.

The ultraviolet is a transition region into the microwave and the even lower-energy RF regions where quantization is unrecognized.

17.05 Generation of Microwaves

An electromagnetic wave in the RF region, say, at a frequency of 10^7 Hz, can be produced as shown schematically in Fig. 17-1A. A tuned circuit energized by a vacuum-tube oscillator feeds an antenna-ground circuit. The oscillator forces electric charges, first of one sign and then of the other, up the antenna. If this has an effective length of $\lambda/4$, a standing-wave pattern of voltage and current will be established along it. Current flow at the outer end will be zero but here the voltage oscillation will have its maximum amplitude. At the base, current flow will be maximal. Oscillating electric and magnetic fields will be set up in the surrounding space and an electromagnetic wave will be generated at the oscillator frequency.

The radio waves leaving the antenna will obey all the general laws of optics. They will be reflected from metal surfaces with only a slight penetration and will be transmitted readily by most *dielectrics*, or electrical insulators. RF frequencies are very low compared to the natural frequencies of atoms and molecules and so absorption will be small. Under favorable conditions a few watts of power suffice to transmit signals for hundreds of miles.

Radio waves can also be diffracted, which is to say that they tend to spread out from a well-defined beam, and can bend around obstacles. The amount of spreading or bending is proportional to the wavelength, which is so long in the RF region that wave propagation is far from linear. Diffraction prevents the casting of deep shadows at RF wavelengths.

Quarter-wavelength antennas become small at microwave frequencies and are generally replaced, perhaps by a dipole located at the focal point of a metal reflector, Fig. 17-1B. The dipole, driven by a vacuum-tube oscillator, will set up an electromagnetic wave in the surrounding space. Diffraction patterns become smaller and the transmitted beam begins to behave like a

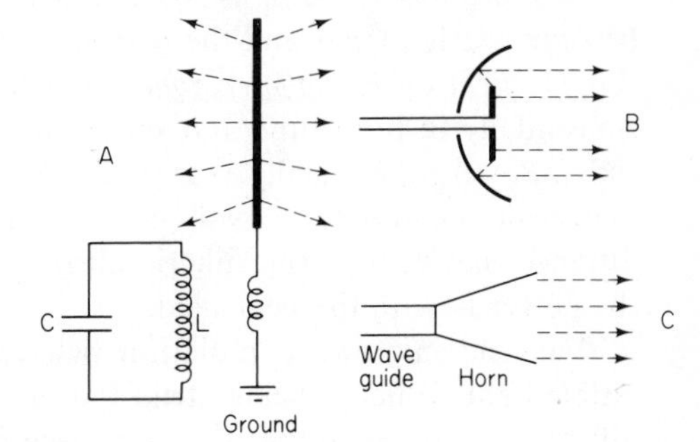

Figure 17-1. (A) A nondirectional radiator for radio frequencies. (B) A center-fed dipole antenna and reflector for microwaves. (C) Wave guide terminating in a horn for microwave transmission.

beam of visible light. Waves propagated backward from the dipole will be reflected by the mirror and a rather sharply defined forward beam will be formed.

At the short-wavelength end of the microwave spectrum it is convenient to generate the electromagnetic waves in a resonant cavity and to conduct them to the point of release in a *wave guide.* A wave guide consists of a rectangular or circular tube made of a good electrical conductor such as copper, perhaps silver-plated on its inner surfaces. When the cross-sectional dimensions of the wave guide are properly matched to the wavelength, electromagnetic waves will be transmitted through the guide with little loss. At the release point the wave guide may be expanded into a *horn*, Fig. 17-1C, in order to achieve a maximum transfer of energy from the guide into free space. Diffraction effects are now so small that the transmission is essentially *line of sight* just as if the beam was at the frequency of visible light.

17.06 Antenna Field Patterns

Near a microwave antenna, be it a dipole reflector or a horn, the electromagnetic field pattern is very complex. Interference patterns are set up between waves reflected from various portions of the antenna and the waves that enter the forward beam without reflection. Intensity nulls and maxima depend in a complex way on the geometrical arrangement of the radiating system.

The Fresnel theory of diffraction, extended by Kirchhoff, applies to the spherical wavefronts that exist near the point of microwave generation. A region known variously as the *Fresnel zone* or the *near field* extends for some distance in front of the radiator. In principle the spatial and temporal distribution of energy in the near field can be calculated from the Fresnel theory. In practice the geometrical relations may be too complex for mathematical treatment. Even the extent of the near field is not exactly calculable. For common types of radiators the near field may persist to a distance of $A/4\lambda$ where A is the cross-sectional area of the horn or reflector.

At a distance of perhaps A/λ in front of the radiator the wavefronts have become nearly planar and the principles of Fraunhofer diffraction apply. In this *far field* or *Fraunhofer zone* the electric and magnetic fields will vary sinusoidally in both time and space and will have constant relative amplitudes. Power flow, usually expressed in terms of milliwatts centimeter^{-2} or in microvolts centimeter^{-1}, will now decrease as the inverse square of the distance (neglecting atmospheric absorption). No simple relation exists for the power flow in the near field.

Far-field microwave radiation behaves in all respects like a beam of visible light. When a beam strikes an interface between two materials with different electrical properties, some energy will be reflected back into the

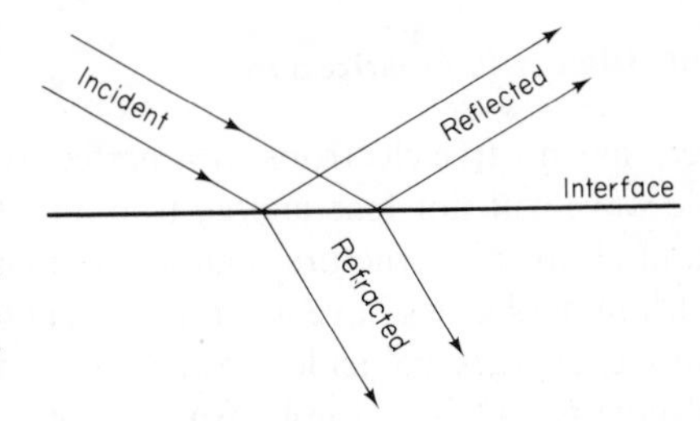

Figure 17-2. A photon beam incident on an interface between two different media will be partially reflected and partially refracted.

first material and some will be *refracted* into the second, Fig. 17-2. The relative amounts of energy reflected and transmitted depend largely on the electrical conductivity of the second medium. If this is a good conductor, reflection will be total and transmission zero. A poor conductor or dielectric will transmit a large fraction of the incident wave, with little reflection. Refraction will take place at the interface because in general the velocity of wave propagation will change at this point. The velocity ratio $v_1/v_2 = n_{21}$, the *index of refraction* of the second medium referred to the first. If the first medium is free space, $v_1 = c$ and the absolute index of refraction is $n = c/v_2$.

When a wave enters a new medium, the frequency and the quantum energy remain constant. The wavelength must, however, change with the change in the velocity of propagation.

The properties of living tissues fall between the conductivity extremes that result in either total reflection or total transmission. In these so-called *lossy* materials an appreciable fraction of the incident energy will be transmitted and absorbed, and an appreciable fraction will be reflected.

17.07 Energy Loss in a Conducting Medium

An electromagnetic wave can interact with matter and hence give up energy, either through its electric or its magnetic component. For materials of biological importance, magnetic interactions are negligible. The magnetic properties of any material are characterized by a *magnetic permeability* μ which has a value in free space of $\mu_0 = 1.26 \times 10^8$ henries m^{-1}. In all materials of present interest, μ has a value so close to μ_0 that the magnetic component of the field is not aware of the material and hence does not interact with it.

Free electric charges exist in any electrical conductor, in particular in the ionized water milieu of living tissues. Any free electrons will be accelerated by the electric field component of a microwave and will then distribute the energy thus acquired in a series of collisions. Energy transferred from the field to free electrons produces the well-known *ohmic* or I^2R heating effect of an electric current. In this case the current does not flow through a completed external circuit but the heating effect arises nevertheless from the collision energy losses of moving charges.

17.08 Dielectric Polarization

There are no free electrons in a perfect dielectric and hence an electromagnetic wave will not lose energy to ohmic heating in it. However, even some bound electronic structures can acquire energy from the wave.

Many molecules have a symmetry such that the "center of gravity" of the negative charges coincides exactly with that of the positives. An oxygen molecule O═O is a simple example of such a *nonpolar* structure. When a nonpolar molecule experiences an electric field, as from a microwave, the electronic structure will be distorted or *polarized.* Positive charges will be displaced in the direction of the field and negative charges will move in the opposite direction to create a temporary *dipole*, Fig. 17-3. The direction of the dipole moment thus created will be exactly parallel to the field. Charge movement will be essentially instantaneous and the induced moment will be in phase with the field.

Except at very high field intensities all the charge distortion in polarization is produced by elastic forces. Energy extracted from the field at charge separation will be recovered exactly as the dipole reverts back to its nonpolar configuration. No net energy will be absorbed from the field by a nonpolar molecule.

An induced dipole moment effectively reduces the strength of the electric field that produces it. The magnitude of the polarization is related to the electric *permittivity* of the material being polarized. Free space has a permittivity $\epsilon_0 = 8.8 \times 10^{-12}$ farad m^{-1}. All ponderable materials will have a permittivity greater than this because of polarization. The ratio $\epsilon/\epsilon_0 = K$ is known as the *dielectric constant* of the polarized material.

Unsymmetrical structures may have a permanent dipole moment because the "centers of gravity" of the two sets of charges do not coincide. A simple example is the keto group >C═O, in which the negative charges are slightly displaced toward the oxygen atom. In the absence of an electric field a group of *polar* molecules will have a random orientation in space, with each in thermal agitation around its equilibrium position, Fig. 17-4A.

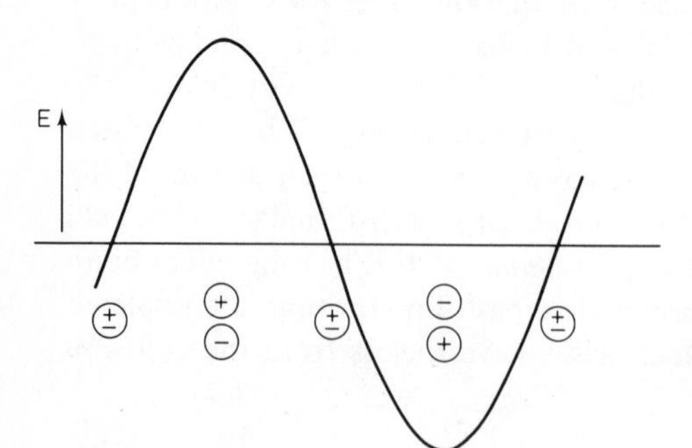

Figure 17-3. A transient dipole moment is induced in a dielectric by the electric component of a passing electromagnetic wave.

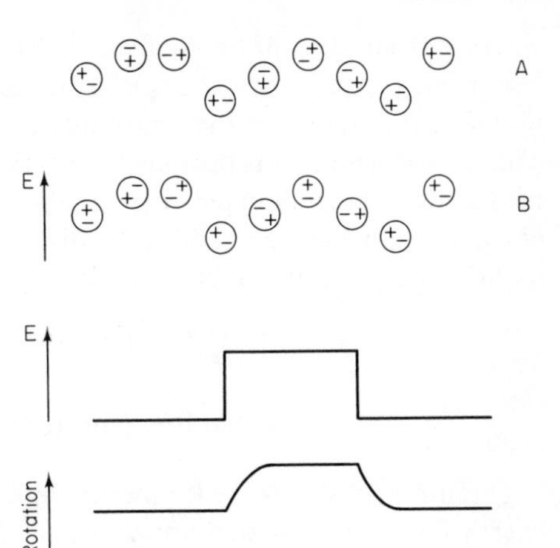

Figure 17-4. A random array of permanent dipoles (A) will be rotated and ordered by an electric field (B).

Figure 17-5. The rotational response of a permanent dipole lags behind a square-wave orienting electric field.

When an external field is applied, each polar molecule will attempt to align itself with the field, Fig. 17-4B. The attempted rotation may be hindered by the viscous forces of a surrounding liquid medium and by the constraints of bonds with adjacent atoms. As a sinusoidal field passes through zero, the oriented structures will relax toward their original positions and will take up new orientations when the field reverses direction. A rotational motion only is produced. A neutral molecule will experience no net translational force.

Dipole orientation is a movement of atomic masses through a viscous medium which cannot take place instantaneously. The temporal relations can be visualized more easily in terms of the response to a square-wave pulse rather than to the sinusoidal field that actually exists. As Fig. 17-5 illustrates, the displaced position is attained somewhat later than the force which produced it and the return to the original random state is also delayed. The time required for the displacement to return to $1/e = 0.37$ of its maximum value is the *relaxation time*.

A sinusoidal field $A \sin \omega t$ will evoke a response of the form $B \sin \omega t + C \cos \omega t$, which indicates that the displacement lags behind the driving force. The electric permittivity will now have a real and an imaginary part, $\epsilon = a\epsilon' + bj\epsilon''$. Nonreturnable energy is now extracted from the field to appear as heat in the medium. In a lossy material such as biological tissues, heat will be produced both by ohmic currents and by dipole orientation.

17.09 Microwave Absorption

A microwave beam will be attenuated exponentially as it gives up energy to free-electron conduction and to dipole rotations. The energy flux at depth x in a homogeneous absorber will be given by

$$\mathcal{E} = \mathcal{E}_0 e^{-\alpha x} \tag{17-2}$$

where α is an absorption coefficient that is a function of photon frequency. Electromagnetic theory shows that the energy in a wave motion is proportional to the square of the amplitude of the electric or magnetic oscillation. The electric field that is responsible for the transfer of energy to the absorbing medium will be attenuated with an absorption coefficient $\alpha/2$. As before, we can define a mean free path as the distance in which the flux is reduced from its original value by a factor of $1/e$.

$$\begin{aligned} \text{mean free path (energy)} &= \frac{1}{\alpha} \\ \text{mean free path (amplitude)} &= \frac{2}{\alpha} \end{aligned} \tag{17-3}$$

Figure 17-6 shows the frequency dependence of the mean free path of the energy flux in a simulated muscle tissue. No resonance absorption bands are to be expected since the frequency range lies well below that of molecular vibrations and rotations. The energy transferred from a microwave beam to dipole rotations should be roughly proprotional to the rate at which the dipoles are oriented, which is just the frequency of the wave. This expectation is reflected in the decreasing mean free path seen in Fig. 17-6.

Conditions in actual living tissues will be far more complicated than would be inferred from Fig. 17-6. Reflection and refraction will take place at

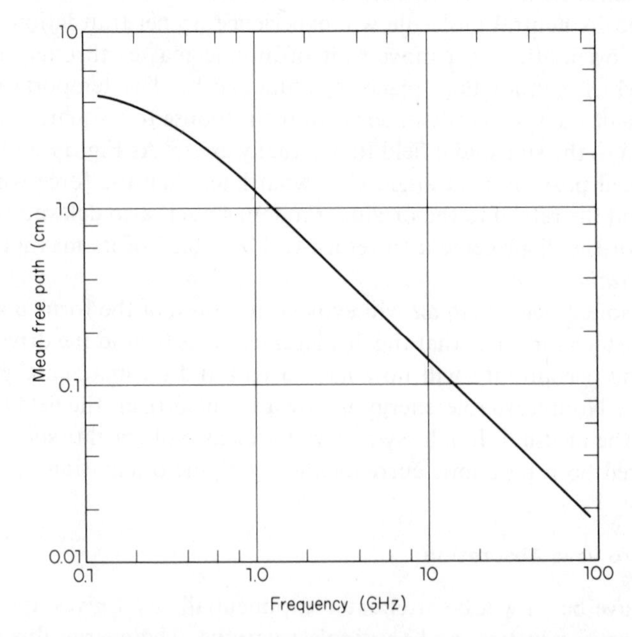

Figure 17-6. The mean free path of photons in a muscle-simulating medium.

each interface, as between fat–muscle and organ–cavity boundaries. Wavelengths are comparable to some body dimensions and so reflections may lead to the establishment of standing-wave patterns with alternating regions of large and zero wave amplitudes. Even in the far field region, energy deposition can be expected to be nonuniform and still greater variations will occur in the near field. On the average, however, it is evident that microwaves are much more penetrating that the infrared and that relatively more energy will be given up to deep tissues.

17.10 Biological Effects of Microwaves

Exposure of mammals to intense microwave fields has produced a variety of injuries including damaged epiphyses of growing bone, testicular degeneration, focal lesions of the nervous system, and lens opacities. None of the effects observed has been demonstrated to be due to a nonthermal injury. Some frequency-dependent effects might be expected because of the possibility of setting up standing-wave patterns in some organs but the damage could still be only thermal rather than specific and nonthermal.

The eye would be expected to be a sensitive organ because of its limited blood supply and the consequential poor conditions for heat removal. Experience with a limited number of human injury cases indicates that a microwave-induced lens opacity forms in a localized region and remains confined to this area when overexposure is terminated.

For the body as a whole any energy absorbed in it will be dissipated by increased heat losses from the surface, within the physiological capabilities of the organism. The basal metabolic rate of an adult human is about 70 watts, which is readily dissipated from the body surface area at a rate of about 4 mwatts cm^{-2}. On a cloudless day in summer, solar radiation delivers energy at the rate of about 70 mwatts cm^{-2}, a value which most sunbathers find excessive for an extended period of time. Skin burns in animals have been seen after exposures to 165 mwatts cm^{-2}.

Only scant evidence is available on which to base allowable exposure levels. In the United States this level has been set at 10 mwatts cm^{-2} with no allowances for particular wavelength ranges or times of exposure. The Soviet Union prefers a much lower limit, 10 μwatts cm^{-2}, in the belief that some nonthermal injuries occur at very low levels. It is also of interest that the Soviet Union has proposed exposure standards for lower frequencies based on the magnitude of the electric component of the electromagnetic field. The values proposed are

100 kHz	3 MHz	20 volts m^{-1}
3 MHz	30 MHz	5
30 MHz	300 MHz	5

Much work remains to be done, particularly in the development of measuring techniques, before generally acceptable exposure levels can be established.*

17.11 Medical Uses of RF and Microwaves

For many years physicians have been inducing hyperthermia either locally or systemically in treating a variety of conditions such as bursitis, sprains, and pelvic inflammatory disease. Heating increases the blood flow through arterial and cappillary dilatation and thus accelerates natural repair processes. No conclusive evidence for nonthermal therapeutic effects has been presented.

The most commonly used source of external heat is infrared from a radiator heated to a dull red glow by an electric current. Radiation from these sources penetrates effectively to 1–3 mm of tissue and consequently only superficial layers are directly heated. Blood flow removes the energy locally deposited for eventual dissipation from the entire body surface.

Three broad classes of more penetrating radiation have been used, Table 17-2.

TABLE 17-2
FREQUENCY RANGES USED IN THERAPY

Classification	*Band in Electromagnetic Spectrum*		*FCC licensed**	
	Wavelength, m	*Frequency, MHz*	*m*	*MHz*
Longwave diathermy	3000–30	0.1–10	22.0	13.66
Shortwave diathermy	30–3	10–100	11.0	27.32
			7.3	40.98
Microwaves	1–0.01	300–3000	0.122	2450

*The demand for usage of the electromagnetic spectrum is so great that the Federal Communications Commission (FCC) allocates and licenses frequencies for specific uses. The 2450-MHz frequency is also available for industrial and experimental use.

Longwave diathermy is now little used, and its replacement, 27.32 MHz is in turn being replaced by 2450 MHz. Diathermy power is usually applied to the treatment area through a pair of condenser plates, to provide a strong electric field for ohmic heating and dipole orientation. Limbs and other selected areas may be treated by wrapping with a few turns of cable through which RF current is passed. Heating in this case comes from electric currents induced in the conducting tissues. Microwave power is directed onto the treatment area from a horn, usually under far field conditions.

*McRee, D. I. and F. T. Pendergrass, "Interaction of a 2450 MHz Microwave Field with Thermocouples and Thermistors." *Health Physics*, **25**, 180, 1973.

17.12 Microwave Heating and Communications

Microwave frequencies are now widely exploited for a variety of purposes. Frequency bands have been assigned and have been given letter designations, Table 17-3.

TABLE 17-3
LETTER DESIGNATIONS OF MICROWAVE BANDS

Designation	*Frequency, MHz*	*Wavelength, cm*
P	225–390	133.3–76.9
L	390–1550	76.9–19.37
S	1550–5200	19.37–5.77
X	5200–10,900	5.77–2.75
K	10,900–36,000	2.75–0.834
Q	36,000–46,000	0.834–0.652
V	46,000–56,000	0.652–0.536

Various industries have taken advantage of the ability of RF and microwave power to produce heat at a depth in a dielectric. For example, high-powered oscillators produce RF energy for heating and setting the adhesives used in many laminating processes. A series of bands has been licensed for use in industrial heating. These bands range from 13.55–5875 MHz and many applications make use of the lower frequencies in order to obtain deeper penetration. Most of the installations are designed for production line use with the high-intensity fields carefully confined for maximum effectiveness and to reduce the chance of human overexposure.

Microwave ovens operating at 2450 MHz are widely used in homes and in public establishments for heating foods. Cooking times can be drastically reduced when heat is supplied throughout a volume rather than being conducted to the interior from the outer surface. These units are operated by people who are relatively unfamiliar with the hazards of radiation overexposure but proper design and construction will ensure that leakage fields do not exceed allowable limits.

High-frequency fields abound as a result of our increasing use of all forms of communication. Radio broadcast and television transmitters regularly operate at power levels of 50 kwatts or more. With some antenna designs the radiated energy will be propagated almost isotropically and will be reduced to biologically negligible levels only a short distance from the radiator. In other applications the radiated power may be concentrated in a beam. Even in the latter case experience suggests that damaging field intensities exist only in close proximity to the source. Maintenance personnel on duty during transmitter operation may through carelessness enter regions of high intensity but there is little chance of overexposure to the general public.

Every living human being is now exposed almost continuously to a wide variety of man-made electromagnetic fields ranging upward in frequency from 60 Hz. Each wave contributes a minute amount of heat in passing without in any way adding to the hazards of contemporary living. Perhaps the greater hazard lies in the information content of some of these radiations rather than in their energy content.

REFERENCES

ANDREWS, C. L., *Optics of the Electromagnetic Spectrum.* Prentice-Hall, Inc., Englewood Cliffs, N.J., 1960.

DANIEL, V. V., *Dielectric Relaxation.* Academic Press, New York, N.Y., 1967.

Electronic Product Radiation and the Health Physicist. BRH/DEP 70.26, 1970, National Technical Information Service, Springfield, Va.

HARRISON, G. R., R. C. LORD, AND J. R. LOOFBOUROW, *Practical Spectroscopy.* Prentice-Hall, Inc., Englewood Cliffs, N.J., 1948.

KRUSEN, F. H., F. J. KOTTKE, AND P. M. ELLWOOD, *Handbook of Physical Medicine and Rehabilitation.* W. B. Saunders Co., Philadelphia, Pa., 1968.

PÜSCHNER, H., *Heating with Microwaves.* Springer Verlag New York Inc., New York, N.Y., 1966.

18

Radiation Health Protection

18.01 Prevention, Not Cure

Although carefully controlled doses of radiation are used in some therapeutic procedures, the general results of radiation absorption must be considered to be deleterious. At least three of the many end results of radiation over-exposure—genetic injury, shortening of life span, and carcinogenesis—are of serious import. This is not to say that every radiation experience, even at a relatively high dose, inevitably leads to an overt injury. The exact site and nature of a radiation absorption are determined by chance, and even a drastic change in a redundant system or one capable of repair may produce no permanent effect on the individual.

Once radiation has been absorbed in living tissue, little can be done to counteract its effects. The early workers with demonstrable signs of injury tried a vast number of procedures in vain attempts to promote healing of ulcers and other lesions. Successive amputations failed to halt the progress of radiogenic malignancies, which metastisized beyond the limits of practical surgery.

The situation is little better today. A very large number of chemical compounds have been investigated as possible radiation protective agents. Some have been found partially effective when they are administered prior to the irradiation. The same substances are almost universally ineffective when given after the radiation has been received.

Some *dose reduction*, not to be confused with the *reduced effectiveness* of a given dose, can be achieved with some radioactive nuclides. For example, a large intake of some stable isotope may increase the rate of excretion of an already ingested radionuclide. In a special case, an intake of a stable iodine compound may saturate the thyroid's ability to accept iodine, and thus the uptake of a subsequent dose of one of its radioactive isotopes may be reduced or blocked. In general, chemical toxicities and the limited physiological responses of living systems restrict even the preirradiation use of many compounds.

Radiation health protection must emphasize dose prevention rather than post facto therapy. Dosage limits must be set for the protection of radiation workers and the general public at levels such that the chance of radiation injury is acceptably low. These levels are subject to change. As new scientific information and operating techniques become available, prescribed limits must be reevaluated and adjusted as necessary. Whatever the levels that are deemed acceptable, they must be considered as upper bounds and every reasonable effort should be made to keep exposures well below them.

18.02 The NCRP

For several years after the hazards of X rays and radium were recognized, the field of radiation protection was in an unorganized state without even a satisfactory, universally accepted unit of radiation dose. In 1928, at the Second International Radiological Congress, held in Stockholm, the International Commission on Radiological Protection, the ICRP, was organized. Early in 1929 the Bureau of Standards joined with the leading radiological socities of the United States in sponsoring the Advisory Committee on X-ray and Radium Protection. This committee effectively promoted the principles and practice of radiation safety in the United States until it was reorganized in 1946.

The large-scale exploitation of nuclear fission increased the hazards from radiation exposures manyfold. To meet the new requirements the Advisory Committee was expanded and renamed the National Committee on Radiation Protection (NCRP). A later reorganization changed the name to the present version, National Council on Radiation Protection and Measurements, with the abbreviation NCRP retained.

The NCRP has no regulatory authority and sets no exposure limits. Its members, selected from the best informed workers in the field, analyze available data on radiation effects, make risk estimates, and promulgate recommendations. None of the recommendations has the force of law, but with the weight of scientific authority behind them they usually form the basis for the standards that are set by regulatory bodies. NCRP published reports now cover a wide variety of radiation exposure situations.

18.03 The Radiation "Mystique"

Immediately after the use of nuclear weapons at Hiroshima and Nagasaki, and the partial lifting of security restrictions, the public was subjected to a barrage of articles and radio programs dramatizing the biological consequences of large radiation doses. Injuries such as cancer and the production of radiation-induced monsters became fixed in the public mind without any clear understanding of the very low probabilities involved in their production. An exaggerated fear of radiation developed; a fear of radiation that was, and still is, out of proportion to the acceptance of the benefits to be derived from its use. This fear is in sharp contrast to the unquestioned acceptance of the many, more common hazards of contemporary living.

In the absence of firm knowledge, many estimates of injury probabilities were made and applied to whole populations and to long time spans. Some of these estimates were based on incorrect, or incorrectly interpreted, data. Simple arithmetic led from the assumptions to some frightening numbers of predicted injuries but in many cases the assumptions were not closely related to reality. Some of the early dire predictions still remain in the minds of the general population.

18.04 Conflict of Interest?

The Atomic Energy Acts of 1946 and 1954 created the United States Atomic Energy Commission (AEC) and charged it with organizing a variety of programs relating to the uses of radiation.One of the charges was "A program to encourage widespread participation in the development and utilization of atomic energy for peaceful purposes to the maximum extent consistent with the common defense and security and with the health and safety of the public."

According to the provisions of the Acts, the AEC became the chief source of radioactive nuclides in the United States. Most of these materials were produced by (n, γ) reactions in nuclear reactors. At the same time other divisions within the AEC were responsible for establishing operating standards which would ensure the safety of radiation workers and the general public. Such a dichotomy was bound to lead to suspicions that in a conflict of interests some safety precautions might be sacrificed to promotional zeal.

The Biological Effects of Atomic Radiation Committee was organized by the National Academy of Sciences–National Research Council to provide an independent evaluation of the hazards associated with the peacetime uses of nuclear energy. This group had no direct vested interest either in promoting or discouraging the use of radiation. Two sets of reports, in

1956 and 1960, were designed to provide the public with an interpretation of the existing scientific data.

The 1956 report of the BEAR Committee recognized the importance of genetic injuries and suggested a *maximum permissible exposure* of 50 R per individual, from all sources, through the age of 30 yr. The corresponding figure for all sources of man-made radiation was set at 10 R. By 1960 the possibility of some genetic repair was noted, and the second report suggested that the earlier levels might have been ultraconservative. An NAS–NRC Advisory Committee on the Biological Effects of Ionizing Radiation issued a comprehensive report in November, 1972, analyzing the effects of low levels of radiation on large populations. Although the NAS–NRC reports were reassuring, they did not completely allay suspicion. Today, with increasing pressures for more and larger nuclear power plants, there are still some who feel that the AEC cannot promote the uses of radiation and at the same time regulate those uses with due regard for public safety.

18.05 The Code of Federal Regulations

Regulations of all sorts adopted by the United States Government appear in the Federal Register, published daily. From time to time the scattered regulations pertaining to a particular area are pulled together and published as a unit, forming a portion of the Code of Federal Regulations, or CFR. Title 10CFR pertains to Atomic Energy, from the formation of the Atomic Energy Commission to all aspects of its widespread operations. We are here particularly concerned with Title 10 Part 20, cited as 10CFR20, entitled "Standards for Protection Against Radiation" and 10CFR30, "Licensing of Byproduct Material." Only brief accounts of the provisions of these regulations can be given here.

The radioactive materials over which the AEC has jurisdiction are divided into three classes:

1. *Source material* is uranium or thorium, or any combination thereof, in any form, and any ores containing by weight 0.05% of these elements, alone or in combination.
2. *Special nuclear material* consists primarily of the fissile materials plutonium, uranium-233, and uranium enriched in either uranium-233 or uranium-235.
3. *Byproduct materials* are any radioactive materials produced or made radioactive by exposure to the radiations incident to the production or utilization of special nuclear materials.

The vast majority of applications of radioactive isotopes involve by-product materials only since these include all active materials produced in nuclear reactors.

It should be noted that the AEC does not have jurisdiction over naturally occurring radioactive materials such as ^{226}Ra and its daughters nor over X-ray generators or other high-energy accelerators. 10CFR applies to ionizing radiations only. Sound or radio waves and visible, infrared, and ultraviolet light are specifically excluded.

All dosage limits specified in 10CFR20 are given in terms of dose equivalents, or rems, which assumes a knowledge of the relative effectiveness of each form of radiation. Exact effectiveness values may not be known but a table of values is given in 10CFR20 so that common factors will be used for all dose–dose equivalent conversions.

In a desire to decentralize authority, the AEC has transferred much of its regulatory power to those states which have acceptable regulations and the organizations to enforce compliance with them. The regulations of these *agreement states* must be at least as stringent as those of the AEC itself. In practice most state regulations closely follow the provisions of 10CFR20.

10CFR30 describes the various classes of licenses issued for the possession and use of by-product materials, together with a list of the possession limits (mostly very small) for which a license is not required. Except for the small *exempt* quantities, some sort of a license is required for the possession or use of any by-product material. An isotope user may be covered by a *general* license issued in the name of the institution which has supervisory authority over him and over the areas in which the by-product materials will be used.

An application for a *specific* license covering an individual must show that

1. The applicant's equipment and and facilities are adequate to protect health and minimize danger to life or property.
2. The application is for a purpose authorized by the Atomic Energy Acts.
3. The applicant is qualified by training and experience to use the requested material for the purposes requested in such manner as to protect health and minimize danger to life and property.
4. The applicant satisfies any special requirements, such, for example, as might apply when the material is to be used in human beings.

Provisions are made in the regulations for inspections of premises and records to check compliance. Failure to comply can mean loss of a license and other penalties.

18.06 Permissible and Acceptable Doses

The early recommendations of the NCRP included numerical limits known as maximum permissible exposures or doses and maximum permissible concentrations of radioactive materials in air or water. Several undesirable connotations are associated with the word *permissible*. For one thing it conveys a sense that some high authority has granted permission for the use of radiation up to the levels specified. This meaning was not the intent of the NCRP recommendations.

The word *permissible* suggests that a radiation dose below the limit is safe and that a greater dose is harmful. As experimental data accumulated, it became evident that the dose-response relationships were smoothly graded functions, with the risk of injury increasing steadily with dose. If there is no dose threshold (and there is no conclusive evidence either for or against one), any dose, no matter how small, carries some possibility of injury. It becomes evident, therefore, that safety regulations must be based on acceptable probabilities rather than on a sharply defined safe/dangerous permissible limit.

The word *permissible* is still used in some of the recommendations pertaining to radiation exposures. Current usage, however, implies an acceptable probability rather than an absolute authorization. In spite of intensive research on the biological effects of radiation it is not possible to quantify precisely the risk associated with a particular dose. As the NCRP reports emphasize, recommended limits must be based on value judgements made by experienced, responsible groups since precise scientifically established values cannot be obtained. Value judgements become even more essential when expected risk/expected benefit balances are being struck.

18.07 Some Numerical Values

As more sophisticated considerations were introduced into radiation safety regulations, the concept of benefit-risk ratios was developed. If the use of a radiation source leads to some personnel exposure, it also presumably leads to some benefit or else the use would not be considered. The benefit derived may be hard to define and is impossible to quantitate. Benefit to a research worker may be the holding of a desirable position and the satisfaction of discovering new scientific facts. A physician treating a patient with radiation benefits from the fees received and from the satisfaction derived from assisting humans to regain health and happiness.

For relatively routine operations, the NCRP, in Report No. 39, issued January 15, 1971, recommends a maximum whole-body dose equivalent, from occupational sources, of 5 rems per year, with a long-term accumulation not to exceed $5(N - 18)$ rems up to age N years. This recommendation emphasizes that persons under the age of 18 years should not be allowed to work in radiation areas.

In an emergency situation, as for the saving of human life, benefit/risk considerations lead the NCRP to consider permissible a single dose, not to be repeated, of 100 rems whole-body radiation.

AEC licensees using by-product material are governed by the requirements of 10CFR20, §20.101. In this section the term *restricted area* refers to an area under the control of the licensee for the purpose of protecting individuals from exposure to radiation and radioactive materials. Control measures might, for example, exclude all individuals not provided with monitoring equipment that is capable of measuring the radiation exposure in the area. Access to an unrestricted area is not controlled by the licensee for the purposes of radiation protection.

§20.101 Exposure of individuals to radiation in restricted areas.

(a) Except as provided in paragraph (b) of this section, no licensee shall possess, use, or transfer licensed material in such a manner as to cause any individual in a restricted area to receive in any period of one calendar quarter from radioactive material and other sources of radiation in the licensee's possession a dose in excess of the limits specified in the following table:

	Rems per calendar quarter
1. Whole body; head and trunk; active blood-forming organs; lens of the eyes; or gonads	$1\frac{1}{4}$
2. Hands and forearms; feet and ankles	$18\frac{3}{4}$
3. Skin of the whole body	$7\frac{1}{2}$

(b) A licensee may permit an individual in a restricted area to receive a dose to the whole body greater than that permitted under paragraph (a) of this section, provided:

(1) During any calendar quarter the dose to the whole body from radioactive material and other sources of radiation in the licensee's possession shall not exceed 3 rems; and

(2) The dose to the whole body, when added to the acculated occupational dose to the whole body, shall not exceed $5(N - 18)$ rems where "N" equals the individual's age in years at his last birthday; . . . .

The requirements of 10CFR20 are in essential agreement with the recommendations of the NCRP except that the AEC has put the accounting on a quarterly rather than on an annual basis.

18.08 The Federal Radiation Council

The regulations of the AEC as set forth in 10CFR20 are intended to cover the occupational exposure of a limited number of workers in restricted areas such as laboratories, medical institutions, and industrial plants. In addition to the licensee, all the workers in the area are supposed to have received some instruction in the hazards of radiation and in the techniques by which exposures can be minimized.

The atmospheric testing of nuclear devices led to the inescapable, involuntary exposure of entire populations to the radiations from fallout debris. Extensive nonoccupational exposures of large populations can also be envisioned as following a nuclear accident such as a massive power reactor failure.

It is obviously impossible to keep an entire population sufficiently informed to permit them to make individual benefit/risk estimates and govern their behavior accordingly. In many cases benefit may accrue to a population as a whole, while risk is an individual, personal matter. Fallout from the testing of nuclear weapons may put at risk each individual in the testing country or perhaps everyone on earth. The benefit is assumed to derive from the development of a strong deterrent to agressive actions by hostile forces. Value judgements of (national benefit)/(individual risk) must be made at a high political level by those whose responsibility for national and international policies transcends that of the individual citizen.

In 1959 the Federal Radiation Council was formed to ". . . advise the President with respect to radiation matters, directly or indirectly affecting health . . . ." Recognizing the semantic difficulties with "tolerance, permissible, and acceptable" the Council introduced the term *Radiation Protection Guide*, or RPG. This is defined as the radiation dose which should not be exceeded without careful consideration of the reasons for doing so.

The RPG values set by the Council for radiation workers followed closely the $5(N - 18)$ rems recommended by the NCRP and adopted by the AEC. For a single individual in the general population the Council recommended an RPG of one-tenth that of the radiation worker, or 0.5 rem per year. In order to limit the total genetic impact on a large population the Council selected an RPG of $0.5/3 = 0.17$ rem per year as the *average* gonadal dose limit in a large group.

The Council also introduced the concept of *action guides* to be applied to various exposure ranges that might be expected to develop after the release of radioactive material to the environment.

Range 1. Calculations indicate that an inappreciable number of individuals will experience exposures of even a large fraction of the RPG. The recommended action is surveillance only, in order to confirm the calculations.

Range 2. Average population exposures are not expected to exceed the RPG. Active surveillance and some routine controls are recommended.

Range 3. Calculations indicate that if continued, exposures will exceed the RPG. Appropriate control measures may be required if conditions appear to exist for some time instead of being transient. In recommending control measures, such as the evacuation of the population from an affected area, any hazards associated with the control measures themselves must be estimated and balanced against the radiation hazards.

The Federal Radiation Council has now been merged with other governmental agencies charged with evaluating many different forms of environmental pollution. Further recommendations governing radiation exposures will be forthcoming as new scientific information becomes available.

18.09 Medical Exposures

In a technically developed country such as the United States, diagnostic and therapeutic procedures recommended by physicians constitute the greatest source of human exposure to man-made radiation. In these human applications the expected benefit is immediately apparent although quantitation is impossible. The risk, either to the patient or to those concerned with the administration of the radiation, may be equally impossible to evaluate.

Exposure to a large field of penetrating radiation, as in an extensive fallout situation, may deliver an essntially whole-body dose even to a body as large as an adult human. Reasonably good dose/risk estimates can be made for this situation, even though they must be based on meager human data and extensive animal experimentation.

Most radiological procedures expose only a small fraction of the body to the primary radiation so that the tissue volume involved is almost unique for each patient. Many procedures are carried out at sites well away from the gonads, for example, which will then receive only scattered radiation. Disparities in size and anatomical differences prevent meaningful transfers of data from animals to humans, beyond the generalization that the effectiveness of radiation is reduced when it is applied to only a portion of an organism.

A considerable amount of information is available on the exposures required, under normal circumstances, for carrying out a variety of radiological procedures. However, only the physician in charge of a case can strike a valid benefit/risk balance, weighing the urgency of the need for the procedure against the possibility of a radiation injury. Organizations such as the NCRP and the AEC have wisely refrained from even suggesting dosage limits for the medical applications of radiation.

Although the physician has wide latitude in prescribing procedures requiring radiation exposures (subject as always to the accusation of malpractice if he deviates too far from accepted standards), he must make every effort to keep exposures to the lowest practicable level. Radiation fields from external sources such as X-ray machines must be kept as small as possible. The amount of any radioactive isotope administered must be sufficient to permit assays with the required precision but must not be excessive. All procedures must be carefully planned and executed to prevent technical failures which might require repeating the radiation experience.

Physicians and dentists are well aware of the desirability of keeping human radiation exposures as low as possible. Their situation has not been improved by at least one court decision which ruled negligence for failure to use all available diagnostic procedures, with no reference to how applicable some of these procedures might be to the case in hand.

18.10 Health Physics

As the use of radiation in industry and medicine increased, a need developed for specialists skilled in measuring radiation exposures and in developing and applying methods for reducing them. These specialists must be hybrids, combining an understanding of the physical nature of the radiations with an appreciation of the biological consequences of their absorption. Some competence in construction engineering is helpful in designing adequate and practical shielding structures.

These hybrids, known as health physicists, form a vital part of any organization using radioactive materials or any radiation generator. Man is indeed his own worst enemy. There are too many cases of well-informed radiation workers who cut corners, took chances, or deliberately violated basic safety principles in order to save time or perhaps the money that was required to provide adequate protective devices. The record keeping alone that is required for the orderly acquisition, use, disposal, or transfer of radioactive materials requires a specialist who is familiar with all the applicable regulations and procedures and whose primary responsibility is to see that they are followed. This specialist must have, in addition to his technical skills, infinite tact and patience in order to persuade some workers to obey regulations that were designed for their safety. When persuasion fails, the health physicist must have the backing of his administration so that he can insist upon compliance.

The Health Physics Society was organized in 1956 to serve as a focus for the growing number of people entering the profession. Growth has been steady and in 1965 the Society served as the nucleus of the International Radiation Protection Association. The term *health physicist* does not translate well into some foreign languages whereas *radiation protection* is universally understood.

The Health Physics Society publishes an official journal and sponsored the American Board of Health Physics. This Board has developed standards of competence and gives examinations which, if passed, lead to the designation Certified Health Physicist. Certification ensures a relatively high level of technical competence and experience and is frequently specified as a job requirement.

18.11 Radiation Protection Methods

Fixed sources with well-defined maximum outputs are usually installed with permanent shielding forming an integral part of the building structure. Once the initial integrity of the installation has been demonstrated, safety depends primarily on the devices designed to prevent access to the shielded space during operation. Routine inspections and checks on the operational procedures usually suffice to ensure personnel safety.

Sources of a less permanent nature are more suspect. Shielding may be breached, new procedures introduced, and the radiation output may be increased. Frequent surveillance and detailed measurements may be required to verify that operational safety is being maintained.

In any external source, exposures can be reduced by increasing the shielding, increasing the distance between the source and the users, and decreasing the exposure time. These three protective measures may not be entirely independent. It is quite possible to increase shielding to the point where increased time requirements offset the expected exposure reduction. Although time is an important factor, operations must not be speeded up to the point where errors in manipulations and perhaps spills of active materials become probable.

Radioactive materials must be confined to designated working areas and must not be allowed to come in contact with the body nor to enter it. Good housekeeping procedures and frequent areal and personnel monitoring are required to ensure that the operations are going as planned.

When the apparently inevitable spills occur, vigorous, prompt decontamination procedures should be started, to minimize the spread of the contamination. A relatively simple cleanup job may expand into an extensive operation unless positive steps are taken to confine the spill.

The health physicist has available a number of sophisticated instruments for detecting and measuring all forms of ionizing radiation. Sensitivities are sufficient to detect activities so low that the chance of a radiation injury is negligibly small. Cleanup at these low levels is desirable if only to remove a nuisance which may be slightly increasing the background of some assay equipment. These sensitive instruments give the health physicist a decided technical advantage over those who are responsible for controlling dangerous chemicals and biologicals, which can seldom be detected in very low concentrations by direct measurement.

18.12 Waste Disposal

The management of radioactive wastes presents the health physicist with one of his most irksome but important problems. These wastes range all the way from a few microcuries in a small laboratory or clinic to the megacurie fission product residues from reactor fuel processing. Concentration and segregation appear to be the only answers to the problem of the ultimate disposal of radioactive wastes. They can be changed in chemical and physical form and can be reduced in volume but no manipulation will alter the rate at which each nuclide decays.

When waste materials leave a restricted area controlled by a licensee, they enter the public domain in which individual radiation doses must not exceed 0.5 rem per year. Disposal methods are rigidly controlled by the provisions of 10CFR20.

Disposal by incineration is forbidden except upon a specific authorization from the AEC, and then only under carefully controlled conditions.

Disposal by burial in soil is limited to very small quantities which may be buried under carefully restricted conditions. Large-scale burials in areas under the direct control of the AEC are authorized in special cases.

Discharge of soluble wastes into municipal sewerage systems is permitted only up to limits specified in 10CFR20. These limits control both the allowable instantaneous concentration and the total annual quantity that can be released.

Allowable limits for airborne contaminants, also listed in 10CFR20, govern the discharges from hoods and other ventilating systems.

Within restricted areas the nearby sink offers the most convenient means of disposal of liquid wastes. Temptation can easily overcome good judgement. The health physicist must develop efficient and convenient waste collection systems if he is to prevent an improper disposal of many wastes. In a large institution where many workers are using active materials, only a single, directly responsible officer such as the health physicist is in a position to estimate the total amount of material that is being put into the municipal sewerage system.

Short half-life materials can be collected and allowed to decay under the control of the health physicist until the activity is reduced to a level acceptable for disposal under the regulations. Space limitations usually prevent storage for decay of isotopes with half-lives longer than a month or so. In any case, possession limits in the license usually interdict long-term storage.

Some commercial firms have been licensed by the AEC to collect and dispose of radioactive wastes in carefully controlled burial sites. This type of service probably represents the best solution to the ultimate disposal problem that is available to the average licensee.

REFERENCES

Advisory Committee on the Biological Effects of Ionizing Radiation, *The Effects on Populations of Exposure to Low Levels of Ionizing Radiation.* National Academy of Sciences–National Research Council, Washington, D.C. 20006, Nov., 1972.

Federal Radiation Council, *Staff Reports* 1–8. U.S. Government Printing Office, Washington, D.C. 20402.

MARTLAND, H. S., *Collection of Reprints on Radium Poisoning*. Technical Information Service, Oak Ridge, Tenn., 1951.

National Academy of Sciences–National Research Council, *The Biological Effects of Atomic Radiation*. Summary Reports, 1960. NAS–NRC, Washington, D.C. 20006.

NCRP Reports. From NCRP Publications, P.O. Box 30175, Washington, D.C. 20014.

United Nations Scientific Committee on the Effects of Atomic Radiation, *Ionizing Radiation Levels and Effects*, Vol 1 and 2. Publication No. E 72, IX 17, United Nations, New York, N.Y., 1972.

United States Environmental Protection Agency, *Estimates of Radiation Doses in the United States*, 1960–2000. U.S. Government Printing Office, Washington, D.C. 20402.

United States Public Health Service, *Radiological Health Handbook, Revised.* U.S. Government Printing Office, Washington, D.C. 20402, 1970.

Appendix

Appendix 1

COMMON PREFIXES

giga-	G-	10^9	milli-	m-	10^{-3}
mega-	M-	10^6	micro-	μ-	10^{-6}
kilo-	k-	10^3	nano-	n-	10^{-9}
			pico-	p-	10^{-12}

TIME CONVERSIONS

Seconds	*Minutes*	*Hours*	*Days*	*Years*
60	1.0			
3.60×10^3	60	1.0		
8.64×10^4	1.44×10^3	24	1.0	
3.15×10^7	5.26×10^5	8.76×10^3	365	1.0

ENERGY CONVERSIONS

Ergs	*Electron volts*	*Grams*	*Mass units*
1.0	6.24×10^{11}	1.1126×10^{-21}	6.701×10^2
1.602×10^{-12}	1.0	1.7826×10^{-33}	1.0736×10^{-9}
8.9875×10^{20}	5.610×10^{32}	1.0	6.0225×10^{23}
1.4923×10^{-3}	9.3148×10^8	1.6604×10^{-24}	1.0

10^7 ergs = 1 joule   4.185 joules = 1 calorie

LENGTH EQUIVALENTS

1 centimeter	cm	10^{-2} meter
1 micrometer	μm (formerly micron)	10^{-6}
1 nanometer	nm (formerly millimicron)	10^{-9}
1 angstrom	Å	10^{-10}

Appendix 2

VALUES OF PHYSICAL CONSTANTS

Symbol	*Quantity*	*Value*
c	Velocity of light in vacuum	2.9979×10^{8} m s^{-1}
e	Elementary charge	1.6021×10^{-19} coulomb 1.6021×10^{-20} emu 4.8029×10^{-10} esu
h	Planck's constant	6.6256×10^{-34} J s
$\hbar$	$h/2\pi$	1.0545×10^{-34} J s
N_a	Avogadro's number	6.02252×10^{23} mole^{-1}
V_0	Volume of 1 mole of ideal gas	2.241×10^{-2} m^{3} mole^{-1}
k	Boltzmann's constant	1.380×10^{-23} J °K^{-1}
a_0	Bohr radius	5.2917×10^{-11} m
λ_c	Compton wavelength of electron	2.426×10^{-12} m
m_e	Electron mass	9.109×10^{-31} kg 5.4859×10^{-4} u 5.1098×10^{5} eV
m_p	Proton mass	1.6725×10^{-27} kg 1.00727 u 9.3855×10^{8} eV
m_n	Neutron mass	1.6748×10^{-27} kg 1.00866 u 9.39767×10^{8} eV

Appendix 3

X-RAY CRITICAL-ABSORPTION AND EMISSION ENERGIES

Calculated from the conversion keV = 12.398/λ (angstroms)

Z	Element	keV			
		K_{abs}	K_{emm}*	L_{abs}	L_{emm}*
13	Aluminum	1.56	1.49	0.087	—
22	Titanium	4.96	4.51	0.53	0.45
23	Vanadium	5.46	4.95	0.60	0.51
24	Chromium	5.99	5.41	0.68	0.57
25	Manganese	6.54	5.90	0.76	0.64
26	Iron	7.11	6.40	0.85	0.71
27	Cobalt	7.71	6.93	0.93	0.78
28	Nickel	8.33	7.48	1.02	0.85
29	Copper	8.98	8.05	1.10	0.93
30	Zinc	9.66	8.64	1.20	1.02
42	Molybdenum	20.00	17.48	2.88	2.29
45	Rhodium	23.22	20.21	3.42	2.69
46	Palladium	24.35	21.18	3.62	2.84
47	Silver	25.52	22.16	3.81	2.98
48	Cadmium	26.71	23.17	4.02	3.13
49	Indium	27.93	24.21	4.25	3.29
50	Tin	29.19	25.27	4.46	3.44
73	Tantalum	67.40	57.52	11.67	8.15
74	Tungsten	69.51	59.31	12.09	8.40
79	Gold	80.71	68.79	14.35	9.71
80	Mercury	83.11	70.82	14.81	9.99
82	Lead	88.00	74.96	15.87	10.55
92	Uranium	115.60	98.43	21.75	13.61

*The most prominent emission line in each group.

Appendix 4

SOME ATOMIC AND NUCLEAR PROPERTIES

Element	*Z*	*A*	*Chem. At. Wgt.*	*At. Mass, MeV*	*Percent Abundance*
Neutron	0	1	—	939.550	—
Hydrogen	1	1	1.008	938.767	99.98
		2		1,876.092	0.02
		3		2,809.384	
Helium	2	3	4.0026	2,809.365	10^{-4}
		4		3,728.337	100
Lithium	3	6	6.939	5,602.956	7.5
		7		6,535.253	92.5
		8		7,472.770	
Beryllium	4	7	9.013	6,536.115	
		9		8,394.653	100
Boron	5	8	10.82	7,474.747	
		10		9,326.832	18.4
		11		10,254.876	81.6
		12		11,191.106	
Carbon	6	11	12.011	10,256.856	
		12		11,177.736	98.892
		13		12,112.339	1.108
		14		13,043.712	
Nitrogen	7	14	14.008	13,043.556	99.64
		15		13,972.270	0.36
Oxygen	8	16	16	14,898.911	99.759
		17		15,834.318	0.037
		18		16,765.822	
Fluorine	9	19	19.00	17,696.596	100
Neon	10	20	20.183	18,662.518	90.9
		21		19,555.308	0.27
		22		20,484.491	8.83
Sodium	11	22	22.990	20,487.334	
		23		21,414.466	100
		24		22,347.054	
Magnesium	12	24	24.32	22,341.539	78.6
		25		23,273.759	10.2
		26		24,202.214	11.2
Aluminum	13	27	26.98	25,132.710	100
Silicon	14	28	28.09	26,059.894	92.27
		29		26,990.968	4.68
		30		27,919.901	3.05
Phosphorous	15	31	30.974	28,851.380	100
		32		29,782.993	
Sulfur	16	32	32.066	29,781.283	95.02
		33		30,712.191	0.75
		34		31,640.318	4.12
		35		32,572.883	

Appendix 4 (cont.)

Element	*Z*	*A*	*Chem. At. Wgt.*	*At. Mass, MeV*	*Percent Abundance*
Chlorine	17	35	35.457	32,572.715	75.4
		37		34,432.921	24.6
Argon	18	36	39.944	33,502.976	0.337
		38		35,361.446	0.063
		40		37,224.082	99.600
Potassium	19	39	39.100	36,293.839	93.08
		40		37,225.587	0.0119
		41		38,155.046	6.91
		42		39,087.056	
Calcium	20	20	40.08	37,224.272	96.97
		44		40,943.572	2.06
Vanadium	23	51	50.95	47,453.179	99.75
Chromium	24	50	52.01	46,523.651	4.31
		51		47,453.931	
		52		48,381.445	83.76
Iron	26	54	55.85	50,243.566	5.84
		55		51,176.816	
		56		52,102.163	91.68
		57		53,034.070	2.17
		58		53,963.577	0.31
		59		54,896.542	
Cobalt	27	59	58.94	54,894.969	100
		60		55,827.029	
Nickel	28	58	58.69	53,965.49	67.76
		60		55,824.21	26.16
Copper	29	63	63.54	58,617.531	69.09
		64		59,549.164	
		65		60,578.80	30.91
Zinc	30	64	65.37	59,548.592	48.87
		65		60,480.15	
		66		61,406.67	27.62
Silver	47	107	107.88	99,579.7	51.35
		109		101,442.3	48.65
Lead	82	204	207.21	189,996.4	1.48
		206		191,860.7	23.6
		207		192,793.6	22.6
		208		193,725.8	52.3
Bismuth	83	209	209.0	194,660.8	100
Uranium	92	234	238.07	218,004.0	0.0058
		235		218,938.2	0.715
		236		219,186.3	
		238		221,739.0	99.28
		239		222,673.8	

Appendix 5

PROPERTIES OF SOME RADIOACTIVE NUCLIDES

Z	*Element*	*A*	*Half-Life*	λ*	$\bar{E}_\beta$ *MeV*	*Rhm*†
1	H	1	12.3 y	0.055	0.0057	—
6	C	14	5730 y	1.22×10^{-4}	0.049	—
11	Na	22	2.6 y	0.266	0.191	1.20
		24	15.0 h	0.046	0.55	1.84
15	P	32	14.3 d	0.0485	0.695	—
16	S	35	88 d	7.9×10^{-3}	0.049	—
17	Cl	36	3×10^5 y	2.3×10^{-6}	0.27	—
19	K	40	1.3×10^9 y	5.3×10^{-10}	0.46	—
		42	12.4 h	0.056	1.43	0.15
20	Ca	45	165 d	4.2×10^{-3}	0.076	—
		47	4.5 d	0.154	0.12	0.64
24	Cr	51	27.8 d	0.0250	EC	0.018
26	Fe	55	2.6 y	0.266	EC	EC
		59	45 d	0.0154	0.12	0.64
27	Co	57	270 d	2.57×10^{-3}	EC	0.09
		58	71 d	9.8×10^{-3}	0.03	0.55
		60	5.26 y	0.132	0.095	1.3
36	Kr	85	10.8 y	0.064	0.25	—
38	Sr	90 }	28.1 y	0.0247	1.13	—
39	Y	90 }				
53	I	131	8.05 d	0.0862	0.183	0.22
		132	2.3 h	0.30	0.59	1.2
55	Cs	137	30 y	0.0232	0.25	0.33
77	Ir	192	74 d	9.4×10^{-3}	0.17	0.59
79	Au	198	2.7 d	0.257	0.33	0.23
80	Hg	203	46.9 d	0.0148	0.098	0.16
88	Ra	226	1602 y	4.34×10^{-4}	In 0.5-mm Pt	0.84

*In reciprocal half-life units, y^{-1}, d^{-1}, ...

†Roentgens per hour per curie at 1 meter.

Appendix 6

CHARACTERISTICS OF THE STANDARD MAN

From recommendation of the International Commission on Radiological Protection, *Brit. J. Radiol.*, Supp. 6, 1955.

Total body 70 kg

Tissue	*Mass, g*	*Percent*
Muscle	30,000	43
Total fat	10,000	14
Bones without marrow	7,000	10
Red marrow	1,500	2.1
Yellow marrow	1,300	2.1
Blood	5,400	7.7
Subcutaneous tissue	4,100	5.8
Skin	2,000	2.9
Gastrointestinal tract (empty)	2,000	2.9
Liver	1,700	2.4
Brain	1,500	2.1
Lungs (total)	1,000	1.4
Lymphoid tissue	700	1.0
Kidneys (total)	300	0.43
Heart	300	0.43
Spleen	150	0.21
Urinary bladder	150	0.21
Pancreas	70	0.10
Testes (total)	40	0.057
Eyes	30	0.043
Thyroid	20	0.029
Adrenals (total)	20	0.029
Thymus	10	0.014
Miscellaneous	500	0.71

Chemical Composition

Element	*Mass, g*	*Percent*
Oxygen	45,500	65.0
Carbon	12,600	18.0
Hydrogen	7,000	10.0
Nitrogen	2,100	3.0
Calcium	1,050	1.5
Phosphorous	700	1.0
Sulfur	175	0.25
Potassium	140	0.2
Sodium	105	0.15
Chlorine	105	0.15
Magnesium	35	0.05
Iron	4	0.006
Copper	0.1	0.0001
Iodine	0.03	0.00004

Index

B

C

F

G

R

S

W

X